Vectorization

IEEE Press
445 Hoes Lane
Piscataway, NJ 08854

Vectorization

A Practical Guide to Efficient Implementations of Machine Learning Algorithms

Edward DongBo Cui
Director of Data Science, DotDash Meredith

IEEE Press
WILEY

Contents

About the Author

Edward Dongbo Cui is a data science and machine learning engineering leader who holds a PhD in neuroscience from Case Western Reserve, USA. Edward served as Director of Data Science at DotDash Meredith, building large-scale search and recommender systems for online digital publications. At NBC Universal, he served as Director of Data Science, building the first recommendation system on the new Peacock streaming platform. Previously, he was lead data scientist at Nielsen Global Media. He is an expert in ML engineering, research, and MLOps to drive data-centric decision-making and enhance product innovation.

Preface

Over the years, I have accrued valuable experiences in implementing machine learning algorithms, and this book is my endeavor to share those insights with you. Machine learning is a vast field, encompassing areas including but not limited to computer vision, speech processing, natural language processing, and recommendation systems. While the diversity of these topics is immense, there are underlying mathematical and computational principles that weave them together.

Structure of the Book: The book is structured into 11 chapters, each dedicated to a pivotal topic in the realm of scientific computation and machine learning:

- **Chapter 1** delves into the power of vectorization, contextualized within parallel computing.
- **Chapter 2** lays the foundation by discussing the basic tensor operations.
- **Chapter 3** details the art of tensor indexing, elucidating how to access individual or subsets of tensor elements.
- **Chapter 4** probes deeper into vectorization in tensor multiplications and common linear algebraic routines, which form the backbone of many machine learning algorithms.
- **Chapter 5** sheds light on the pivotal concepts of masking and padding, which come into play when handling data of nonuniform sizes.
- **Chapter 6** touches upon string processing techniques, invaluable for natural language processing (NLP) tasks.
- **Chapter 7** introduces the realm of sparse matrices, discussing their data structures and integral operations.
- **Chapter 8** takes a closer look at a unique kind of sparse tensor – ragged or jagged tensors – and the nuances of processing them.
- **Chapter 9** breaks down the data wrangling pattern of groupby-apply-aggregate, which is crucial for data summarization.

- **Chapter 10** explores the topic on sorting, reordering, sampling, and batch processing through a series of case studies.
- **Chapter 11** concludes the book with a review of the Transformer++ architecture (a.k.a LLaMA, a foundational large language model architecture from Meta), weaving together the principles discussed in the preceding chapters.

Audience: While I have tailored this book for those embarking on their journey in machine learning algorithm implementation, it is not solely for beginners. Having a basic understanding of machine learning and deep learning will certainly be an advantage, but every concept introduced is supplemented with sufficient background to make it comprehensible to all readers.

Prerequisites:

- **Familiarity with the Python programming language:** Readers can find online resources to learn Python or through classic books such as "Think Python."
- **Undergraduate linear algebra:** To this, I highly recommend Gilbert Strang's lectures on Linear Algebra at MIT, or his book "Introduction to Linear Algebra."
- **Foundational knowledge of data science and machine learning:** The goal of this book is not to introduce machine learning, but rather to guide the reader in solving real-world problems encountered in machine learning (oftentimes deep learning). Basic knowledge of data science and machine learning can be gained from either previous more foundational courses or practical knowledge gained by doing machine learning projects hands-on. Courses such as Andrew Ng's CS229 Machine Learning can serve as a foundation.

Features:

- Every chapter is punctuated with numerous case studies to ensure the concepts are grounded in practical scenarios. Accompanying these are detailed code implementations with line-by-line explanations. Most of the implementations will use only the knowledge acquired within the chapter. Certainly, as the reader learns along the way, one may come up with better implementations by leveraging more advanced concepts.
- The book covers vectorization techniques using NumPy, Tensorflow, and PyTorch (and for certain operations, Pandas as well). When introducing the operations, I will include tables that enumerate equivalent operations in these three libraries, in the hope that readers who are familiar with one of the libraries can also learn to translate algorithms and models implemented in another library. Most case study code examples will be implemented in one of the libraries (often NumPy), but readers are encouraged to implement the same functionality using alternative libraries by referencing equivalent operations.

- All code examples are available at github.com/sapphire008/vectorization-book-supplement. Note that some code examples are simulations using random inputs, so rerunning the notebooks will not necessarily get the same results as the ones provided in the book.
- To further clarify many of the vectorization operations, illustrative figures have been incorporated to visually represent the logic behind them, often accompanied by detailed explanations either within the text body or in the caption.
- The principles discussed in this book should be agnostic to specific versions of Python, NumPy, TensorFlow, or PyTorch. By default, my code examples were tested using `python==3.10.6`, `numpy==1.23.4`, `pandas==2.1.1`, `scikit-learn==1.3.0`, `scipy==1.11.1`, `tensorflow==2.13.0`, `keras==2.13.1`, and `torch==2.1.0`. However, I will specify the library versions alongside code examples when the versions used are not the default, or whenever relevant or necessary to emphasize the versions.

Limitations:

- Note that the carriage return symbol "↪" within the code block (with shaded background) is used to indicate a line break to accommodate formatting of long lines.
- Code examples and case studies implemented to solve certain problems may not always be the most optimal solution, though it is honestly my best effort to come up with these solutions when writing this book. Sometimes, I also specifically made a decision to implement certain things not as efficiently as possible but to illustrate the practical application of the concepts introduced in the chapter. Because the materials of each chapter in the book build upon each other, I have made an explicit choice not to use a functionality that would be introduced in a later chapter. Still, this demonstrates the beauty of vectorization and coding in general: there is not a single solution to a specific problem in practice. Nonetheless, I encourage readers to find more efficient ways to solve the case study problems. On the other hand, if the readers found a bug in one of the implementations, please feel free to open an issue at github.com/sapphire008/vectorization-book-supplement.
- When comparing the vectorization libraries, it is possible that I may be ignorant of certain functionalities specific to the library even though the information contained in this book is to the best of my current knowledge. I may sometimes claim that one library does not implement functionality that another library implements at the time of writing, but if readers found the other library indeed implemented that functionality, I stand corrected and will make the best effort to update this information in future editions of the book.

My hope is that by the end of this book, the readers will be well-versed in the foundational and advanced vectorization techniques of implementing scalable machine learning algorithms and will be ready to tackle real-world challenges.

November 29, 2023
New York

Edward DongBo Cui
Director of Data Science
DotDash Meredith, New York

Acknowledgment

I express my deepest gratitude to Ryan Beauchemin and Kendall Meyer for their invaluable reviews and contributions, which have greatly enhanced the quality of this book. I also thank Ilmir Moussikaev for believing in my potential and providing me with the opportunity to create something remarkable on the Peacock streaming platform. This specific experience was the cornerstone of my career, leading me to acquire the knowledge necessary to write this book. I also want to extend my heartfelt appreciation to my students. Teaching and mentoring them on various techniques in machine learning and data science provided me with the insights and material needed to write this book. A special thanks to Chris Haroun, whose inspiration and guidance motivated me to embark on this writing endeavor. Lastly, I am immensely thankful to my wife for her unwavering support throughout this journey.

1

Introduction to Vectorization

1.1 What Is Vectorization

Vectorization is a type of parallel computing paradigm that performs arithmetic operations on an array of numbers [Intel, 2022] within a single data processing unit (either CPU or GPU). Modern CPUs can perform between 4 and 16 single precision (float32) parallel computations, depending on the type of instructions [Intel, 2022] (see Figure 1.1).

Here, SSE, or streaming SIMD (single-instruction-multiple-data) extension, has 128 registers; each register is used to store 1-bit of data for computations. This is equivalent to performing four single-precision (float32) operations or two double-precision (float64) operations. Intel's AVX512, or Advanced Vector Extension 512, on the other hand, has 512 registers and can therefore process 512 bits of data simultaneously. This is equivalent to performing 16 single-precision or 8 double-precision operations.

The term **"Vectorization"** also refers to the process of transforming an algorithm computable on one data point at a time to an operation that calculates a collection of data simultaneously [Khartchenko, 2018]. A classic example is the dot product operation,

$$\sum_{i}^{n} a_i b_i = \mathbf{a} \cdot \mathbf{b}$$

where $\mathbf{a} := [a_1, a_2, a_3, \dots, a_n]$, and $\mathbf{b} := [b_1, b_2, b_3, \dots, b_n]$.

The dot product operation on the right-hand side of the equation is the vectorized form of the summation process on the left. Instead of operating on one element of the array at a time, it operates on the collections or arrays, **a** and **b**.

Vectorization: A Practical Guide to Efficient Implementations of Machine Learning Algorithms, First Edition. Edward DongBo Cui.

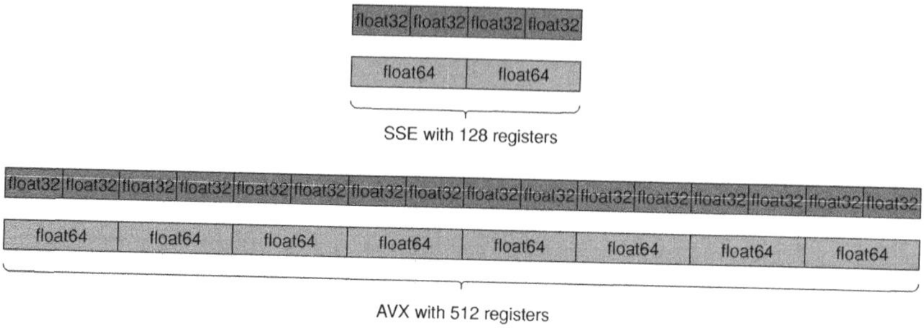

Figure 1.1 CPU vector computations.

> **Note**
>
> Another definition of the term "vectorization" is related to the more recent advances in large language models: using a vector or an array of values to represent a word. This is more commonly known as **embedding** in the literature which originated from the word2vec model in natural language processing. In Chapters 6 and 7, we will introduce the idea of representing string or categorical features as numerical values, i.e. integer indices, sparse one-hot encoding vectors, and dense embedding vectors.

1.1.1 A Simple Example of Vectorization in Action

Suppose we want to take the dot product of 1 million random numbers drawn from the uniform distribution. Using Python's `for loop`, we would typically implement this algorithm as the follows:

```
import random

def sqrt_of_n_random_numbers(n=1_000_000):
    result = 0
    for _ in range(n):
        a = random.uniform(0, 1)
        b = random.uniform(0, 1)
        result += a * b
    return result
```

On my M1 MacBook Pro under Python 3.11.4, this is the runtime of the above function:

```
294 ms ± 1.39 ms per loop (mean ± std. dev. of 7 runs, 1 loop each)
```

If we implement the algorithm using `numpy` with vectorization

```
import numpy as np

def sqrt_of_n_random_numbers_np(n=1_000_000):
    a = np.random.rand(n)
    b = np.random.rand(n)
    result = np.dot(a, b)
    return result
```

Under the same setup, we have the following runtime of the vectorized implementation.

```
7.9 ms ± 119 µs per loop (mean ± std. dev. of 7 runs, 100 loops each)
```

This is a ~37× speed-up by vectorization!

1.1.2 Python Can Still Be Faster!

However, the above example is not to discount that certain native Python operations are still faster than NumPy's implementations, despite NumPy being often used for vectorization operations to simplify for-loops in Python.

For example, we would like to split a list of strings delimited by periods ".". Using Python's list comprehension, we have

```
1 X = ["foo.bar", "bar.foo", "x.y", "z.k"] * 1000
2 [x.split(".") for x in X]
```

Line 2 runs for

```
299 µs ± 5.98 µs per loop (mean ± std. dev. of 7 runs, 1,000
  ↪ loops each)
```

However, in comparison, if we use NumPy's "vectorized" split operation

```
1 Y = ["foo.bar", "bar.foo", "x.y", "z.k"] * 1000
2 Y = np.array(Y)
3 np.core.defchararray.split(Y, ".")
```

Line 3 runs for

```
1.18 ms ± 4.37 µs per loop (mean ± std. dev. of 7 runs, 1,000
↪ loops each)
```

which is about ~4.0× slower than native Python.

1.1.3 Memory Allocation of Vectorized Operations

When using vectorized operations, we need to consider the trade-off between memory and speed. Sometimes, vectorized operations would need to allocate more memory for results from intermediate steps. Therefore, this has implications on the overall performance of the program, in terms of space vs. time complexity. Consider a function in which we would like to compute the sum of the squares of the elements in an array. This can be implemented as a for-loop, where we iterate through each element, take the square, and accumulate the square values as the total. Alternatively, the function can also be implemented as a vectorized operation where we take the square of each element first, store the results in an array, and then sum the squared vector.

```
import numpy as np

def sum_of_square_vectorized(X):
    x = np.asarray(X)
    x_sq = np.square(x) # x ** 2
    res = x_sq.sum()
    return res


def sum_of_square_for_loop(X):
    res = 0
    for x in X:
        res += x ** 2
    return res
```

Let us use the `memory_profiler` package to examine the memory usage of each line of the two functions. We can first install the package by `pip` `install memory_profiler==0.61.0` and load it as a plugin in a Jupyter Notebook.

```
!pip install memory_profiler==0.61.0
%load_ext memory_profiler
```

We also save the above script into a file (which we call `memory_allocation_example.py`), then we measure the memory trace of each function in the notebook as follows:

```
from memory_allocation_example import sum_of_square_vectorized
X = np.linspace(-50, 50, 100_000).tolist()
%mprun -f sum_of_square_vectorized sum_of_square_vectorized(X)
```

This gives us the following line-by-line memory trace of the function call:

```
Line # Mem usage Increment Occurrences  Line Contents
=====================================================================
     4                                 def sum_of_square_vectorized(X):
     5 104.3 MiB 104.3 MiB          1      x = np.asarray(X)
     6 105.1 MiB 0.8 MiB            1      x_sq = np.square(x)  # x ** 2
     7 105.4 MiB 0.3 MiB            1      res = x_sq.sum()
     8 105.4 MiB 0.0 MiB            1      return resx
```

We can also use `%timeit` magic function to measure the speed of the function:

```
%timeit sum_of_square_vectorized(X)
```

which gives

```
3.91 ms ± 64.7 µs per loop (mean ± std. dev. of 7 runs, 100
↪  loops each)
```

Similarly, let us also look at the for-loop implementation

```
from memory_allocation_example import sum_of_square_for_loop
X = np.linspace(-50, 50, 100_000).tolist()
%mprun -f sum_of_square_for_loop sum_of_square_for_loop(X)
```

```
Line #    Mem usage    Increment  Occurrences   Line Contents
=============================================================
    11     68.2 MiB     68.2 MiB            1   def sum_of_square_for_loop(X):
    12     68.2 MiB      0.0 MiB            1       res = 0
    13     68.2 MiB      0.0 MiB       100001       for x in X:
    14     68.2 MiB      0.0 MiB       100000           res += x ** 2
    15     68.2 MiB      0.0 MiB            1       return res
```

And we also test the run time as

```
%timeit sum_of_square_for_loop(X)
```

which gives

```
5.72 ms ± 39 µs per loop (mean ± std. dev. of 7 runs, 100 loops each)
```

We can see that although the for-loop function is slower than the vectorized function, the for-loop had almost no memory increase. In contrast, the vectorized solution has a ~1M increase in memory usage while reducing the run time by about 30%. If `X` is several magnitudes larger than our current example, then we can see that even more memory will be consumed in exchange for speed. Hence, we need to consider carefully the trade-off between speed and memory (or time vs. space) when using vectorized operations.

1.2 Case Study: Dense Layer of a Neural Network

Vectorization is a key skill for implementing various machine learning models, especially for deep learning algorithms. A simple neural network consists of several feed-forward layers, as illustrated in Figure 1.2.

A feed-forward layer (also called dense layer, linear layer, perceptron layer, or hidden layer, as illustrated in Figure 1.3) has the following expression:

$$z = \sigma(xw + b)$$

Notice that $y = xw + b$ is a linear model. Here,

- x is the input, and typically of shape `(batch_size, num_inputs)`
- w is the weight (a.k.a. kernel) parameter matrix of shape `(num_inputs, hidden_size)`
- b is the bias parameter vector of shape `(1, hidden_size)`

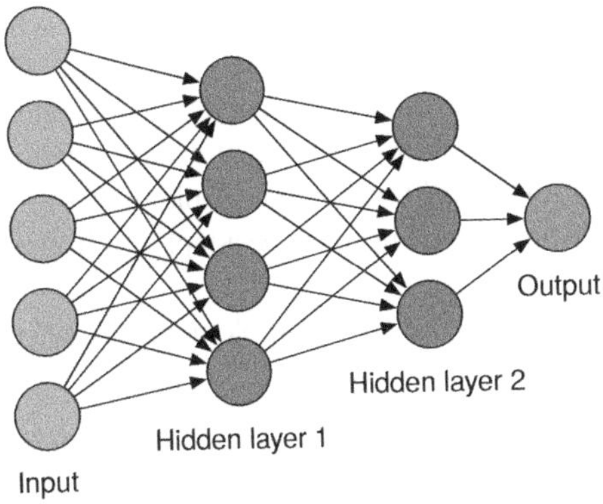

Figure 1.2 Multilayer perceptron.

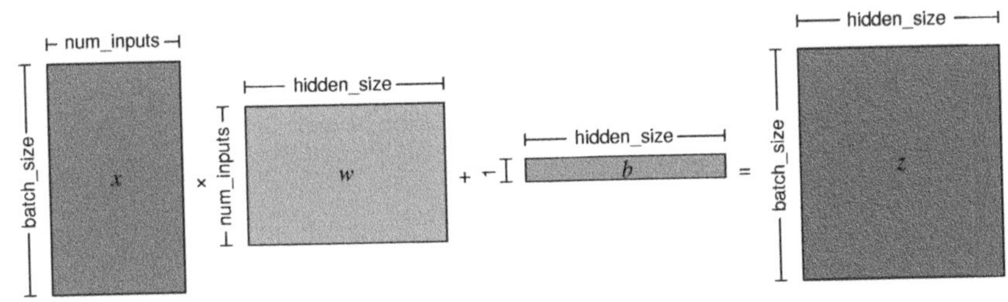

Figure 1.3 Feed-forward layer.

- σ is an activation function that is applied to every element of the resulting linear model operation. It can be, e.g., `relu`, `sigmoid`, or `tanh`.
- z is the output of shape `(batch_size, hidden_size)`

Stacking multiple feed-forward layers creates the **multilayer perceptron model**, as illustrated by our example neural network above.

If we were to implement a feed-forward layer using only Python naively, we would need to run thousands to millions of iterations of the `for-loop`, depending on the `batch_size`, `num_inputs`, and `hidden_size`. When it comes to large language models like generative pretrained transformer (GPT) which has hundreds of billions of parameters to optimize, using `for loops` to make such computations can become so slow that training or making an inference using the model is infeasible. But if we take advantage of vectorization on either CPU or GPU, we would see significant gains in the performance of our model. Furthermore, certain operations like matrix multiplication are simply inefficient to perform if implemented naively by simply parallelizing a `for-loop`, so in practice must rely on special algorithms such as Strassen's Algorithm [Strassen, 1969] (or more recently, a new algorithm discovered by AlphaTensor [Fawzi et al., 2022]). Implementing these algorithms as an instruction on either CPU or GPU is a non-trivial task itself. Practitioners of machine learning and data science should therefore take advantage of pre-existing libraries that implement these algorithms. But to use these libraries efficiently, one needs to be adept in thinking in terms of vectorization.

In the following, let's take a look at an example of how to implement the above perceptron layer using `numpy`.

```
1  import numpy as np
2
3  def sigmoid(x):
4      """Implement sigmoid activation function."""
5      return 1. / (1 + np.exp(-x))
6
7
8  def dense_layer(x, w, b):
9      # linear model: (batch_size, hidden_size)
10     y = x @ w + b
11     # activation
12     y = sigmoid(y)
13     return y
```

- Lines 3–5 define the sigmoid activation function, $f(x) = \frac{1}{1+e^{-x}}$
- Line 10 implements the linear model
  - Here, the @ operator is the matrix multiplication operator that is recognized by typical vector libraries like `numpy`, `tensorflow`, and `torch`. We will discuss more on matrix operations in Chapter 4.
  - The result of multiplying x and w should be a matrix of shape `(batch_size, hidden_size)`, but in this implementation, we have b of shape `(1, hidden_size)`. How can a vector be added to a matrix? In Chapter 2, we will explore the concept called **broadcasting**.
- Line 12 applies the sigmoid activation we defined previously to compute the final output of this dense layer. This is an element-wise operation, where we apply the sigmoid function to each element of the matrix. We will explore the idea of element-wise operations in Chapter 2.

As you can see, vectorization can greatly simplify the representation of neural networks operating on high-dimensional arrays with a large number of elements or data points.

1.3 Vectorization vs. Other Parallel Computing Paradigms

Knowing that vectorization is a type of parallel computing paradigm, it is natural to ask the question: how does vectorization compare to other parallel computing paradigms such as multithreading, multiprocessing, or multiworker distributed computing?

1.3.1 Multithreading

Like vectorization, multithreading operates on a single processing unit, with shared memory. This can be a single core of a CPU or GPU. Vectorization processes multiple data inputs within the register in parallel, given a single instruction. In contrast, multithreading can only process one stream of data at a time, instead of processing data in parallel. It can switch between different threads of the same program (or sometimes different threads of different programs), with each thread representing a different part of the process. This is referred to as "concurrency." Of course, multithreading can be used in combination with vectorization. In fact, modern GPUs operate by combining vectorization, multithreading, and multiprocessing to increase their processing powers [NVIDIA, 2012].

The multithreading paradigm is quite advantageous in tasks like reading data from remote servers. For example, a database has 10 parts. If we have only a single thread, then for each part of the data to be read, we need to wait for one minute until the data is downloaded before reading the next part. Then, it will take 10 minutes in total to download the entire database. However, if we have 10 threads, each making a request to read one part of the data. Since the time to make a request to read the data is negligible compared to the time to download the data, each thread takes one minute to download the data. However, since the downloading process is in parallel with each other, it will take only one minute in total to download the entire database. See Figure 1.4 for an illustration.

1.3.2 Multiprocessing

Multiprocessing, on the other hand, operates on multiple processing units, e.g. multi-core CPUs or GPUs. Each core of the processing unit can access data through a shared memory (the RAM). Unlike vectorization and multithreading, multiprocessing does not share the same data between the processes. Therefore, to process the same data in parallel using different instructions, multiprocessing needs to duplicate the data across the cores. Unlike multithreading, multiprocessing will not be efficient if the process is bounded by I/O (i.e. data reading and writing). Conversely, multiprocessing is a great choice if the process is bounded

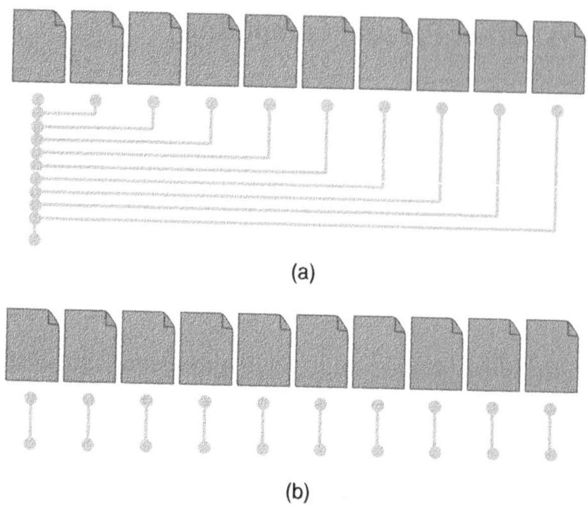

Figure 1.4 Reading remote database (a) with single thread and (b) with multithreading.

by the processing power (a.k.a. CPU bound) and can greatly speed up the process by distributing the computational load to multiple cores of the processing unit.

A typical use of multiprocessing is to perform in-memory batch data processing. Suppose we want to compute the min–max normalized score for a bunch of users:

```
import pandas as pd

df = pd.DataFrame({
    "id": [
        "A", "A", "A", "A", "B", "B", "B", "B", "B",
        "C", "C", "C", "C", "C", "C", "C", "C", "C"
    ],
    "scores": [
        98, 97, 100, 80, 78, 80, 82, 81, 61, 51, 72,
        70, 82, 80, 75, 62, 67, 53
    ],
})
```

We can define the following user-defined function (UDF) to be used to compute each user's normalized score:

```
def apply_func(pdf):
    """Compute the min-max normalized score"""
    min_val = pdf["scores"].min()
    max_val = pdf["scores"].max() + 1e-6
    pdf["scores"] = (pdf["scores"] - min_val) / (max_val - min_val)
    return pdf
```

Each grouped batch based on the user's id will be transformed using the `apply_func`:

```
df_out = df.groupby(
    by="id", as_index=False
).apply(apply_func)
```

This takes about 1.5 ms on my MacBook Pro M1 machine with 8 cores, yielding the following results in Table 1.1.

We can perform the same transformation using multiprocessing, where each process will operate on each of the grouped batches of data. We first define the helper function `apply_parallel` which will use Python's `joblib` to create the multiprocessing environment.

Table 1.1 Min–Max normalized scores

Id	Scores
A	0.9
A	0.85
A	1
A	0
B	0.809524
B	0.904762
B	1
B	0.952381
B	0
C	0
C	0.677419
C	0.612903
C	1
C	0.935484
C	0.774194
C	0.354839
C	0.516129
C	0.0645161

```python
from tqdm import tqdm
import multiprocessing
from joblib import Parallel, delayed
import pandas as pd

def apply_parallel(df_grouped, func):
    """
    Apply a function to grouped pandas dataframe using multiprocessing.
    Usage:
    def apply_func(pandas_df):
        ...

    df = apply_parallel(
        df.groupby(
            by=grouped_by_columns,
            as_index=False,
        ), apply_func)
    """
    ret = Parallel(n_jobs=multiprocessing.cpu_count())(
        delayed(func)(group) for name, group in tqdm(df_grouped)
    )  # enumerate(tqdm(dfGrouped))
    return pd.concat(ret)
```

We then use this helper function to perform our batchwise transformations

```python
df_out_parallel = apply_parallel(
    df.groupby(by="id", as_index=False), apply_func
)
```

This takes 677 ms, which is much slower than using the native `pandas.DataFrame.apply` function. This is because setting up multiprocessing has an overhead, so most of the process is used to set up the multiprocessing environment instead of performing the actual computation. However, when the size of the data and the number of batches are large, we can expect that the additional overhead can be worthwhile for the overall speed gain from multiprocessing. Let's now instead randomly generate 100,000 rows of data for our transformation.

```python
import string
import random

df = pd.DataFrame({
    "id": [
        "".join(
            random.choices(string.ascii_letters, k=3)
        ) for _ in range(100_000)
    ],
    "scores": [
        random.randint(1, 100) for _ in range(100_000)
    ],
})
```

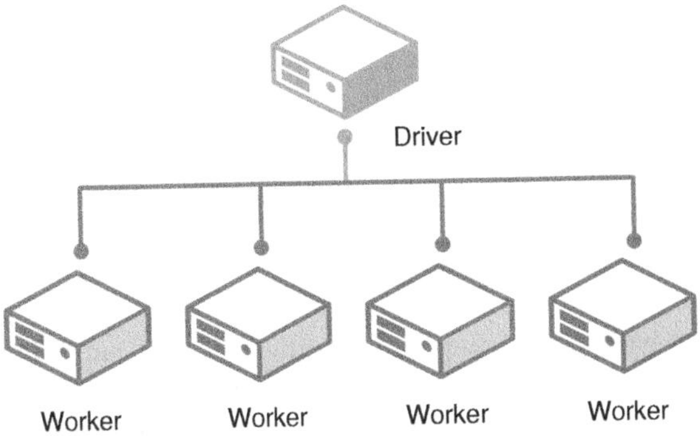

Figure 1.5 Driver and workers in distributed computing. A driver (or main) node issues a series of commands to create workers and assigns each worker a task to process a portion of the data. Once each worker completes their own task, the worker will send the results back to the driver for aggregation. The driver then waits for all the workers to finish their tasks before returning the results to the user.

If we use `pandas.DataFrame.apply`, the transformation will take about 19 seconds. If we use `apply_parallel`, it now takes 16 seconds to perform the transformation, a ~16% gain in processing speed.

1.3.3 Multiworker Distributed Computing

In modern big data analytics, practitioners often need to process data ranging from anywhere between several gigabytes and hundreds of terabytes in size (and this is often only the compressed size on disk; the size of the data occupying the memory can be much larger). No single consumer-grade laptop or desktop computer can hold hundreds of gigabytes to terabytes of data in memory, nor is any single CPU or GPU fast enough to process this much data without the user waiting hours to days (well, at least no such powerful processing unit exists yet as I am writing this book). Data analysts, data scientists, and machine learning engineers instead rely on cloud technologies to host such a large amount of data on distributed file storage systems and process the data using distributed computing. The term "distributed" here simply means multiple machines are storing, accessing, and processing the data, whereas machines communicate with each other via networking. Figure 1.5 illustrates a multiworker distributed computing system.

When dealing with the task of processing a large amount of data, a "divide and conquer" approach is key. The idea is if it takes 100 hours for a single computer to process our data, then 100 computers can finish the same task in about one hour. If we can somehow divide such a large amount of data into smaller pieces, let each computing node processes that small piece of data, and finally aggregate the results at the end from all the parallel workers, it can then greatly reduce the physical (or wall) time we need to wait for our data processing task to finish. In fact, this is the core idea behind MapReduce [Dean and Ghemawat, 2008].

A "Hello World" example often used to illustrate MapReduce is the task of counting the number of occurrences of each word from a large body of documents. In this example, we will use Apache Beam [Apache Foundation, 2023a] to illustrate how to write a MapReduce algorithm for word count. We can install `apache-beam` with `pip install apache-beam` (I am using `apache-beam==2.49.0`. Note that Apache Beam currently does not support `pandas >=2.0.0`[Apache Foundation, 2023b] at the time of writing, so I am using `pandas==1.5.3` in this example).

```
1  import apache_beam as beam
2  from apache_beam.options.pipeline_options import PipelineOptions
3
4  # Setting up running environment
5  pipeline_args = [
6      "--job_name=word_count",
7      "--runner=DirectRunner", # local runner
8  ]
9  pipeline_options = PipelineOptions(pipeline_args)
10
11 # Run the beam pipeline
12 with beam.Pipeline(options=pipeline_options) as p:
13     _ = (
14         p
15         # Creating the example document
16         # This can also be read from files
17         | "Create texts" >> beam.Create([
18             "lorem ipsum dolor sit amet",
19             "sit amet dolor lorem",
20             "ut enim ad minim veniam",
21             "ut enim ad dolor lorem",
22         ])
23         # Splitting the documents into words
24         | "Split" >> beam.FlatMap(lambda x: x.split())
25         # Convert to a tuple (key, value) = (word, 1)
26         | "Pair with 1" >> beam.Map(lambda x: (x, 1))
27         # Summing the value by each key/word
28         | "Count" >> beam.CombinePerKey(sum)
29         # Print the output
30         | "Print" >> beam.Map(print)
31     )
```

- Lines 1–2 import the `apache-beam` library and the `PiplineOptions` class to configure the running environment.
- Lines 5–9 set up the pipeline running environment. Beam can automatically figure out when to use multithreading, multiprocessing, or distributed

computing depending on the specified runner type. In this small example, since we are running using `DirectRunner`, Beam assumes this is running on a single machine, so it will find the opportunity to use multiprocessing whenever appropriate. But, if we run the pipeline on a distributed environment such as Apache Flink or Apache Spark, Beam will use multiworker processing, while each worker will then also use multiprocessing if multiple cores are also available on each worker. This small technical detail, however, should not prevent us from understanding how MapReduce works in a distributed, parallel environment.

- Line 12 starts the context of the Beam pipeline, with the options we specified.
- Lines 14–22 initialize the pipeline by creating some mock documents. In this small example, we assume we have four documents. Each document contains words separated by spaces.
- Line 24 splits the documents into individual words. Here, `FlatMap` means applying the word splitting function to each document and then flat the results. So instead of returning the list of lists, it returns a flattened list.
- Line 26 performs the classic Map step of MapReduce, where it creates the tuple `(word, 1)` for each word in the flattened list. This is often described as the `<key, value>` map format, where the `word` is the key and the static value `1` is the `value`.
- Line 28 performs the classic Reduce step of the MapReduce, where it aggregates the tuples from the previous step by each key, summing together the values. This is equivalent to `.groupby(...).sum()` operations in `pandas.DataFrame`.
- Line 30 prints the results of the pipeline.

After executing the above pipeline, we have the following printout that counts the number of occurrences of each word:

```
('lorem', 3)
('ipsum', 1)
('dolor', 3)
('sit', 2)
('amet', 2)
('ut', 2)
('enim', 2)
('ad', 2)
('minim', 1)
('veniam', 1)
```

If the number and size of documents are both very large, we could imagine that each document will be read, split, mapped, aggregated, and returned as the above summary by each worker node. Then, the driver node can do further merging of the aggregated results from each worker to output the final results.

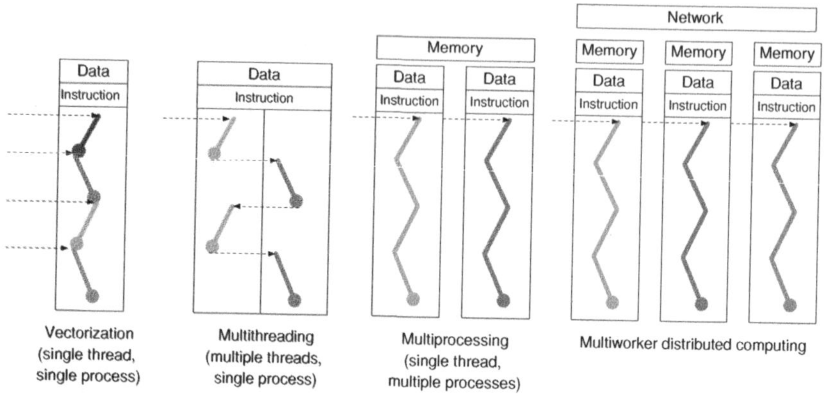

Figure 1.6 Comparison of parallel computing paradigms. Vectorization stores data within a single registry of the processing unit and processes multiple data representations on the registry using a single instruction simultaneously. Multithreading needs to switch between different threads of the processing unit and instructions are executed concurrently. Multiprocessing initiates multiple processes across different computing cores or different processing units on the machine, where data are communicated through shared memory. Multiworker distributed computing uses multiple machines to process a small piece of data, where the data are hosted on distributed file systems and communicated through networking.

Figure 1.6 summarizes and compares different paradigms of parallel computing we have discussed so far.

I hope this section provides enough information on different paradigms of parallel computing. It can be deduced that multithreading, multiprocessing, and multiworker paradigms are agnostic to the exact instructions of the program, whereas vectorization manipulates specific data using dedicated algorithms. Therefore, it is correct to expect that when combining vectorization with other types of parallel computing techniques such as multiprocessing and multiworker distributed computing, we can maximize our computing efficiency given limited resources.

Bibliography

Apache Foundation. Apache Beam. https://beam.apache.org/, 2023a.

Apache Foundation. Support Pandas==2.x in Apache Beam. #27221. https://github.com/apache/beam/issues/27221, June 2023b.

Jeffrey Dean and Sanjay Ghemawat. MapReduce: Simplified data processing on large clusters. *Communications of the ACM*, 51(1):107–113, January 2008. ISSN 0001-0782, 1557-7317. doi: 10.1145/1327452.1327492.

Alhussein Fawzi, Matej Balog, Aja Huang, Thomas Hubert, Bernardino Romera-Paredes, Mohammadamin Barekatain, Alexander Novikov, Francisco J. R. Ruiz, Julian Schrittwieser, Grzegorz Swirszcz, David Silver, Demis Hassabis, and Pushmeet Kohli. Discovering faster matrix multiplication algorithms with reinforcement learning. *Nature*, 610(7930):47–53, October 2022. ISSN 1476-4687. doi: 10.1038/s41586-022-05172-4.

Intel. Get Outstanding Computational Performance without a Specialized Accelerator. *Intel Technology Brief*, July 2022.

Evgueny Khartchenko. Vectorization: A Key Tool To Improve Performance On Modern CPUs. https://www.intel.com/content/www/us/en/developer/articles/technical/vectorization-a-key-tool-to-improve-performance-on-modern-cpus.html, January 2018.

NVIDIA. NVIDIA's Next Generation CUDATM Compute Architecture. Technical report, 2012.

Volker Strassen. Gaussian elimination is not optimal. *Numerische Mathematik*, 13:354–356, 1969.

2

Basic Tensor Operations

Numerical libraries like NumPy [Harris et al., 2020], Tensorflow [Abadi et al., 2015], and PyTorch [Paszke et al., 2019] implement various types of vectorized operations to facilitate scientific computing. Throughout this book, we define the following set of terminologies: if a variable is called a

- **scalar**, then the variable has zero dimensions and represents a single numerical value.
- **array**, or **vector**, then the variable has one dimension and represents a list of values.
- **matrix**, then the variable has two dimensions and represents a typical two-dimensional matrix.
- **tensor**, then the variable is a generic collection of values of N-dimension(s), where $N \in \{0, 1, 2, \dots\}$ is a natural number. More often, I will use the term "tensor" to indicate the high-dimensional collection of values where $N > 2$.

In this chapter, we will examine various operations that are used on a single tensor or a pair of tensors.

2.1 Tensor Initializers

Tensor initializers provide quick and convenient ways to pre-allocate some spaces in memory to store values to be filled in later on. It can also be used for masking and indexing operations, which we will see in later chapters. To initialize a list of values, we would typically use a for loop like the following example:

```
# Initialize an array of zeros
value = []
for _ in range(1_000_000):
    value.append(0)
```

Vectorization: A Practical Guide to Efficient Implementations of Machine Learning Algorithms,
First Edition. Edward DongBo Cui.

which seems to be slow (on my M1 MacBook Pro)

```
32.3 ms ± 390 µs per loop (mean ± std. dev. of 7 runs, 10
 ↪ loops each)
```

Using list comprehension

```
[0 for _ in range(1_000_000)]
```

seems to improve the performance

```
17.3 ms ± 586 µs per loop (mean ± std. dev. of 7 runs, 100
 ↪ loops each)
```

We can also use a one-liner with Python's syntax candy operator `*` on the list

```
[0] * 1_000_000
```

which is much faster

```
2.69 ms ± 66.6 µs per loop (mean ± std. dev. of 7 runs, 100
 ↪ loops each)
```

Still, none of the above beats the performance of the NumPy function `np.zeros`

```
np.zeros(1_000_000)
```

which is the fastest

```
537 µs ± 39.8 µs per loop (mean ± std. dev. of 7 runs, 1,000
 ↪ loops each)
```

By using `np.zeros`, we can also easily create a multi-dimensional tensor (not just a 1-D vector), which then helps us perform additional downstream vectorized operations.

But why pre-allocate or initialize a vector at all? This again has something to do with performance. It can be said that extending an array on the fly is more costly than pre-allocating a portion of the memory for our computations. Let's show this using an example where we compute the square root of values between 0 and 1 million

```
import math
def sqrt_iter():
    values = []
    for ii in range(1_000_000):
        values.append(math.sqrt(ii))
    return values
```

Extending the array `values` while iterating over our examples took some time

```
87.5 ms ± 1.27 ms per loop (mean ± std. dev. of 7 runs, 10
 ↪  loops each)
```

Alternatively, we can pre-allocate an array of zeros to store the results

```
import math
def sqrt_iter_preallocate():
    values = [0] * 1_000_000
    for ii in range(1_000_000):
        values[ii] = math.sqrt(ii)
```

This reduced the runtime by about 10%

```
78.4 ms ± 828 µs per loop (mean ± std. dev. of 7 runs, 10
 ↪  loops each)
```

i Anecdote

I first started doing vectorized programming in MATLAB. Pre-allocating a vector before starting a for-loop is a classic lesson on MATLAB performance. Although in the above small Python example, we were only seeing about 10% improvement, in the case MATLAB, this can be a 10–50× performance gain! See for MATLAB' official documentation on this: https://www.mathworks.com/help/matlab/matlab_prog/preallocating-arrays.html. *Anecdotes* in real-world software development have also suggested that preallocation is an effective strategy for better performance when dealing with large arrays in many other languages such as JavaScript [Nenashev, 2024], R [Sellors, 2024], and C++ [Usi, 2023].

Table 2.1 List of tensor initializers in Python, NumPy, Tensorflow, and PyTorch.

Init Operation	Python	NumPy	Tensorflow	PyTorch
Zeros[a]	`[0] * n`	`np.zeros`	`tf.zeros`	`torch.zeros`
Ones[a]	`[1] * n`	`np.ones`	`tf.ones`	`torch.ones`
Empty[a]	`[[]] * n`	`np.empty`	`tf.experimental.numpy.empty`	`torch.empty`
Full[a]	`[value] * n`	`np.full`	`tf.fill`, `tf.experimental.numpy.full`[b]	`torch.full`
Random uniform[a]	`random.uniform`	`np.random.rand`	`tf.random.uniform`	`torch.rand`
Random normal[a]	`random.normal`	`np.random.randn`	`tf.random.normal`	`torch.randn`
Random int[a]	`random.randint`	`np.random.randint`	`tf.experimental.numpy. random.randint`[c]	`torch.randint`
Range	`range`	`np.arange`	`tf.range`	`torch.arange`[d]
Linspace	N/A, use `range`	`np.linspace`	`tf.linspace`	`torch.linspace`
Identity matrix	N/A	`np.eye`[d]	`tf.eye`	`torch.eye`
Diagonal matrix	N/A	`np.diag`	`tf.linalg.diag`	`torch.diag`

Note:

a) Many of these tensor initializers have corresponding `_like` methods where the shape of the initialized tensor can be inferred from the shape of another provided tensor. These methods are convenient when trying to initialize tensors with the same shape as the existing tensors. In other scenarios, where a tensor's shape may not be known until runtime, the `_like` methods allow the user to initialize the tensor dynamically by inferring its shape from the runtime result.
- `zeros`: `np.zeros_like`, `tf.zeros_like`, `torch.zeros_like`
- `ones`: `np.ones_like`, `tf.ones_like`, `torch.ones_like`
- `empty`: `np.empty_like`, `tf.experimental.numpy.empty_like`, `torch.empty_like`
- `full`: `np.full_like`, `tf.experimental.numpy.full_like`, `torch.full_like`
- `uniform`: `torch.rand_like`
- `normal`: `torch.randn_like`
- `randint`: `torch.randint_like`

b) Tensorflow implements an experimental NumPy module `tf.experimental.numpy`. Oftentimes, many NumPy operations equivalent or alternative to the Tensorflow operations (e.g. `full` vs. `fill`) can be found in this module. This module does not include all NumPy operations; some operations have limited capabilities compared to NumPy (most likely due to the need for efficient graph tracing [Abadi et al., 2023a] in Tensorflow).

c) Alternatively `tf.random.uniform(start, end, shape, dtype=tf.int32)` can also be used to draw integers given a range.

d) PyTorch also has the method `torch.range`, which is end inclusive. This means, `torch.range(1, 5) = [1, 2, 3, 4, 5]`, while all other range methods are end-exclusive, i.e. `range(1, 5) = [1, 2, 3, 4]`.

e) An alternative method to `np.eye` is `np.identity`. `np.identity` only returns matrices with the entries at the main diagonal to be 1s. In contrast, `np.eye` can return matrices with off-diagonal entries to be 1s. The corresponding methods in Tensorflow `tf.eye` and PyTorch `torch.eye` are actually much more similar to `np.identity` where the matrices returned only have the main diagonal entries set to 1s. To illustrate

```
import numpy as np
np.eye(2, dtype=int)
# array([[1, 0],
#        [0, 1]])
np.eye(3, k=1, dtype=int)
# array([[0, 1, 0],
#        [0, 0, 1],
#        [0, 0, 0]])
np.identity(3)
# array([[1, 0, 0],
#        [0, 1, 0],
#        [0, 0, 1]])
```

Table 2.1 provides a summary of a list of tensor initializers implemented by native Python, NumPy, Tensorflow, and PyTorch. Most of these initializers are used in similar ways, where a `shape` and optionally, a `dtype` (discussed in the following section) arguments are needed to construct the tensor. There are four basic categories of tensor initializers:

1. **Static value initializers**: Where all the entries of the initialized tensors are filled with a single static value. They include "zeros," "ones," "empty," and "full" initializers.
2. **Random value initializers**: Where the entries of the initialized tensors are filled with random values drawn from some distribution, e.g. random uniform, random normal, or random int.
3. **Range value initializers**: Where the entries of the initialized tensors are ordered and are within a predefined range of values. They include range and `linspace` initialization operations.
4. **Special matrix initializers**: Where they initialize special matrices such as diagonal or identity matrices of various sizes.

2.2 Data Type and Casting

In many areas of scientific computing and machine learning applications, we need to manipulate the underlying data stored in the tensor. The value of the data may be represented and interpreted differently on the machine. I will not go into detail on how a computer represents information at the fundamental level. Interested readers can pick up an introductory computer science book or a C [Prinz and Crawford, 2005] or Java programming book to gain more understanding of computational data types. Although Python is not a strongly type-checked language like C or Java, a good understanding of fundamental data types within the libraries we use for our tasks (which often interface with numerical libraries written in C) would help us call methods and functions that efficiently implement dedicated algorithms specifically optimized for certain data types. Table 2.2 provides a list of data types implemented by Python, NumPy, Tensorflow, and PyTorch.

Table 2.2 List of data types in Python, NumPy [Harris et al., 2023a], Tensorflow [Abadi et al., 2023b], and PyTorch [Paszke et al., 2023b].

dtype	Python	NumPy	Tensorflow	PyTorch
Boolean	`bool`	`np.bool_`	`tf.bool`	`torch.bool`
8-bit int[a),b)]	N/A	`np.int8`	`tf.int8`	`torch.int8`
16-bit int (short)[a)]	N/A	`np.int16` (`np.short`)	`tf.int16`	`torch.int16` (`torch.short`)
32-bit int (int)[a),b)]	`int`	`np.int32` (`np.intc`)	`tf.int32`	`torch.int32` (`torch.int`)
64-bit int (long)	N/A	`np.int64` (`np.int_`)	`tf.int64`	`torch.int64` (`torch.long`)
8-bit unsigned int[a),b)]	N/A	`np.uint8`	`tf.uint8`	`torch.uint8`
16-bit unsigned int[a)]	N/A	`np.uint16` (`np.ushort`)	`tf.uint16`	N/A
32-bit unsigned int	N/A	`np.uint32` (`np.uintc`)	`tf.uint32`	N/A

Notes:

a) Tensorflow implements a quantized version of these integer dtypes, which are used for model size compression.

- `tf.int8`: `tf.qint8`
- `tf.int16`: `tf.qint16`
- `tf.int32`: `tf.qint32`
- `tf.uint8`: `tf.quint8`
- `tf.uint16`: `tf.quint16`

b) PyTorch implements a quantized version of these integer dtypes, which are used for model size compression. It is interesting to note that PyTorch does not implement `torch.qint16` and `torch.quint16` like Tensorflow but implements an additional quantized 4-bit integer type (`torch.quint4x2`, which is an 8-bit integer)

- `torch.int8`: `torch.qint8`
- `torch.uint8`: `torch.quint8`
- `torch.int32`: `torch.qint32`
- _: `torch.quint4x2`

(continued)

Table 2.2 (Continued)

dtype	Python	NumPy	Tensorflow	PyTorch
64-bit unsigned int	N/A	`np.uint64` (`np.uint`)	`tf.uint64`	N/A
float16 (half)	N/A	`np.float16` (`np.half`)	`tf.float16`	`torch.float16`
float32 (single)	N/A	`np.float32` (`np.single`)	`tf.float32`	`torch.float32` (`torch.float`)
float64 (double)	`float`	`np.float64` (`np.double`)	`tf.float64`	`torch.float64` (`torch.double`)
complex32	N/A	N/A	N/A	`torch.complex32` (`torch.chalf`)
complex64	N/A	`np.complex64`	`tf.complex64`	`torch.complex64` (`torch.cfloat`)
complex128	`complex`	`np.complex128` (`np.complex_`)	`tf.complex128`	`torch.complex128` (`torch.cdouble`)
complex256	N/A	`np.complex128` (`np.longcomplex`)	N/A	N/A
bytes	`bytes`	`np.bytes_`	`tf.string`[c)]	N/A[d)]
str	`str`	`np.str_`	`tf.string`[c)]	N/A[d)]

c) Strings in Tensorflow will be converted into bytes. It is common when casting from Tensorflow back into NumPy arrays, we do `x.numpy().astype(str)` to cast an array of bytes into an array of strings.

d) Unlike NumPy and Tensorflow, PyTorch does not support string and bytes Tensors. Instead, it relies on NumPy and pure Python to handle string processing. In Chapter 6, we will introduce the topic of converting strings to numerical representations via indexing and embedding lookup.

2.2.1 Tips on Specifying the dtypes During Tensor Initialization

Most tensor initializers listed in the previous section should accept an argument dtype where the user can specify the dtypes of the tensors after construction. Using the zeros initializer as an example:

- For NumPy, x = np.zeros(5, dtype=np.float32) will initialize an array of np.float32 zeros.
- For Tensorflow, x = tf.zeros(5, dtype=tf.float32) will initialize an array of tf.float32 zeros.
- For PyTorch, x = torch.zeros(5, dtype=torch.float32) will initialize an array of torch.float32 zeros.
- Both Tensorflow and PyTorch dtypes are compatible with native Python dtypes. So it is also possible to specify the dtypes using native Python dtypes, although care needs to be taken to make sure desired precision is used. For example, x = tf.zeros(5, dtype=float) is equivalent to x = tf.zeros(5, dtype=tf.float32), which is equivalent to x = tf.zeros(5, dtype=np.float32) (yes, Tensorflow dtypes is also compatible with NumPy's dtypes in many cases), even though native Python's float represents 64-bit floats. However, in PyTorch, x = torch.zeros(5, dtype=float) is equivalent to x = torch.zeros(5, dtype=torch.float64) (using np.float64 will, however, throw an error).

2.2.2 Tips on Casting

Casting means to convert one data type to another data type.

- To cast between data types in NumPy, use x<np.ndarray>.astype(np.<dtype>). For example, x = np.array([1, 2, 3]).astype(np.float32) gives array([1., 2., 3.], dtype=float32)
- To cast between data types in TensorFlow, use tf.cast(<tf.Tensor>x, tf.<dtype>)
- To cast between data types in PyTorch, use x<torch.Tensor>.type(torch.<dtype>)

2.3 Mathematical Operations

Various scientific computing libraries (NumPy, MATLAB, Mathematica, etc.) are built specifically for the purpose of doing mathematical operations. Table 2.3 shows a list of typical mathematical operations that can be used on each element of the tensor in NumPy, Tensorflow, and PyTorch. Note that this table is not meant to be exhaustive of the library.

Table 2.3 List of common mathematical operations in Python, NumPy, Tensorflow, and PyTorch.

Math Operation	Python	NumPy	Tensorflow	PyTorch
Addition	+	+(np.add)	+(tf.add)	+(torch.add)
Subtraction	-	-(np.subtract)	-(tf.subtract)	-(torch.subtract)
Multiplication	*	*(np.multiply)	*(tf.multiply)	*(torch.multiply)
Division	/	/(np.divide)	/(tf.divide)	/(torch.divide)
Safe division	N/A	np.divide[a)]	tf.math.divide_no_nan	N/A[b)]
Integer division	//	np.floor_divide	tf.math.floordiv	torch.floor_divide
Reciprocal	N/A	np.reciprocal	tf.math.reciprocal	torch.reciprocal
Safe reciprocal	N/A	np.reciprocal[c)]	tf.math.reciprocal_no_nan	N/A[d)]
Remainder	%(math.remainder)	%(np.remainder, np.mod)	tf.math.floormod	torch.remainder
Power	**(math.pow)	np.power	tf.math.pow	torch.pow
Exponential	math.exp	np.exp	tf.exp	torch.exp
$\exp(x) - 1$[e)]	math.expm1	np.expm1	tf.math.expm1	torch.expm1
log base e	math.log	np.log	tf.math.log	torch.log
log base 2	math.log2	np.log2	N/A[f)]	torch.log2
log base 10	math.log10	np.log10	N/A[f)]	torch.log10
$\log(1+x)$[g)]	math.log1p	np.log1p	tf.math.log1p	torch.log1p
$\log(\exp(x_1) + \exp(x_2))$[h)]	N/A	np.logaddexp	tf.experimental.numpy. logaddexp	torch.logaddexp
$\log(\sum_i \exp(x_i))$[i)]	N/A	scipy.special.logsumexp	tf.math.reduce:logsumexp	torch.logsumexp

Notes:

a) NumPy does not implement a safe divide operation, but one can use `np.divide(a, b, out=np.zeros_like(b), where=b!=0)` to achieve the same effect as `tf.divide_no_nan`

b) PyTorch does not implement the safe divide operation, but one can use `torch.nan_to_num` to fill the NaNs or `infs` in the quotient to zeros. NumPy also has a similar function `np.nan_to_num`.

c) Like the `divide` method, NumPy does not implement the safe reciprocal operation, but one can use `np.reciprocal(x, out=np.zeros_like(x), where=x!=0)` to achieve the same effect.

d) Like the `divide method`, PyTorch also does not implement the safe reciprocal operation, but one can use `torch.nan_to_num` to fill the NaNs or `infs`.

e) `expm1(x)` is useful for computing the expression $e^x - 1$ more accurately comparing to `exp(x) - 1` when $x \to 0$. A well-known example is to compute annualized interest rate, $(1 + x)^n - 1$, which can be expressed as $\exp(n \log(1 + x)) - 1$, and implemented as `exp1m(n * log1p(x))` (see notes on `log1p` below).

f) Tensorflow does not implement log bases other than e, but one can use `tf.math.log(x) / tf.math.log(base)` to convert bases.

g) `log1p(x)` is useful for computing the expression $\log(1 + x)$ more accurately comparing to `log(1+x)` when $x \to 0$. Well-known examples include computing annualized interest rates (see notes on `expm1`) and implementation of inverse hyperbolic functions. For example, $\text{arctanh}(x) = \frac{1}{2}\log(\frac{1+x}{1-x})$ can be implemented as `0.5 * (log1p(x) - log1p(-x))`.

h) `logaddexp` is useful for computing the expression $\log(\exp(x_1) + \exp(x_2))$ more accurately when x is extremely large or extremely small. This is a special case of `logsumexp` (see notes below).

i) `logsumexp` is a more general implementation of the above `logaddexp`, which can operate on more than two numbers, computing the expressions $\log\left(\sum_i \exp(x_i)\right)$. Notice that the summation expression is the normalization term for the softmax loss function in maximum-likelihood estimation $p_i = \text{softmax}(x_i) = \frac{\exp(x_i)}{\sum_j \exp(x_j)}$, and the log probabilities then have the form $\log(p_i) = x_i - \log\left(\sum_j \exp(x_j)\right)$. NumPy does not have a native implementation of this function, but it can be found in Scipy's `scipy.special.logsumexp`, using the so-called log-sum-exp trick [Wikipedia, 2023]:

$$\begin{aligned} y &= \log\left(\sum_i \exp(x_i)\right) \\ &= \log\left(\sum_i \exp(x_i - x_{\max}) \cdot \exp(x_{\max})\right) \\ &= \log\left(\sum_i \exp(x_i - x_{\max})\right) + \log\exp(x_{\max}) \\ &= \log\left(\sum_i \exp(x_i - x_{\max})\right) + x_{\max} \end{aligned}$$

where $x_{\max} = \max(x_0, x_1, \ldots, x_n)$. A naive implementation is

```
import numpy as np
def logsumexp(x):
x_max = np.max(x)
    return np.log(np.exp(x - x_max).sum()) + x_max
```

(continued)

Table 2.3 (Continued)

Math Operation	Python	NumPy	Tensorflow	PyTorch
Square root	`math.sqrt`	`np.sqrt`	`tf.sqrt`	`torch.sqrt`
Square[j)]	N/A	`np.square`	`tf.square`	`torch.square`
Absolute value	`abs`	`np.abs`	`tf.abs`	`torch.abs`
sine	`math.sin`	`np.sin`	`tf.sin`	`torch.sin`
cosine	`math.cos`	`np.cos`	`tf.cos`	`torch.cos`
tan	`math.tan`	`np.tan`	`tf.tan`	`torch.tan`
tanh (hyperbolic tan)	`math.tanh`	`np.tanh`	`tf.tanh`	`torch.tanh`
Rounding	`round`	`np.round`	`tf.round`	`torch.round`
Rounding up	`math.ceil`	`np.ceil`	`tf.ceil`	`torch.ceil`
Rounding down	`math.floor`	`np.floor`	`tf.floor`	`torch.floor`
Real part of complex	`(complex).real`[k)]	`np.real`	`tf.real`	`torch.real`
Imaginary part of complex	`(complex).img`[k)]	`np.img`	`tf.imag`	`torch.img`
Complex conjugate	`(complex).conjugate`	`np.conj` (`np.conjugate`)	`tf.math.conj`	`torch.conj`

j) You may wonder why we even need to have a specific function fcr computing the square of a number, instead of simply using `pow` or `**`. This is all about performance. It turns out that it is faster to compute `x * x` than `pow(x, 2)`. Having a specialized square function is convenient for implementing something like `this_is_a_variable_with_super_long_name * this_is_a_variable_with_super_long_name`, e.g. See this issue [Paszke et al., 2023a] in PyTorch's repository.

k) Python's complex numbers can be constructed as `<real> + <img>j`, e.g. `1 + 2j` (notice no spaces between `2` and `j`). This is why I typically do not use `j` as indices when writing a for-loop since it may be confused with Python's reserved variable for the imaginary number.

> **Note**
>
> You may notice that, for Tensorflow operations, some functions are under the `tf.` module and some are under the `tf.math.` module. This is because, although all are implemented under the `tf.math` module, some frequently used functions are aliased under the main `tf` module for ease of access. You can verify this by printing out the module path. For example, `tf.math.abs` and `tf.abs` both refer to the operation `<function tensorflow.python.ops.math_ops.abs(x, name=None)>`.

PyTorch has many of these operations implemented as methods on its `torch.Tensor` class. Therefore, instead of using the function from the main module, one can call the corresponding methods on the array/tensor instances. For example, instead of `torch.abs(x)`, we can do `x.abs()`. This makes PyTorch more object-oriented.

2.4 Reduction Operations

Reduction (or reduce) operations are mathematical operations that summarize a set of values into a single value. It "reduces" the size of the tensor, from higher-dimensional to lower-dimensional, or single scalars. Table 2.4 lists examples of reduction operations.

NumPy and PyTorch also have many of these reduction operations implemented as methods on their respective `np.ndarray` and `torch.Tensor` classes. Therefore, instead of using the function from the main module, one can call the corresponding methods on the array/tensor instances. For example, instead of `np.sum(x)` or `torch.sum(x)`, we can do `x.sum()`. Tensorflow is more function-oriented and does not have these reduction operations implemented on its `tf.Tensor` class, likely because it needs to support the construction of computational graphs by tracing through function calls.

2.5 Value Comparison Operations

Value comparison operations compare the values stored within the tensor. Table 2.5 provides a short list of the operations.

Table 2.4 List of tensor reduction operations in Python, NumPy, Tensorflow, and PyTorch.

Reduce Operation	Python	NumPy	Tensorflow	PyTorch
Sum[a]	sum	np.sum	tf.reduce:sum	torch.sum
Mean[a]	N/A	np.mean	tf.reduce:mean	torch.mean
Variance[a]	N/A	np.var	tf.math.reduce:variance	torch.var
Standard deviation[a]	N/A	np.std	tf.math.reduce:std	torch.std
Median[a]	N/A	np.median	N/A[b]	torch.median
Product[a]	N/A	np.prod	tf.reduce:prod	torch.prod
Maximum[a]	max	np.max	tf.reduce:max	torch.max
Minimum[a]	min	np.min	tf.reduce:min	torch.min
Any true	any	np.any	tf.reduce:any	torch.any
All true	all	np.all	tf.reduce:all	torch.all
Tensor norm	N/A	np.linalg.norm	tf.norm	torch.linalg.norm

Notes:
a) If the tensor has any NaN ("Not a Number") value, then the reduce operations will always return NaN. Fortunately, many reduce operations have corresponding versions that can handle missing values filled with NaNs:
   1. sum: np.nansum, tf.experimental.numpy.nansum, torch.nansum
   2. mean: np.nanmean, tf.experimental.numpy.nanmean, torch.nanmean
   3. variance: np.nanvar
   4. std: np.nanstd
   5. median: np.nanmedian, torch.nanmedian
   6. prod: np.nanprod, tf.experimental.numpy.nanprod
   7. max: np.nanmax
   8. min: np.nanmin
b) Technically, one can also use a Tensorflow derivative library, tensorflow_probability or tfp to obtain the median, i.e. tfp.stats.percentile(x, 50, interpolation='midpoint'), where x is an array of float numbers.

Table 2.5 List of value comparison and thresholding operations in Python, NumPy, Tensorflow, and PyTorch.

Operation	Python	NumPy	Tensorflow	PyTorch
max of 2	max	np.maximum	tf.minimum	torch.maximum
min of 2	min	np.minimum	tf.maximum	torch.minimum
Clip values	N/A	np.clip	tf.clip_by_value	torch.clip, torch.clamp

2.6 Logical Operations

Logical operations are also known as boolean operations, where operations like AND, OR, NOT, and XOR are computed. Table 2.6 lists these four major boolean operations.

Table 2.6 List of logical or boolean tensor operations in Python, NumPy, Tensorflow, and PyTorch.

Operation	Python	NumPy	Tensorflow	PyTorch
Logical not	`not`, `~`	`~`[a)], `np.logical_not`	`tf.logical_not`	`torch.logical_not`
Logical and	`and`, `&`	`&`[a)], `np.logical_and`	`tf.logical_and`	`torch.logical_and`
logical or	`or`, `|`	`|`[a)], `np.logical_or`	`tf.logical_or`	`torch.logical_or`
logical xor	`^`	`^`[a)], `np.logical_xor`	`tf.math.logical_xor`	`torch.logical_xor`

Notes:

a) The symbols `~`, `&`, `|`, and `^` represent bit-wise operations, "NOT," "AND," "OR," and "XOR," respectively. When using these symbols with pure Python's boolean variables or NumPy's boolean arrays, the effect is the same as the function operations. However, if we are applying it on integers or floats, e.g. `np.array(1) & np.array(2) = 0` is not the same as `np.logical_and(1, 2) = True`.

Table 2.7 List of adjacent element cumulative sum and discrete differences tensor operations in Python, NumPy, Tensorflow, and PyTorch.

Operation	Python	NumPy	Tensorflow	PyTorch
Cumulative sum[a)]	N/A	`np.cumsum`	`tf.cumsum`	`torch.cumsum`
Cumulative product[b)]	N/A	`np.cumprod`	`tf.math.cumprod`	`torch.cumprod`
Adjacent diff[c)]	N/A	`np.diff`	`tf.experimental.numpy.diff`	`torch.diff`

Notes:

a) Suppose we start with a vector $x = [1, 2, 3, 4]$. The cumulative sum of the vector $y_k = \sum_{i=1}^{k} x_i, \forall k = 1, 2, \dots, n$ results $y = [1, 1+2, 1+2+3, 1+2+3+4] = [1, 3, 6, 10]$.

b) Suppose we have a vector $x = [1, 2, 3, 4]$. Then, the cumulative product of the vector $y_k = \prod_{i=1}^{k} x_i, \forall k = 1, 2, \dots, n$ results $y = [1, 1 \times 2, 1 \times 2 \times 3, 1 \times 2 \times 3 \times 4] = [1, 2, 6, 24]$.

c) Suppose we have a vector $x = [1, 3, 6, 7]$. Then, the adjacent difference of the vector $y_k = x_k - x_{k-1}, \forall k = 2, 3, \dots, n$ results $y = [3-1, 6-3, 7-6] = [2, 3, 1]$. This is equivalent of doing `x[1:] - x[:-1]`. Notice that the resulting array y has a length of 1 less than the length of x.

2.7 Ordered Array-Adjacent Element Operations

Array-adjacent element operations perform computations between consecutive elements. See Table 2.7 for examples of cumulative and difference operations.

2.8 Array Reversing

Array reversing operations reverse the array or tensors along a certain dimension. Table 2.8 provides a list of array reversing operations.

Table 2.8 List of array/tensor reversing operations in Python, NumPy, Tensorflow, and PyTorch.

Operations	Python	NumPy	Tensorflow	PyTorch
Reverse	`list[::-1]`	`np.flip`	`tf.reverse`	`torch.flip`
Reverse left-right (axis=0)	N/A	`np.fliplr`	`tf.experimental.numpy.fliplr`	`torch.fliplr`
Reverse up-down (axis=1)	N/A	`np.flipud`	`tf.experimental.numpy.flipud`	`torch.flipud`
Reverse padded sequence[a]	N/A	N/A	`tf.reverse_sequence`	N/A

Notes:

a) The padded sequence reverse in Tensorflow `tf.reverse_sequence` is not the same as simple `tf.reverse`. It will only reverse part identified as part of the sequence and leave the padding on the right as is. For example, suppose we have a zero-padded sequence `s=[1, 2, 3, 4, 0, 0, 0]`. `tf.reverse` will reverse the entire sequence including the padding, as `[0, 0, 0, 4, 3, 2, 1]`, whereas `tf.reverse_sequence` can reverse the sequence at the non-padding entries as `[4, 3, 2, 1, 0, 0, 0]`, when given expected sequence length. At the time of writing this book, Tensorflow is the only library that implements this padded sequence reverse method. In Chapter 10, we will see how to implement this method using NumPy as well.

2.9 Concatenation, Stacking, and Splitting

Tensors can be constructed by combining with other tensors or by splitting larger tensors into smaller tensors. Table 2.9 lists functions to perform tensor combination and splitting operations along various dimensions.

2.10 Reshaping

Reshaping operations keep the underlying data of the tensor identical to the inputs but represent the tensor in a different format. Being able to trace the evolution of the tensor shapes through a computational graph is probably one of the most important skills to master before being able to successfully implement a deep learning algorithm. Oftentimes, the outputs of a previous operation may not be directly usable in the subsequent operations. For example, a feed-forward layer may output an array of 9 elements, but the subsequent operation requires us to take an average of 3 groups of 3 elements. One way we can implement this is by reshaping the 1D array of 9 elements into a two-dimensional, 3×3 matrix,

Table 2.9 List of adjacent element cumulative sum and discrete differences tensor operations in Python, NumPy, Tensorflow, and PyTorch.

Operation	Python	NumPy	Tensorflow	PyTorch
concatenate[a]	`+(list.__add__)`	`np.concatenate`	`tf.concat`	`torch.cat`
stacking[a]	N/A	`np.stack`	`tf.stack`	`torch.stack`
row stacking[b]	N/A	`np.vstack`, `np.row_stack`	N/A	`torch.vstack`, `torch.row_stack`
column stacking[b]	N/A	`np.hstack`, `np.column_stack`	N/A	`torch.hstack`, `torch.column_stack`
Depth stacking[b]	N/A	`np.dstack`	N/A	`torch.dstack`

Notes:

a) Concatenation is sometimes similar to stacking. However, concatenation will always combine the tensors along existing axes and will throw an error if the axis does not exist. In contrast, stacking will combine the tensors along either an existing or a new axis and will change the dimensions of the tensors sometimes. When operating on 1D vectors, the horizontal stack operation (`hstack`) is identical to the concatenation (`concatenate`). Figure 2.1 illustrates the concatenation and stacking operations on 1D vectors. When operating on higher dimensions matrices or tensors, stacking can act similarly to concatenation. For example, `vstack` is equivalent to concatenating along the 1st axis (`concatenate(..., axis=0)`) while `hstack` is equivalent to concatenating along the second axis (`concatenate(..., axis=1)`). `dstack` will stack along the third dimension (and will create it if not already existing). Figure 2.2 illustrates the concatenation and stacking operations when applied to a 2D matrix.

b) `row_stack` or `vstack` always stacks the tensors along the first dimension, equivalent to `stack(..., axis=0)`; `column_stack` or `hstack` always stacks the tensors along the second dimension, equivalent to `stack(..., axis=1)`; `dstack` always stacks the tensors along the third dimension, equivalent to `stack(..., axis=2)`.

(continued)

Table 2.9 (Continued)

Operation	Python	NumPy	Tensorflow	PyTorch
Append[c)]	`<list>.append`	`np.append`	`tf.experimental.numpy.append`	N/A

c) The append operation is similar to concatenation. When `axis=None`, then the inputs are first flattened before being concatenated.

```
# one dimensional vectors
a = np.array([1, 2, 3])
b = np.array([4, 5, 6])
c = np.append(a, b, axis=None)
# results
# array([1, 2, 3, 4, 5, 6])
```

```
# two dimensional matrices
x = np.array([[1, 2, 3], [4, 5, 6]])
y = np.array([[-1, -2, -3], [-4, -5, -6]])
z = np.append(x, y, axis=None)
# results
# array([ 1,  2,  3,  4,  5,  6, -1, -2, -3, -4, -5, -6])
```

However, if `axis` is specified (e.g. `axis=0`), then this function may act as `concatenation`.

```
# one dimensional vectors
a = np.array([1, 2, 3])
b = np.array([4, 5, 6])
# specifying axis=1 will throw AxisError:
# axis 1 is out of bounds for an array of dimension 1
c = np.append(a, b, axis=0)
# results
# array([1, 2, 3, 4, 5, 6])
```

```
# two dimensional matrices
x = np.array([[1, 2, 3], [4, 5, 6]])
y = np.array([[-1, -2, -3], [-4, -5, -6]])
# concatenate along axis=0
z = np.append(x, y, axis=0)
# results
# array([[ 1,  2,  3],
#        [ 4,  5,  6],
#        [-1, -2, -3],
#        [-4, -5, -6]])
# concatenate along axis=1
z = np.append(x, y, axis=1)
# results
# array([[ 1,  2,  3, -1, -2, -3],
#        [ 4,  5,  6, -4, -5, -6]])
```

Table 2.9 (Continued)

Operation	Python	NumPy	Tensorflow	PyTorch
Split[d)]	N/A (use slicing)	`np.split`, `np.array_split`[e)]	`tf.split`[e)]	`torch.split`
Row split[f)]	N/A	`np.hsplit`	N/A	`torch.hsplit`
Column split[f)]	N/A	`np.vsplit`	N/A	`torch.vsplit`
Depth split[f)]	N/A	`np.dsplit`	N/A	`torch.dsplit`

d) Splitting operation is the reverse of stacking. Splitting will always keep the resulting tensors the same number of dimensions as the input tensor. Figure 2.3 illustrates examples of split operations.

e) When specifying the second argument of the split methods (e.g. `num_or_size_splits`) as integers, `np.split` and the subsequent alias methods `np.vsplit, np.hsplit, np.dsplit`, as well the corresponding Tensorflow method `tf.split`, will divide the tensor into equal sized pieces along a certain dimension; they will therefore throw an error if the tensor is not evenly divisible. `np.array_split` is more general and is equivalent to `torch.split`, which both allow a tensor to be divided into unequal pieces.

f) These additional split operations can also be achieved by simply using `split` (which may be why Tensorflow does not implement them natively).

- `vsplit` always splits the tensors along the first dimension, equivalent to `split(..., axis=0)`, and can be used as the reverse of `vstack`.
- `hsplit` always splits the tensors along the second dimension, equivalent to `split(..., axis=1)`, and can be used as the reverse of `hstack`.
- `dsplit` always splits the tensors along the third dimension, equivalent to `split(..., axis=2)`, and can be used as the reverse of `dstack`.

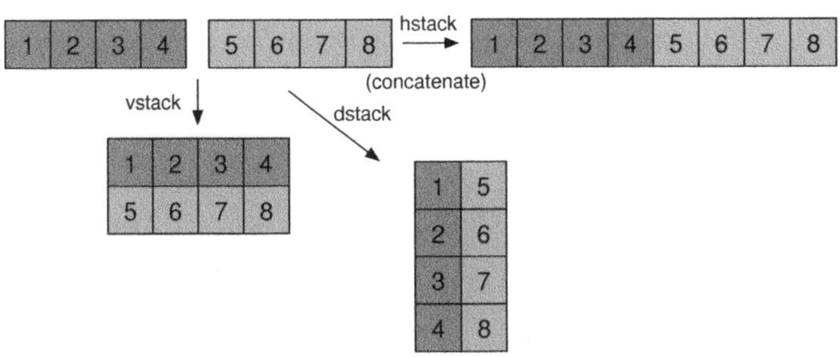

Figure 2.1 Concatenation and stacking in 1D vectors.

Figure 2.2 Concatenation and stacking in 2D matrices.

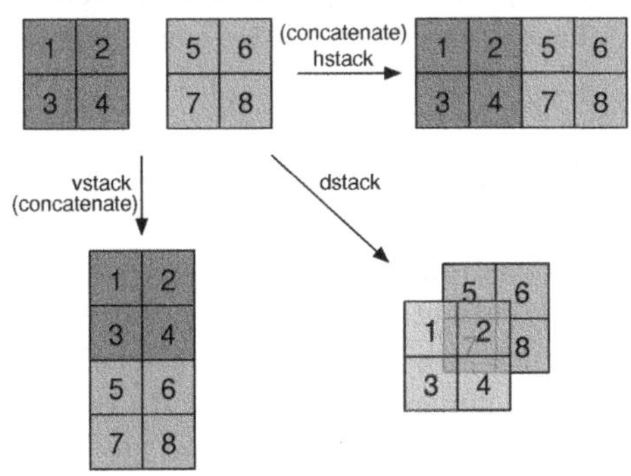

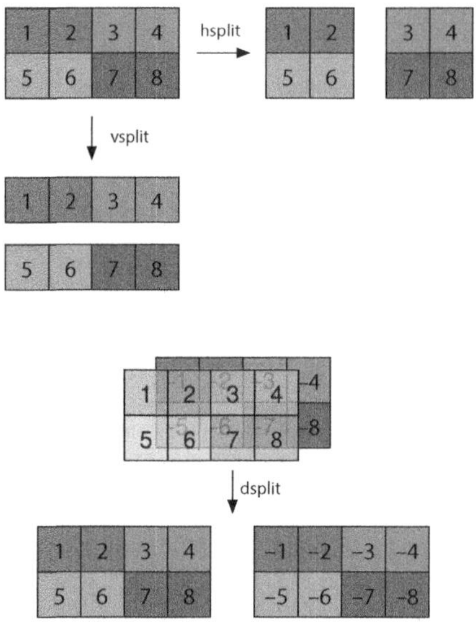

Figure 2.3 Splitting operations.

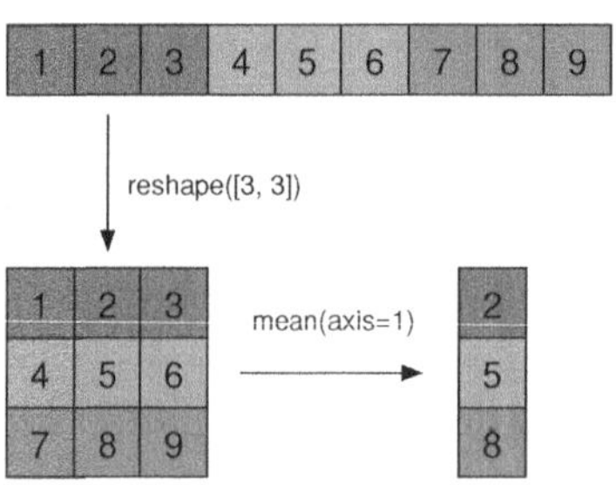

Figure 2.4 Reshaping a vector to a 3 × 3 matrix before taking averages.

then taking the mean by each row (see Figure 2.4). Table 2.10 provides a list of reshaping operations that attempt to achieve similar purposes of making inputs and outputs of the operations compatible with each other.

Table 2.10 provides a list of commonly used reshaping operations.

2.11 Broadcasting

Broadcasting is an element-wise operation on a pair of tensors of different dimensions [Harris et al., 2023c]. The tensor of the lower dimension is as if repeated many times to match the size of the higher dimensional tensor. For a simple example, a scalar (0-dimensional tensor) and a vector (1D tensor) can be added together via broadcasting, as shown in Figure 2.6.

Table 2.10 List of tensor reshaping operations in Python, NumPy, Tensorflow, and PyTorch.

Operation	Python	NumPy	Tensorflow	PyTorch
Reshape[a)]	N/A	`np.reshape`	`tf.reshape`	`torch.reshape`, `torch.Tensor.view`
Add singleton dim[b)]	N/A	`np.expand_dims`	`tf.expand_dims`	`torch.unsqueeze`
Remove singleton dim[b)]	N/A	`np.squeeze`	`tf.squeeze`	`torch.squeeze`

Notes:

a) There are two important concepts when working with reshaping operations.

- **Views**: An important concept in NumPy's reshape is the concept of "views." Instead of making a copy of the array with the new shape, NumPy can figure out a smart way to map the elements from the original array to the new array via "views" [Harris et al., 2023b]. This reduces the memory trace of the program by storing the data only once. Well-known operations that return views instead of copies include `np.reshape` and `np.transpose` (as well as slicing operations which we will introduce in Chapter 3). Tensorflow does not specifically mention the concept of view, but the `tf.reshape` operation also reuses the underlying data buffer rather than making a copy of the data according to their documentation [Abadi et al., 2023c]. PyTorch also has similar concepts about views and copies. Therefore, when appropriate, it is more preferred to use `Tensor.view(shape)` method to reshape the tensor rather than using `torch.reshape` (in the latter case, a copy of the tensor with the new shape is returned, contrasting `np.reshape`).
- **Order**: Another important concept, `order`, is related to how the underlying tensor data is represented in the memory (RAM). There are typically two conventions in data ordering in memory: row-major (`order="C"` as in the C language) or column-major (`order="F"` for Fortran). Consider, for a 2D matrix, the row-major representation will store elements along the rows of the matrix next to each other, while the column-major representation will store elements along the columns of the matrix next to each other (see Figure 2.5); typically, RAM does not have the concept of shape when it stores data. This matters for the sake of computational efficiency. Suppose we would like to sum together all the elements in a single row. It would be faster to look up the values needed to be combined by reading contiguous elements along the registers; in contrast, it would be slow to sum these numbers together if it needs to constantly skip elements and read more elements along the memory register until all needed elements are gathered. More concretely, in the example provided in Figure 2.5, it would be faster to compute the sum of the rows in a row-major (C) representation of the matrix (i.e. summing together, `[1, 2, 3]`, `[4, 5, 6]`, `[7, 8, 9]`) than summing together the rows in a column-major (F) representation of the matrix (i.e. summing together `[1, 4, 7]`, `[2, 5, 8]`, `[3, 6, 9]`).

b) `squeeze` and `expand_dims` (a.k.a `unsqueeze`) are a pair of reciprocal operations. `squeeze` removes the singleton dimension (dimension or axis of size 1), e.g. $(3, 1, 2) \xrightarrow{\text{squeeze}} (3, 2)$. In contrast, `expand_dimsor` `unsqueeze` adds a singleton dimension, e.g. $(3, 2) \xrightarrow{\text{unsqueeze(dim=1)}} (3, 1, 2)$. Note that NumPy and Tensorflow call the dimensions `axis` (or `axes` when used plurally), while PyTorch calls the dimensions `dim` (or `dims` when used plurally). Using slicing notations, we can also expand the dimensions using `None` or `np.newaxis`. For example, a tensor of shape `(3, 2)` to be converted to shape `(3, 1, 2)` can be implemented in two ways:

```
import numpy as np
X = np.random.rand(3, 2)
# using expand dims
y1 = np.expand_dims(X, axis=1)
# using None slicing
y2 = X[:, None, :]
# using np.newaxis
y3 = X[:, np.newaxis, :]
```

Similarly, squeezing a dimension can also be implemented by slicing

```
import numpy as np
X = np.random.rand(3, 1, 2)
# using squeeze
y1 = np.squeeze(X, axis=1)
# using slicing
y2 = X[:, 0, :]
```

We will discuss more on slicing operations in Chapter 3. Note that slicing is less preferred in Tensorflow implementations than the `squeeze` and `expand_dims` operations, as it sometimes may not be compatible with the graph execution.

(continued)

Table 2.10 (Continued)

Operation	Python	NumPy	Tensorflow	PyTorch
Ensure vector shape[c]	N/A	`np.atleast_1d`	`tf.experimental.numpy.atleast_1d`	`torch.atleast_1d`
Ensure matrix shape[c]	N/A	`np.atleast_2d`	`tf.experimental.numpy.atleast_2d`	`torch.atleast_2d`
Ensure 3D tensor shape[c]	N/A	`np.atleast_3d`	`tf.experimental.numpy.atleast_3d`	`torch.atleast_3d`
Transpose[d]	N/A	`np.transpose`	`tf.transpose`	`torch.transpose`
Swap axes[d]	N/A	`np.swapaxes`	`tf.transpose`	`torch.transpose`, `torch.swapaxes`, `torch.swapdims`

c) The `atleast_nd` functions help ensure the inputs are of the expected shape. These operations are often helpful when we need to perform broadcasting operations (i.e. element-wise operations between lower and higher dimensional tensors), which we will discuss in the upcoming section. They are somewhat equivalent to and are often used as shortcuts for the `expand_dims` operations. Suppose we would like a scalar value to be at least three-dimensional. Then, we can either implement using `expand_dims` as

```
import numpy as np
x = np.array(6)
y = np.expand_dims(x, axis=0) # np.array([6])
y = np.expand_dims(y, axis=0) # np.array([[6]])
y = np.expand_dims(y, axis=0) # np.array([[[6]]])
```

Or simply as

```
import numpy as np
x = np.array(6)
y = np.atleast_3d(x) # np.array([[[6]]])
```

One nice feature about these functions is that we do not need to know beforehand how many `expand_dims` operations we need to make before reaching the desired number of dimensions. So in the above example, by using `np.atleast_3d`, I can also let the input `x = np.array([6])` without modifying any subsequent code to ensure the shape of y to be always at least three-dimensional, which gives the input shape a lot of flexibility. On the other hand, the name `atleast` suggests that, when the input x is of a higher dimension, all the dimensions will be kept as is:

```
import numpy as np
x = np.array([[[[[6]]]]]) # 5-dimensional
y = np.atleast_3d(x) # still 5-dimensional
```

d) A matrix transpose operation can be seen as swapping the x and y coordinates. For example, the transpose of a matrix of shape (3, 4) is a matrix of shape (4, 3). It can be deduced that for higher dimensional tensors, we can also transpose or swap the axes to facilitate different subsequent operations. NumPy implements several functions for changing the order of the dimensions of the multi-dimensional tensor. By contrast, Tensorflow implements only a single but more general transpose operation that

Table 2.10 (Continued)

Operation	Python	NumPy	Tensorflow	PyTorch
Flatten to array[e)]	N/A	`np.ndarray.flatten`, `np.ravel`	`tf.experimental.numpy.flatten`	`torch.flatten`

covers a variety of use cases. This is a common operation in many machine learning applications such as convolutional neural networks. In image classification using a convolutional neural network, it is typical to represent the input images either as `(batch_size, height, width, channels)` (Tensorflow convention) or `(batch_size, channels, height, width)`. Notice that the two conventions represent a set of images, but the dimensions of the tensors are not in the same order. To convert from the Tensorflow convention to the PyTorch convention, we can leverage the transpose operations.

```
import numpy as np
image_tf = np.random.rand(8, 32, 15, 3)
image_torch = np.transpose(image_tf, axes=[0, 3, 1, 2])
```

Though it takes two lines, we can also use `np.swapaxes`

```
image_torch = np.swapaxes(image_tf, 1, 3) # (8, 3, 15, 32)
image_torch = np.swapaxes(image_torch, 2, 3) # (8, 3, 32, 15)
```

Note PyTorch `torch.Tensor` has `torch.Tensor.t()` method for matrix transpose (<=2 dimensions). NumPy's `np.ndarray` has `np.ndarray.T` for tensor transpose, which essentially reverses the ordering of the dimensions. This probably corresponds to the typical mathematical notation for transpose, i.e. $\cdot^T$. The transpose of a 1D array is still a 1D array, as there are no additional dimensions to swap with.

e) NumPy's `np.ndarray.flatten` method will always return a copy of the data. To return a view (if possible), use `np.ravel` instead. PyTorch's `torch.flatten` is more similar to `np.ravel` which will return a view if possible. Tensorflow's `tf.experimental.numpy.flatten` method will also return a view if possible as its underlying implementation uses `tf.reshape`. Note that it is also important to pay attention to the concept of `order` when flattening arrays.

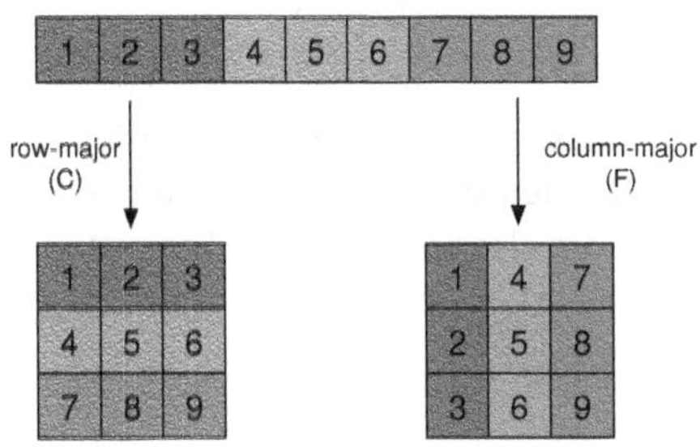

Figure 2.5 Concept of order in reshaping.

Figure 2.6 Scalar to vector broadcasting.

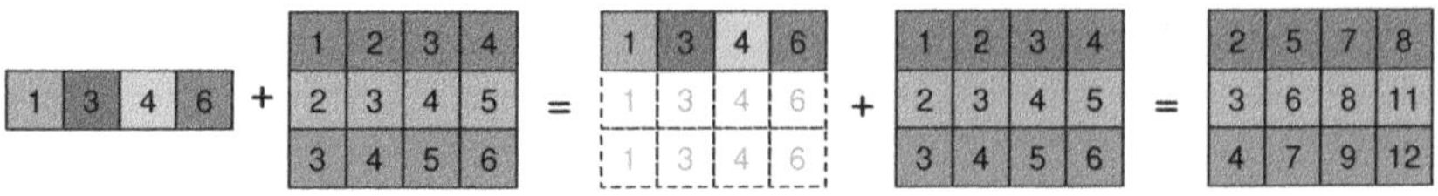

Figure 2.7 Vector to matrix broadcasting.

Note that the mathematical operation between a scalar and an array makes no sense,

$$1 + \begin{bmatrix} 2 & 4 & 6 \end{bmatrix}$$

However, after broadcasting, the equivalent mathematical operation becomes

$$\begin{bmatrix} 1 & 1 & 1 \end{bmatrix} \oplus \begin{bmatrix} 2 & 4 & 6 \end{bmatrix} = \begin{bmatrix} 3 & 5 & 7 \end{bmatrix}$$

where $\oplus$ denotes element-wise addition operation. Similarly, I will use

- $\ominus$ to denote element-wise subtraction
- $\odot$ to denote element-wise multiplication
- $\oslash$ to denote element-wise division

Likewise, a vector can also be broadcasted to a matrix (Figure 2.7). The operation between a vector and a matrix also makes no mathematical sense,

$$\begin{bmatrix} 1 & 3 & 4 & 6 \end{bmatrix} + \begin{bmatrix} 1 & 2 & 3 & 4 \\ 2 & 3 & 4 & 5 \\ 3 & 4 & 5 & 6 \end{bmatrix}$$

The equivalent broadcasting operations become

$$\begin{bmatrix} 1 & 3 & 4 & 6 \\ 1 & 3 & 4 & 6 \\ 1 & 3 & 4 & 6 \end{bmatrix} \oplus \begin{bmatrix} 1 & 2 & 3 & 4 \\ 2 & 3 & 4 & 5 \\ 3 & 4 & 5 & 6 \end{bmatrix} = \begin{bmatrix} 2 & 5 & 7 & 8 \\ 3 & 6 & 8 & 11 \\ 4 & 7 & 9 & 12 \end{bmatrix}$$

A tensor with singleton dimensions (`.shape[axis]=1`) is broadcastable to another tensor with non-singleton dimensions, as long as the rest of the dimensions are also broadcastable. For example, broadcasting between a column vector of shape $(1, N)$ and a matrix of size (M, N) (Figure 2.8),

$$\begin{bmatrix} 1 \\ 3 \\ 6 \end{bmatrix} + \begin{bmatrix} 1 & 2 & 3 & 4 \\ 2 & 3 & 4 & 5 \\ 3 & 4 & 5 & 6 \end{bmatrix}$$

has the equivalent broadcasting operations

$$\begin{bmatrix} 1 & 1 & 1 & 1 \\ 3 & 3 & 3 & 3 \\ 6 & 6 & 6 & 6 \end{bmatrix} \oplus \begin{bmatrix} 1 & 2 & 3 & 4 \\ 2 & 3 & 4 & 5 \\ 3 & 4 & 5 & 6 \end{bmatrix} = \begin{bmatrix} 2 & 3 & 4 & 5 \\ 5 & 6 & 7 & 8 \\ 9 & 10 & 11 & 12 \end{bmatrix}$$

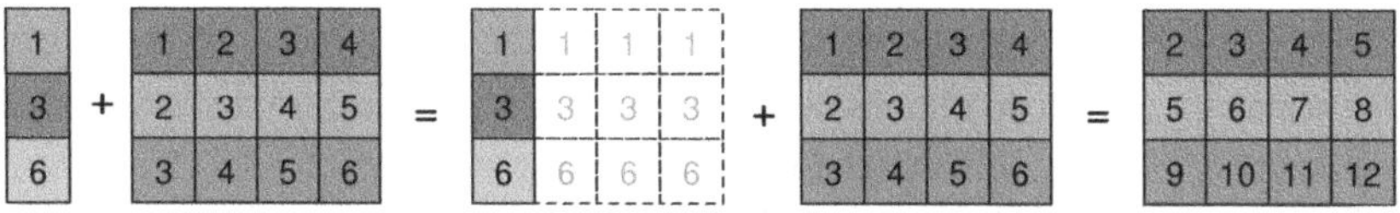

Figure 2.8 Broadcasting between tensor with singleton dimension and non-singleton dimensions.

In general, a scalar (`ndim=0`) can always be broadcasted to any higher-dimensional tensors. A lower dimensional tensor (`ndims > 0`) can be broadcasted to a higher dimension tensor as long as the size of some or all trailing (right-aligned) axes of the lower dimensional tensors is either a singleton (i.e. `.shape[axis]=1`) or match the size of the higher dimensional tensors at the corresponding trailing axes. To define this more concretely, suppose the lower dimensional tensor X has shape $(a_1, a_2, a_3, \dots, a_n)$, $a_i \neq 0, \forall i$ and the higher dimensional tensor Y has shape $(b_1, b_2, \dots, b_m)$, $b_j \neq 0, \forall j$, where $n \leq m$, then for X to be broadcasted to Y, we necessarily need to have

$$\forall k \begin{cases} a_n \in \{b_m, 1\}, \\ a_{n-1} \in \{b_{m-1}, 1\}, \\ \quad \dots \\ a_{n-k} \in \{b_{m-k}, 1\}, \\ \quad \dots \\ a_1 \in \{b_{m-n+1}, 1\} \end{cases}$$

The broadcasted operation will result in a tensor Z of shape $(c_1, c_2, \dots, c_m)$ that takes the larger size of either X or Y, and the number of dimensions will be `Y.ndim` which is larger than `X.ndim`. That is,

$$\forall k, p \begin{cases} c_m &= \max(a_n, b_m) \\ c_{m-1} &= \max(a_{n-1}, b_{m-1}) \\ &\dots \\ c_{n-k} &= \max(a_{n-k}, b_{m-k}) \\ &\dots \\ c_{m-n+1} &= \max(a_1, b_{m-n+1}) \\ c_{m-n} &= b_{m-n} \\ c_{m-n-1} &= b_{m-n-1} \\ &\dots \\ c_{m-n-p} &= b_{m-n-p} \\ &\dots \\ c_1 &= b_1 \end{cases}$$

For examples,

- A scalar, e.g. 5 with no dimensions (`ndim=0`), can be broadcasted to a tensor of shape `(3, 6, 5, 7)`, which produces a tensor of shape `(3, 6, 5, 7)`.
- A tensor of shape X of shape `(5, 7)` can be broadcasted to another tensor Y of shape `(3, 6, 5, 7)`, since the last two dimensions of X and Y are matching. The resultant tensor Z has shape `(3, 6, 5, 7)`.
- A tensor X of shape `(5, 7)` cannot be broadcasted to another tensor Y of shape `(7, 6, 2)`, since the last two dimensions of X and Y are not matching, nor any of the axes of X is a singleton.
- A tensor X of shape `(1, 7)` can be broadcasted to another tensor Y of shape `(3, 6, 5, 7)`, since the last dimension of the two tensors are matching, and the second to last dimension of X is a singleton. The resultant tensor Z has shape `(3, 6, 5, 7)`.
- A tensor X of shape `(6, 1, 7)` can be broadcasted to another tensor Y of shape `(3, 1, 5, 7)`, since the last dimension of the two tensors are matching, the second to last dimension of X is a singleton, and the third to last dimension of Y is a singleton. The resultant tensor Z will have a shape of `(3, 6, 5, 7)`.
- A tensor X of shape `(0, 7)` cannot be broadcasted to another tensor Y of shape `(3, 6, 5, 7)`, since the first axis of X is `0`, which is neither a singleton nor matching the second to last dimension of Y, i.e. `5`.
- Here is an odd example for 0-size tensors. When attempting to broadcast tensor X of shape `(0, 7)` (i.e. empty, 0-size tensor) to another tensor of shape `(1, 7)`, the operation will not throw an error. However, the resulting tensor Z will also be of shape `(0, 7)`, i.e. an empty tensor like X. This is why we need to avoid operations with empty or 0-size tensors in our broadcasting rule.

i Note

Note that the underlying mechanism of broadcasting is not in reality repeating the elements to be broadcasted many times by copying the same data over and over again, as it will not be memory efficient.

2.12 Case Studies

We will close the chapter with the following set of case studies using element-wise operations, reshaping, and broadcasting to solve several practical problems.

2.12.1 Image Normalization

An example application of scalar to tensor broadcasting is the min–max normalization of images. Given image I,

$$\text{Normalize}(I_i) = \frac{I_i - \min(I)}{\max(I) - \min(I)}, \forall i \tag{2.1}$$

where i is the index of the image pixels. We can implement the above as

```
1  import numpy as np
2
3  def normalize_image(image):
4      """Normalize image pixel values to be between 0 and 1"""
5      # image shape (height, width, channel)
6      min_val = np.min(image) # scalar
7      max_val = np.max(image) # scalar
8      normalized_img = (image - min_val) / (max_val - min_val + 1e-6)
9
10     return normalized_img
```

- Lines 6 and 7 compute the global minimum and maximum values, which are scalars.
- Line 8 implements Equation 2.1. The denominator adds a small $\epsilon = 10^{-6}$ to prevent division by zero error. If minimum and maximum values are zeros, then all pixels in the image should have values of zero. In this case, the normalized image will also be all zeros.

If we have a batch of images, we use vector broadcasting, which is implemented as

```
import numpy as np

def normalize_image_batch(images):
    """Normalize image pixel values to be between 0 and 1"""
    # image shape (batch_size, height, width, channel)
    # min / max (batch_size, 1, 1, 1)
    min_val = np.min(images, axis=(1, 2, 3), keepdims=True)
    max_val = np.max(images, axis=(1, 2, 3), keepdims=True)
    normalized_img = (images - min_val) / (max_val - min_val + 1e-6)

    return normalized_img
```

- Lines 7 and 8 compute the minimum and maximum values for each image in the batch. `axis=(1, 2, 3)` specifies the computation is within the batch, but reduce the values globally across all the other dimensions within the input. `keepdims=True` will prevent the output to be squeezed across the reduced dimensions. This will result in the shape of the output tensor being `(batch_size, 1, 1, 1)` which will help us broadcast the min/max values back into the input tensor. If `keepdims=False` (default), then the output tensor will be of shape `(batch_size,)`, which cannot be broadcasted to the input tensor `(batch_size, height, width, channel)`.
- Line 9 implements Equation 2.1 but applies broadcasting between the min tensor with the input tensor (as well as scalar to multidimensional tensor broadcasting, with the term + `1e-6`) to normalize the batch of images in a single expression.

2.12.2 Pearson's Correlation

Pearson's correlation measures the linear relationship between two variables X and Y and is defined as

$$
\begin{aligned}
r &= \frac{\mathrm{COV}(X, Y)}{\sigma(X)\sigma(Y)} \\
&= \frac{\sum_i (x_i - \bar{x}) \cdot (y_i - \bar{y})}{\sqrt{\sum_i (x_i - \bar{x})^2} \cdot \sqrt{\sum_i (y_i - \bar{y})^2}} \\
&= \frac{\sum_i \tilde{x}_i \cdot \tilde{y}_i}{\sqrt{\sum_i \tilde{x}_i^2} \cdot \sqrt{\sum_i \tilde{y}_i}} \\
&= \sum_i \left(\frac{\tilde{x}_i}{\sqrt{\sum_p \tilde{x}_p^2}} \cdot \frac{\tilde{y}_i}{\sqrt{\sum_q \tilde{y}_q^2}} \right)
\end{aligned}
$$

where $\tilde{x}_i = x_i - \bar{x}$ and $\tilde{y}_i = y_i - \bar{y}$

We use vectorization to implement Pearson's correlation between multiple pairs of variables in the following:

```
import numpy as np

def faster_corr(X, Y):
    """
    A faster way to compute Pearson's correlation
    between corresponding rows of two matrices
```

```
7       """
8       # zero means
9       X = X - np.mean(X, axis=1, keepdims=True)
10      Y = Y - np.mean(Y, axis=1, keepdims=True)
11      # L2 normalization
12      X = X / np.sqrt(np.sum(X**2, axis=1, keepdims=True))
13      Y = Y / np.sqrt(np.sum(Y**2, axis=1, keepdims=True))
14      R = np.sum(X*Y, axis=1)
15
16      return R
```

- Lines 9 and 10 remove the mean of each variable. We set `keepdims=True` to result vectors of shape `(batch_size, 1)` to help with subsequent broadcasted subtraction with the original input of shape `(batch_size, num_variables)`. Note that since we are treating each row as each variable, we should set the `order=C` when initializing the two matrices so that each row is contiguously stored in the memory.
- Lines 12 and 13 perform L2 normalization but dividing the mean-removed variables by their respective L2 norm. We again set `keepdims=True` to result in a vector of shape `(batch_size, 1)` to help with subsequent broadcasted division with the mean-removed variables of shape `(batch_size, num_variables)`. This step should output a set of unit-length vectors along each row of the matrix.
- Line 14 implements the last step of the Pearson correlation equation, which essentially performs dot products between corresponding pairs of vectors from each normalized input.

2.12.3 Pair-wise Difference

```
import numpy as np

def pairwise_diff(A, B):
    """
    Calculate the pair-wise difference between
    each element of the two vectors of length N and M,
    respectively, and return a matrix of shape M x N.

    Parameters:
    * A: vector of length M
    * B: vector of length N

    Returns:
```

```
14        * A matrix of shape M x N where the element at (i, j) is A[i] - B[j]
15        """
16        A = np.asarray(A).flatten()[:, np.newaxis]  # shape (M, 1)
17        B = np.asarray(B).flatten()[np.newaxis,:]  # shape (1, N)
18        return A - B  # Resulting shape (M, N)
```

- Lines 16 and 17 prepare the broadcasting in Line 18. `np.asarray` makes sure that the inputs are NumPy arrays; this means, the function can also accept lists or tuples as inputs. `flatten` makes sure the inputs are one-dimensional arrays. This implies that if the input `A` is of shape `(1, M)`, we will first convert it to shape `(M,)`. This then would allow us to use `[:, np.newaxis]` to add a singleton dimension at the second dimension, making it of shape `(M, 1)`. Similarly, when we make sure `B` is of shape `(N,)`, we then use `[np.newaxis,:]` to add a singleton dimension along the first dimension, making it of shape `(1, N)`.
- Line 18 computes the pair-wise difference via broadcasting, i.e. tensor of shape `(M, 1)` broadcasted to a tensor of shape `(1, N)`, resulting in a tensor of shape `(M, N)`.

2.12.4 Construction of Magic Squares

A magic square is a square matrix of size $n \times n$, with its entries filled with numbers between 1 and n^2, such that each row, column, and the main diagonal have identical sums [Arroyo, 2018]. We can easily verify that when $n = 2$, this matrix is impossible to construct: given numbers $1, 2, 3, 4$, if we want to make sure each row of the matrix to have identical sum, then the sum needs to be 5, and the two ways we can construct them are with $(1, 4)$ and $(2, 3)$. However, regardless of how we arrange them, we cannot make the columns sum to 5 (see two examples in Figure 2.9a,b). Similarly, if we want to make sure each column of the matrix has an identical sum, we again use $(1, 4)$ and $(2, 3)$ pairing for each column. But regardless of how we arrange them, we cannot make sure the rows can be summed to 5 (see two examples in Figure 2.9c,d).

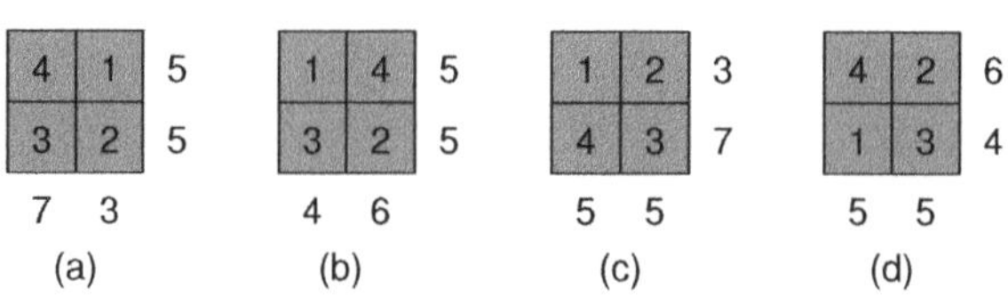

Figure 2.9 2x2 magic squares cannot be constructed.

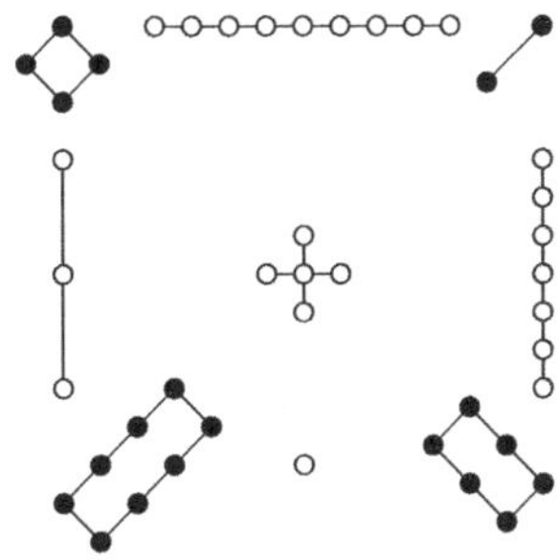

4	9	2	15
3	5	7	15
8	1	6	15
15	15	15	

Figure 2.10 Luo Shu, the earliest known third-order magic square.

A magic square can be classified into three categories: odd ($n = 2k + 1$), doubly even ($n = 2(2k) = 4k$), and singly even ($n = 2(2k + 1) = 4k + 2$) [Kraitchik, 1942]. These three categories of magic squares use three different algorithms to construct. The magic constant of the magic square is the row or column sum of the entries, given by the formula $\sigma(M) = \frac{n(n^2+1)}{2}$. For a 3×3 magic square, the magic constant is

$$\sigma(3) = \frac{3 \times (3^2 + 1)}{2} = 15$$

The oldest third-order magic square is probably the one from "Luo Shu," an ancient Chinese book used for fortune-telling, as shown in Figure 2.10. It also came with a pithy formula that goes like the following: "2 (right) and 4 (left) on the shoulders, 6 (right) and 8 (left) on the feet, 3 on the left, 7 on the right, 9 be the hat, 1 be the shoe, and 5 at the center" (and I am mostly remembering this from the novel "The Legend of the Condor Heroes" written by Jin Yong who cited this pithy formula for the magic square in "Luo Shu").

Odd-Order Magic Squares Construction by Siamese Method

Siamese method is probably the best-known method for constructing magic squares of odd orders [Kraitchik, 1942; De La Loubere, 1693]. The rules for constructing a magic square are the following:

1. Start at the position of the topmost row and central column of the matrix. Fill this position with 1.
2. We then continue filling the matrix by going diagonally upward and right.
   1. If we run out of rows, we go back to the bottom-most row and fill in the values in the corresponding column.
   2. If we run out of columns, we go back to the leftmost column and fill in the values in the corresponding row.
   3. If we run out of both rows and columns (i.e. when going from the upper right corner), or the subsequent position is already occupied, we then fill the position right below the current element by going down 1 row, keeping the same column.

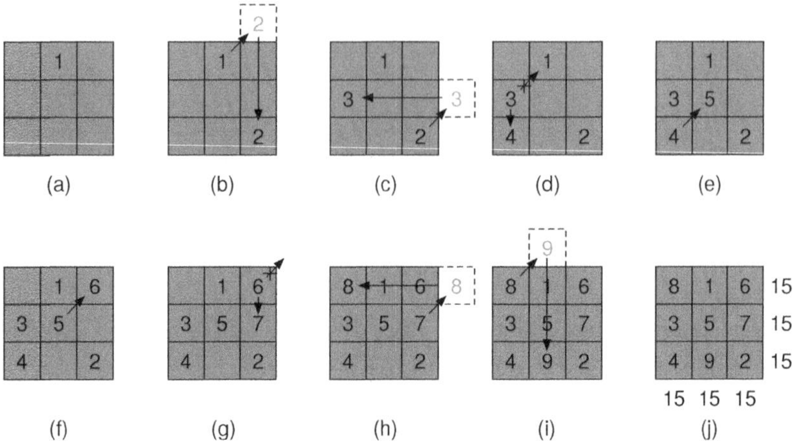

Figure 2.11 Example of constructing a third-order magic square using the Siamese method. (a) Start at the top-center grid and fill with 1 (Rule 1); (b) go up and right 1 grid. Since this grid goes above the rows of the matrix, we use the bottom row, but still fill at column 3 (Rule 2.1); (c) we continue to go diagonally up and right. Since this grid goes beyond the columns of the matrix, we use the left-most column, but still fill at row 2 (Rule 2.2); (d) continuing going diagonally up and right hits the top-center grid, which is already filled with 1. We therefore fill the grid with 4 in the next row at the same column (Rule 2.3); (e) go diagonal up and right to fill 5; (f) go diagonal up and right to fill 6; (g) we can no longer go diagonal since we will run out of both rows and columns. We therefore fill the grid below the current grid at the same column with 7 (Rule 2.3); (h) continue diagonally, we run out of columns again, so we move to the left-most column on row 1 and fill the grid with 8 (Rule 2.2); (i) continue diagonally, we run out of rows again, so we fill the bottom row at column 2 with 9 (Rule 2.1); (j) final completed third-order magic square. We can verify that each row and column, diagonal, and anti-diagonal all summed to 15.

Figure 2.11 illustrates an example of filling out a 3×3 magic square using the Siamese method. Notice that the resulting magic square can be seen as a new magic square constructed by simply swapping rows 1 and 3 from the example in "Luo Shu."

How do we implement the Siamese method using vectorization? We need to make some observations on the patterns of the constructed magic squares. Based on this work [Arroyo, 2018], we can see that Siamese magic squares are composed of two orthogonal Latin squares L_b and L_u and a unitary square matrix J (square matrix filled with 1s), i.e.

$$S = nL_b + L_u + J \tag{2.2}$$

as illustrated in Figure 2.12. Let $b \in \{0, 1, 2, \dots, n-1\}$ be the entries of L_b and $u \in \{0, 1, 2, \dots, n-1\}$ be entries of L_u. Then, the indices of the magic square (i, j)

$$\begin{bmatrix}8&1&6\\3&5&7\\4&9&2\end{bmatrix} = 3\times\begin{bmatrix}2&0&1\\0&1&2\\1&2&0\end{bmatrix} + \begin{bmatrix}1&0&2\\2&1&0\\0&2&1\end{bmatrix} + \begin{bmatrix}1&1&1\\1&1&1\\1&1&1\end{bmatrix}$$

$$\begin{bmatrix}17&24&1&8&15\\23&5&7&14&16\\4&6&13&20&22\\10&12&19&21&3\\11&18&25&2&9\end{bmatrix} = 5\times\begin{bmatrix}3&4&0&1&2\\4&0&1&2&3\\0&1&2&3&4\\1&2&3&4&0\\2&3&4&0&1\end{bmatrix} + \begin{bmatrix}1&3&0&2&4\\2&4&1&3&0\\3&0&2&4&1\\4&1&3&0&2\\0&2&4&1&3\end{bmatrix} + \begin{bmatrix}1&1&1&1&1\\1&1&1&1&1\\1&1&1&1&1\\1&1&1&1&1\\1&1&1&1&1\end{bmatrix}$$

$$\begin{bmatrix}30&39&48&1&10&19&28\\38&47&7&9&18&27&29\\46&6&8&17&26&35&37\\5&14&16&25&34&36&45\\13&15&24&33&42&44&4\\21&23&32&41&43&3&12\\22&31&40&49&2&11&20\end{bmatrix} = 7\times\begin{bmatrix}4&5&6&0&1&2&3\\5&6&0&1&2&3&4\\6&0&1&2&3&4&5\\0&1&2&3&4&5&6\\1&2&3&4&5&6&0\\2&3&4&5&6&0&1\\3&4&5&6&0&1&2\end{bmatrix} + \begin{bmatrix}1&3&5&0&2&4&6\\2&4&6&1&3&5&0\\3&5&0&2&4&6&1\\4&6&1&3&5&0&2\\5&0&2&4&6&1&3\\6&1&3&5&0&2&4\\0&2&4&6&1&3&5\end{bmatrix} + \begin{bmatrix}1&1&1&1&1&1&1\\1&1&1&1&1&1&1\\1&1&1&1&1&1&1\\1&1&1&1&1&1&1\\1&1&1&1&1&1&1\\1&1&1&1&1&1&1\\1&1&1&1&1&1&1\end{bmatrix}$$

Figure 2.12 Deconstructing odd-order Siamese magic squares into a sum of two Latin squares and a unitary square matrix.

has the following relationship with b and u:

$$\begin{aligned} b &= \operatorname{mod}\left(i + j - \frac{n-1}{2} + n,\ n\right) \\ u &= \operatorname{mod}\left(i + 2j - n + 1,\ n\right) \end{aligned} \tag{2.3}$$

The formula indicates that we need to iterate over all potential combinations of i and j. This can be done via broadcasting between a row vector and a column vector, both initialized by `np.arange`. Based on Equation 2.3, we can implement the Siamese method as the following:

```
def _odd_order_magic(n: int):
    """Make Siamese odd-order magic squares via Latin
    ↪ squares."""
    # Initialize p = i / j, used for
    # brodcasting later -> shape n x 1
    p = np.arange(n)[:, None]
    # make L_b
    L_b = np.mod(p + p.T - (n - 1) // 2, n)
    # make L_u
    L_u = np.mod(p + 2 * p.T - n + 1, n)
    # make the magic square
    S = n * L_b + L_u + np.ones_like(L_b)

    return S
```

- Line 5 initializes a column vector using `np.arange`. We are trying to use broadcasting to construct a mesh that helps us iterate over all pairs of (i,j) $[(0,0),(0,1),(0,2),\ldots,(n-1,n-1)]$.
- Line 7 makes the Latin square L_b using the formula in Equation 2.3. Notice that `(n - 1) // 2` is doing an integer division, since we know that input `n` is already an odd integer. This will keep all the results as integers. Otherwise, NumPy will cast everything into `float64`. `np.mod(L_b, n)` will make sure the resulting value of `b` is an integer between 0 and $n-1$ (inclusive). You may notice that the values from line 10 may sometimes be negative. For example, given a third-order Siamese magic square ($n = 3$), when $i = 0, j = 0$, the corresponding $b = -1, u = -2$. However, $\mod(-1, 3) = 2$, $\mod(-2, 3) = 1$. The modulus operation helps cast the negative values to a value between 0 and 2 (inclusive).
- Line 9 implements the formula in Equation 2.3 to make Latin square L_u.
- Line 11 implements the definition of the Siamese odd-order magic squares (Equation 2.2).

Doubly-Even Order Magic Squares

Manuel Moschopoulos in the 1300s illustrated a very elegant way of constructing doubly even magic squares, which I will call the "X" method:

1. Fill the square matrix from n to 1 row-wise.
2. Draw Xs through the main diagonals of each 4×4 block.
3. For each grid where the Xs touch, replace the number with the complement $n^2 + 1 - x_{i,j}$.

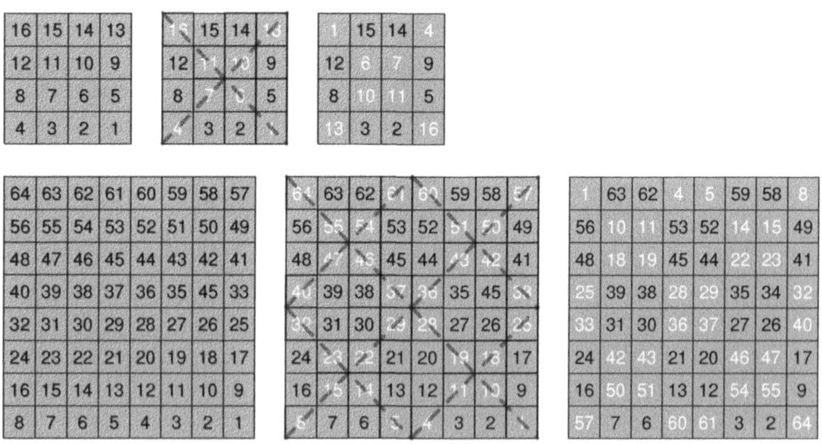

Figure 2.13 The "X" method for constructing doubly even order magic squares.

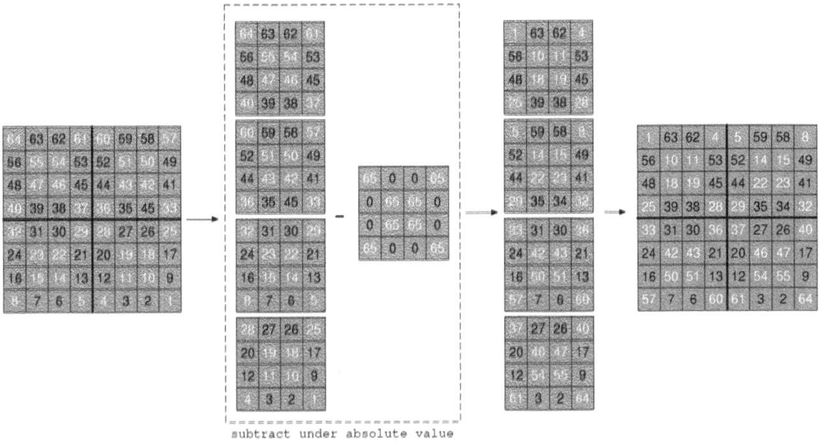

Figure 2.14 Steps of implementing the X-method for constructing doubly even magic squares.

Figure 2.13 illustrates an order-4 and order-8 magic squares constructed using the X-method.

Note that Wolfram Alpha [Weisstein, 2023] illustrated a similar version of the X-method for constructing doubly even magic squares, whose source is unknown. Google search results seem to suggest this method was due to Philippe de la Hire's 1705 Latin translation of Manuel Moschopoulos' work on magic squares. This complementary method simply fills the square matrix from 1 to n instead of n to 1, which yields a complementary magic square. It also appears MATLAB implements this method.

We can again use the idea of reshaping and broadcasting to implement this algorithm. Figure 2.14 illustrates the process step by step.

```
def _doubly_even_order_magic(n: int):
    """Use X method to construct doubly even magic
     ↪ squares."""
    # e.g. n=8, (64,)
    S = np.arange(n**2, 0, -1)
    # Split into 4x4 blocks (2, 4, 2, 4)
    S = S.reshape((n//4, 4, n//4, 4))
    # Create the X mask
    X = np.eye(4, dtype=S.dtype) * (n**2 + 1)
    X = X + np.fliplr(X)
    # (1, 4, 1, 4)
    X = X[None, :, None, :]
    # Subtract under absolute value
```

```
13        S = np.abs(S - X)
14        # Reshape (8, 8)
15        S = S.reshape((n, n))
16
17        return S
```

1. Line 4 initializes the values of the square matrix in descending order, from n^2 to 1.
2. Line 6 splits the matrix into blocks of shape $(4, 4)$, for each row (`axis=0`) and column (`axis=3`). This forms a tensor of size $(k, 4, k, 4)$, where $k = n//4$.
3. Lines 8 and 9 create the 4×4 matrix with diagonal and anti-diagonal entries filled with $n^2 + 1$, which forms the X-shape. This is used to compute the complement values against each block of 4×4.
4. Line 11 adds dimensions at `axis=0` and `axis=3`, making the tensor shape $(1, 4, 1, 4)$. This is used for broadcasting in the next step.
5. Line 13 takes the difference under the absolute value. This replaces the entries touched by the "X" with the complementary values.
6. Line 15 reshapes the tensor back into a matrix of shape (n, n).

Singly Even Order Magic Squares

Lastly, we will construct singly even order magic squares via the "LUX" method due to John Horton Conway [Erickson, 2009; Weisstein, 2023]. Through its implementation in NumPy, I am hoping the readers of the book can appreciate the efficiency and power of this library in handling complex data manipulations. "LUX" is referring to the order of filling out a 2×2 grid, as illustrated in Figure 2.15. The three variants of filling order form the shape of "L," "U," and "X," respectively. Let $n = 2m = 2(2k + 1)$, where $k \geq 1$, then,

1. Starting with an odd-order square matrix $m \times m$, we can determine what shape to fill for each of the grids that represents a 2×2 block:
   - Fill k rows of the matrix with L shapes.
   - Fill the $k + 1$th row with L shapes, except the middle one with an U shape.
   - Fill the $k + 2$th row with U shapes, except the middle one with an L shape.
   - Fill the rest of the rows (if any left) with X shapes
2. The order of the numbers to be filled is determined by the Siamese method, which starts at the top-center grid as the first block to fill. We can derive the mapping between the values in the Siamese odd-order magic matrix and the values within the 2×2 matrix to be filled by "LUX" as $u = 4(v - 1) + i$, where v is the corresponding Siamese odd order magic square and i is the indexed value of the 2×2 block determined by one of the three variants of the filling order.

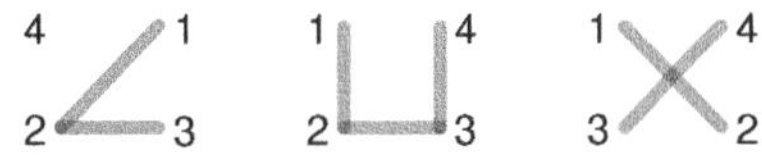

Figure 2.15 LUX: order of filling out a 2 × 2 grid.

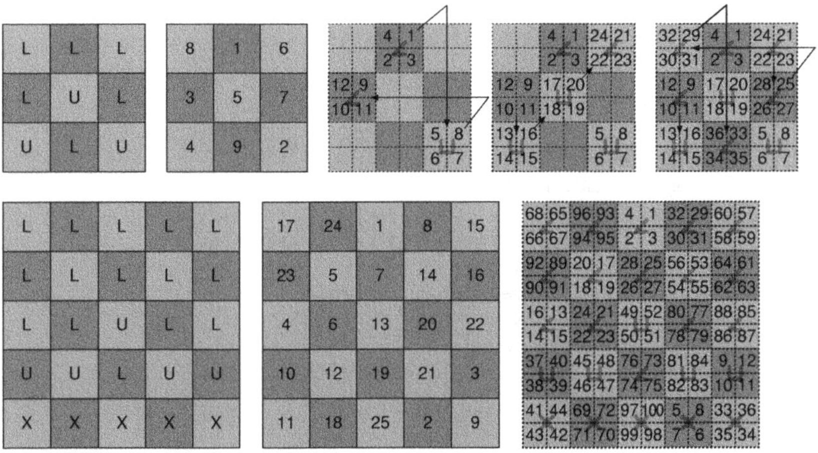

Figure 2.16 Examples of constructing order 6 and 10 magic squares using the LUX method.

Figure 2.16 illustrates examples of constructing order 4 and order 6 magic squares using the "LUX" method.

```
def _singly_even_order_magic(n: int):
    """Use LUX method to construct singly even magic
    ↪ squares."""
    # Compute size constants
    m = n // 2 # odd
    k = (m - 1) // 2
    # Initialize the 3 filling variants
    L = np.array([[4, 1], [2, 3]])
    U = np.array([[1, 4], [2, 3]])
    X = np.array([[1, 4], [3, 2]])
    # Create k-rows of L-filled blocks
    M_L = np.full((k, m, 2, 2), L)
    # Create a row of L-filled blocks,
    # except the middle column is filled with U
    M_LU = np.concatenate([
        np.full((1, k, 2, 2), L),
        U[None, None,:,:],
        np.full((1, k, 2, 2), L),
      ], axis=1)
    # Create a row of U-filled blocks,
```

```
20        # except the middle column is filled with L
21        M_UL = np.concatenate([
22            np.full((1, k, 2, 2), U),
23            L[None, None,:,:],
24            np.full((1, k, 2, 2), U),
25        ], axis=1)
26        # Fill the rest the k-1 rows with X
27        M_X = np.full((k-1, m, 2, 2), X)
28        # Combine all the sub-blocks (p, p, 2, 2)
29        M = np.concatenate([
30            M_L, M_LU, M_UL, M_X
31        ], axis=0)
32        # Initialize the Simaese odd order magic square
33        S = _odd_order_magic(m)
34        # Apply the Siamese mapping formula
35        M += 4 * (S[:,:, None, None] - 1)
36        # Reshaping back to n x n
37        M = np.swapaxes(M, 1, 2).reshape((n, n))
38
39        return M
```

1. Lines 4 and 5 compute some constants that we will use later on to construct the matrices.
2. Lines 7–9 initialize the L, U, X ordering as 2×2 matrix blocks.
3. Line 11 creates k rows of L filled blocks. We can treat it as a block of $k \times m$ matrix, with each matrix filled with 2×2 block, resulting in a tensor of shape $k \times m \times 2 \times 2$.
4. Lines 14–18 create 1 row of L-filled blocks except with the middle entry being U-filled. We do this via concatenation of Concat([$[L, \ldots]$, $[U]$, $[L, \ldots]$]). This results in a tensor of shape $1 \times m \times 2 \times 2$.
5. Lines 21–25 create 1 row of U-filled blocks except with the middle entry being L-filled. Similar to the above step, we do this via concatenation of Concat([$[U, \ldots]$, $[L]$, $[U, \ldots]$]). This results in a tensor of shape $1 \times m \times 2 \times 2$.
6. Line 27 creates $k - 1$ rows of X-filled blocks. When $n = 6, k = 1$, then $k - 1 = 0$, which means we do not add the X blocks below. `np.full` can handle zero dimensions while keeping the tensor to the desired shape. This creates a tensor with shape $(k - 1) \times m \times 2 \times 2$.
7. Lines 29–31 combine the sub-blocks of tensors into a single tensor via concatenation along the first dimension. This creates a tensor of shape $m \times m \times 2 \times 2$.
8. Line 33 initializes the $m \times m$ odd-order Siamese magic matrix, which we need in order to apply the value mapping formula.

9. Line 35 applies the value mapping formula we defined previously, by broadcasting the $m \times m$ Siamese magic matrix to the $m \times m \times 2 \times 2$ tensor M we obtained from Step 7 above. Notice that we expanded two additional dimensions at `axis=2` and `axis=3`.
10. Line 37 reshapes the tensor back into the $n \times n$ matrix. We first swap the second and third axes, making $m \times m \times 2 \times 2$ becomes $m \times 2 \times m \times 2$. Then, we reshape to $n \times n$.

Finally, we can put the three cases of constructing magic squares under a single function:

```
import numpy as np

def magic(n: int):
    """Construct magic squares of order n."""
    n = int(n) # make sure n is integer
    if n < 3: # check value of n
        raise(ValueError("n must be > 2."))
    if n % 2 == 1: # odd case
        return _odd_order_magic(n)
    elif n % 4 == 0: # doubly even
        return _doubly_even_order_magic(n)
    else: # assume singly even
        return _singly_even_order_magic(n)
```

Bibliography

Martín Abadi, Ashish Agarwal, Paul Barham, Eugene Brevdo, Zhifeng Chen, Craig Citro, Greg S. Corrado, Andy Davis, Jeffrey Dean, Matthieu Devin, Sanjay Ghemawat, Ian Goodfellow, Andrew Harp, Geoffrey Irving, Michael Isard, Yangqing Jia, Rafal Jozefowicz, Lukasz Kaiser, Manjunath Kudlur, Josh Levenberg, Dan Mane, Rajat Monga, Sherry Moore, Derek Murray, Chris Olah, Mike Schuster, Jonathon Shlens, Benoit Steiner, Ilya Sutskever, Kunal Talwar, Paul Tucker, Vincent Vanhoucke, Vijay Vasudevan, Fernanda Viegas, Oriol Vinyals, Pete Warden, Martin Wattenberg, Martin Wicke, Yuan Yu, and Xiaoqiang Zheng. TensorFlow: Large-scale machine learning on heterogeneous distributed systems. November 2015.

Martín Abadi, Ashish Agarwal, Paul Barham, Eugene Brevdo, Zhifeng Chen, Craig Citro, Greg S. Corrado, Andy Davis, Jeffrey Dean, Matthieu Devin, Sanjay Ghemawat, Ian Goodfellow, Andrew Harp, Geoffrey Irving, Michael Isard, Yangqing Jia, Rafal Jozefowicz, Lukasz Kaiser, Manjunath Kudlur, Josh Levenberg, Dan Mane, Rajat Monga, Sherry Moore, Derek Murray,

Chris Olah, Mike Schuster, Jonathon Shlens, Benoit Steiner, Ilya Sutskever, Kunal Talwar, Paul Tucker, Vincent Vanhoucke, Vijay Vasudevan, Fernanda Viegas, Oriol Vinyals, Pete Warden, Martin Wattenberg, Martin Wicke, Yuan Yu, and Xiaoqiang Zheng. Introduction to graphs and tf.function. https://www.tensorflow.org/guide/intro_to_graphs, 2023a.

Martín Abadi, Ashish Agarwal, Paul Barham, Eugene Brevdo, Zhifeng Chen, Craig Citro, Greg S. Corrado, Andy Davis, Jeffrey Dean, Matthieu Devin, Sanjay Ghemawat, Ian Goodfellow, Andrew Harp, Geoffrey Irving, Michael Isard, Yangqing Jia, Rafal Jozefowicz, Lukasz Kaiser, Manjunath Kudlur, Josh Levenberg, Dan Mane, Rajat Monga, Sherry Moore, Derek Murray, Chris Olah, Mike Schuster, Jonathon Shlens, Benoit Steiner, Ilya Sutskever, Kunal Talwar, Paul Tucker, Vincent Vanhoucke, Vijay Vasudevan, Fernanda Viegas, Oriol Vinyals, Pete Warden, Martin Wattenberg, Martin Wicke, Yuan Yu, and Xiaoqiang Zheng. TensorFlow Dtypes. https://www.tensorflow.org/api_docs/python/tf/dtypes, 2023b.

Martín Abadi, Ashish Agarwal, Paul Barham, Eugene Brevdo, Zhifeng Chen, Craig Citro, Greg S. Corrado, Andy Davis, Jeffrey Dean, Matthieu Devin, Sanjay Ghemawat, Ian Goodfellow, Andrew Harp, Geoffrey Irving, Michael Isard, Yangqing Jia, Rafal Jozefowicz, Lukasz Kaiser, Manjunath Kudlur, Josh Levenberg, Dan Mane, Rajat Monga, Sherry Moore, Derek Murray, Chris Olah, Mike Schuster, Jonathon Shlens, Benoit Steiner, Ilya Sutskever, Kunal Talwar, Paul Tucker, Vincent Vanhoucke, Vijay Vasudevan, Fernanda Viegas, Oriol Vinyals, Pete Warden, Martin Wattenberg, Martin Wicke, Yuan Yu, and Xiaoqiang Zheng. TensorFlow tf.reshape. https://www.tensorflow.org/api_docs/python/tf/reshape, 2023c.

Joshua Arroyo. A proof of the “Magicness” of the Siam construction of a magic square. *RoseHulman Undergraduate Mathematics Journal*, 19(1):3, 2018.

Simon De La Loubere. *A New Historical Relation of the Kingdom of Siam*. F.L. For Tho. Home at the Royal Exchange, Francis Saunders at the New Exchange, and Tho. Bennet at the Half-Moon in St. Pauls Church-yard:London 1693.

Martin Erickson. Magic squares. In *AHA! Solutions*. Mathematical Association of America, 2009.

Charles R. Harris, K. Jarrod Millman, Stéfan J. van der Walt, Ralf Gommers, Pauli Virtanen, David Cournapeau, Eric Wieser, Julian Taylor, Sebastian Berg, Nathaniel J. Smith, Robert Kern, Matti Picus, Stephan Hoyer, Marten H. van Kerkwijk, Matthew Brett, Allan Haldane, Jaime Fernández del Río, Mark Wiebe, Pearu Peterson, Pierre Gérard-Marchant, Kevin Sheppard, Tyler Reddy, Warren Weckesser, Hameer Abbasi, Christoph Gohlke, and Travis E. Oliphant. Array programming with NumPy. *Nature*, 585(7825):357–362, September 2020. ISSN 1476-4687. doi: 10.1038/s41586-020-2649-2.

Charles R. Harris, K. Jarrod Millman, Stéfan J. van der Walt, Ralf Gommers, Pauli Virtanen, David Cournapeau, Eric Wieser, Julian Taylor, Sebastian Berg, Nathaniel J. Smith, Robert Kern, Matti Picus, Stephan Hoyer, Marten H. van Kerkwijk, Matthew Brett, Allan Haldane, Jaime Fernández del Río, Mark Wiebe, Pearu Peterson, Pierre Gérard-Marchant, Kevin Sheppard, Tyler Reddy, Warren Weckesser, Hameer Abbasi, Christoph Gohlke, and Travis E. Oliphant. NumPy Dtypes. https://numpy.org/doc/stable/user/basics.types.html, 2023a.

Charles R. Harris, K. Jarrod Millman, Stéfan J. van der Walt, Ralf Gommers, Pauli Virtanen, David Cournapeau, Eric Wieser, Julian Taylor, Sebastian Berg, Nathaniel J. Smith, Robert Kern, Matti Picus, Stephan Hoyer, Marten H. van Kerkwijk, Matthew Brett, Allan Haldane, Jaime Fernández del Río, Mark Wiebe, Pearu Peterson, Pierre Gérard-Marchant, Kevin Sheppard, Tyler Reddy, Warren Weckesser, Hameer Abbasi, Christoph Gohlke, and Travis E. Oliphant. NumPy Copies and views. https://numpy.org/doc/stable/user/basics.copies.html, 2023b.

Charles R. Harris, K. Jarrod Millman, Stéfan J. van der Walt, Ralf Gommers, Pauli Virtanen, David Cournapeau, Eric Wieser, Julian Taylor, Sebastian Berg, Nathaniel J. Smith, Robert Kern, Matti Picus, Stephan Hoyer, Marten H. van Kerkwijk, Matthew Brett, Allan Haldane, Jaime Fernández del Río, Mark Wiebe, Pearu Peterson, Pierre Gérard-Marchant, Kevin Sheppard, Tyler Reddy, Warren Weckesser, Hameer Abbasi, Christoph Gohlke, and Travis E. Oliphant. NumPy Broadcasting. https://numpy.org/doc/stable/user/basics.broadcasting.html, 2023c.

Maurice Kraitchik. Chapter 7 Magic squares. In *Mathematical Recreation*. W. W. Norton & Company, Inc.:New York, pages 148–152. 2nd edition, 1942.

Alexander Nenashev. Pre-allocate arrays to make Javascript fast, May 2024.

Adam Paszke, Sam Gross, Francisco Massa, Adam Lerer, James Bradbury, Gregory Chanan, Trevor Killeen, Zeming Lin, Natalia Gimelshein, Luca Antiga, Alban Desmaison, Andreas Köpf, Edward Yang, Zach DeVito, Martin Raison, Alykhan Tejani, Sasank Chilamkurthy, Benoit Steiner, Lu Fang, Junjie Bai, and Soumith Chintala. PyTorch: An imperative style, high-performance deep learning library, December 2019.

Adam Paszke, Sam Gross, Francisco Massa, Adam Lerer, James Bradbury, Gregory Chanan, Trevor Killeen, Zeming Lin, Natalia Gimelshein, Luca Antiga, Alban Desmaison, Andreas Köpf, Edward Yang, Zach DeVito, Martin Raison, Alykhan Tejani, Sasank Chilamkurthy, Benoit Steiner, Lu Fang, Junjie Bai, and Soumith Chintala. Lack of Square Function #2618. https://github.com/pytorch/pytorch/issues/2618, 2023a.

Adam Paszke, Sam Gross, Francisco Massa, Adam Lerer, James Bradbury, Gregory Chanan, Trevor Killeen, Zeming Lin, Natalia Gimelshein, Luca Antiga, Alban Desmaison, Andreas Köpf, Edward Yang, Zach DeVito, Martin Raison, Alykhan Tejani, Sasank Chilamkurthy, Benoit Steiner, Lu Fang, Junjie Bai, and

Soumith Chintala. PyTorch Dtypes. https://pytorch.org/docs/stable/tensors.html, 2023b.

Peter Prinz and Tony Crawford. *C In a Nutshell – A Desktop Quick Reference*. O'Reilly, 1st edition, 2005.

Mark Sellors. Vector Pre Allocation in R – A Simple Example, May 2024.

Usi. Using std::vector::reserve whenever possible, May 2023.

Eric W. Weisstein. Magic Square. https://mathworld.wolfram.com/, September 2023.

Wikipedia. LogSumExp. https://en.wikipedia.org/wiki/LogSumExp, 2023.

3

Tensor Indexing

An **index** is a reference that points to the location of the value in an arrangement of an N-dimensional tensor. **Indexing** is the operation to select or manipulate an element or a group of elements within a tensor using such a reference system. In this chapter, we will introduce various ways of indexing. We will also illustrate how indexing can be used to manipulate elements in a tensor through several case studies.

3.1 Get Values at Index

3.1.1 Integer Indexing

In Python, elements of a list (e.g. `l`) at a specific index can be accessed by the syntax `l[index]`, where `index` is an integer starting from 0 (i.e. 1st index). For example, let `l = [2, 3, 5, 7, 11]` be a list of integers. Then, the third item (`index=2`) can be retrieved by `l[2]`, returning `5`. Here, the variable `index` can also be negative numbers, which access the element starting from the end of the list. For example, `l[-1]` accesses the last element in the list and `l[-2]` accesses the second to last element. In general, given a list of length `N` and a negative integer `index`, then `l[index] == l[N+index]`. See Figure 3.1 for an illustration. 1D arrays in NumPy (`np.ndarray`), Tensorflow (`tf.Tensor`), and PyTorch (`torch.Tensor`) all follow such indexing conventions as well.

For 2D matrices, the indices of the elements are represented by a pair (tuple) of integers, in the format of `(row, column)`. As illustrated in Figure 3.2, the value `7` resides in the second row and fourth columns of the 3×4 matrix and can be referenced by the index `(1, 3)`. It is worthwhile to point out that this convention is different from the Cartesian coordinates `(x, y)`, where `x` is the horizontal position, and `y` is the vertical position. To reference the same value `7` in this matrix, we would instead use `(3, 1)` using the Cartesian coordinate system. (We may also treat `(x, y)` as the `(column, row)` convention.)

Vectorization: A Practical Guide to Efficient Implementations of Machine Learning Algorithms,
First Edition. Edward DongBo Cui.

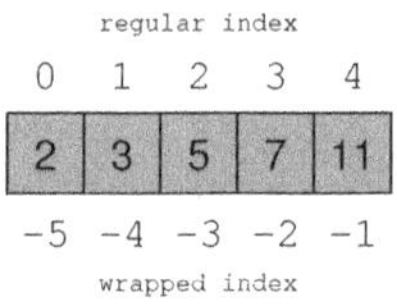

Figure 3.1 Illustration of integer indexing using positive (regular) and equivalent negative (wrapped) index.

We can represent a 2D matrix as a Python list of lists, e.g. X = [[5, 2, 8, 3], [4, 9, 6, 7], [8, 11, 12, 10]]. If we want to retrieve the element 4 from the matrix, then we need to use the X[1][0] syntax. If X is a NumPy, Tensorflow, or PyTorch tensor, then we should instead use X[1, 0] syntax to retrieve the same element.

To reference an element in higher-dimensional tensors, we can use a list of integers (i_0, i_1, ..., i_d, ..., i_D), where i_d is the *d*th dimension/axis of the tensor X. We can then retrieve the element at a specific index using the list-of-integers indexing syntax, e.g. X[3, 2, 5, 8]. Note that it is also legitimate to use negative integers for indexing in this case.

Furthermore, using the integer indexing system, we can also reference/select multiple elements in a tensor, using the syntax [[axis_0_index_0, axis_0_index_1, ..., axis_0_index_k], [axis_1_index_0, axis_1_index_1, ..., axis_1_index_k], ..., [axis_d_index_0, axis_d_index_1, ..., axis_d_index_k]]. Notice that indices of the same axis/dimension are grouped together and specified as a list; the order of the indices follows the corresponding order of the elements to be retrieved. For example, we would like to select elements both at (1, 3) and at (2, 2) from X we defined in Figure 3.2, we can do X[(1, 2), (3, 2)]. Note that indexing of multiple elements simultaneously is only allowed in NumPy and PyTorch, but not in Tensorflow or base Python lists. If X is a tf.Tensor, then X[(1, 2), (3, 2)] will throw the following error:

```
TypeError: Only integers, slices (':'), ellipsis ('...'),
 ↪  tf.newaxis ('None') and scalar tf.int32/tf.int64 tensors
 ↪  are valid indices, got (1, 2)
```

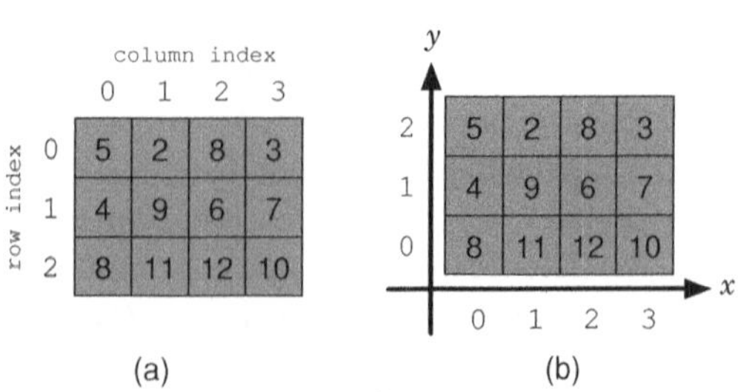

Figure 3.2 Indexing a matrix. (a) Row-column index. Here, "row" refers to the first dimension, and "column" refers to the second dimension of the matrix. (b) Cartesian XY coordinate of the matrix.

It is also interesting to see how Tensorflow handles (or perhaps is not able to handle) the multi-element indexing when `X` is a 1D vector. For example, if `X = np.array([1, 2, 3, 4])` is a 1D vector, and we would like to retrieve the first and third elements, then we can do `X[[0, 2]]` which results `[1, 3]`. This works in both NumPy and PyTorch, but not in TensorFlow. If `X` is a `tf.Tensor`, then `X[[0, 2]]` throws the following error:

```
InvalidArgumentError: {{function_node
↪   __wrapped__StridedSlice_device_/job:
↪   localhost/replica:0/task:0/device:CPU:0}} Index out of
↪   range using input dim 1; input has only 1 dims
↪   [Op:StridedSlice] name: strided_slice/
```

Since Tensorflow considers `[0, 2]` an index of a single element from a 2D matrix, rather than selecting multiple elements from a 1D vector. As we will see later in the chapter, in Tensorflow, we can instead use `tf.gather` to get multiple elements from a tensor.

3.1.2 Flat Index vs. Multi-index

For matrices (2D) or higher-dimensional tensors, there are in general two conventions to represent the indices of the elements: **flat index** (a.k.a. linear index) and **multi-index**. The examples we have seen in Section 3.1.1 were using multi-index, which points to the spatial coordinate of the elements across multiple dimensions and represents the indices as a list of integers (each one corresponding to each dimension of the tensor). In contrast, flat indices enumerate the elements from 0 up to the number of elements in the tensor and represent each index as a single integer. As illustrated in Figure 3.3, suppose we have a 3×4 2D matrix, which has 12 elements. The flat index enumerates the elements one by one using indices from 0 to 11. In Chapter 2, we briefly discussed the "`C`" order and "`F`" order conventions when introducing reshaping operations. So depending on the ordering convention used by the matrix, there are two ways to specify the indices, either along the rows (row-major "`C`" ordering) or along the columns (column-major "`F`" ordering). In the case of "`C`" ordering, flat indices `[0, 1, 2, ..., 11]` correspond to `(0, 0), (0, 1), (0, 2), ..., (1, 0), (1, 1), ..., (2, 3)`, where the indices along the second dimension increase first; in the case of "`F`" ordering, flat indices `[0, 1, 2, ..., 11]` correspond to `(0, 0), (1, 0), (2, 0), ..., (0, 1), (1, 1), ..., (2, 3)`, where the indices along the first dimension increase first. Another way to examine the differences between the two conventions is to look at the multi-index at a specific flat index. For example, the element at flat index `6` corresponds to

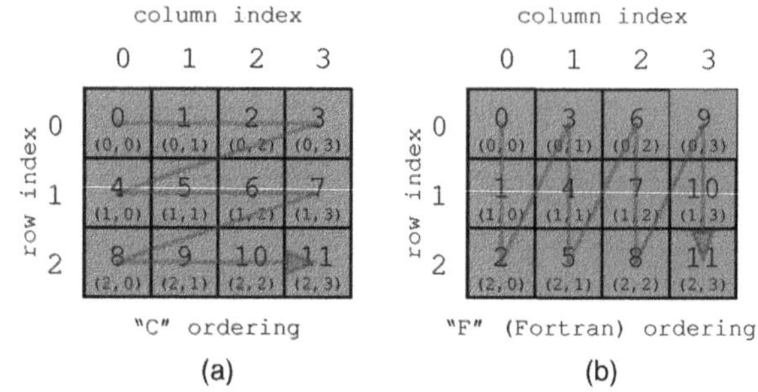

Figure 3.3 Flat and multi(dimensional) indices. There are two conventions to order the elements of a matrix in memory, "C" and "F" ordering. (a) In "C" ordering, the matrix is stored in the row-major convention. In this convention, values along the flat index move along each row of the matrix, or when indexed by multi-index, from $(0,0),(0,1),\ldots,$ $(1,0),(1,1),\ldots$ to $(n,0),(n,1),\ldots,(n,m)$. (b) In "F" or Fortran order, the matrix is stored in the column-major convention. In this convention, values along the flat index move along each column of the matrix, or when indexed by multi-index, from $(0,0),(1,0),\ldots,(0,1),$ $(1,1),\ldots$ to $(0,m),(1,m),\ldots,(n,m)$. See also the arrowed lines that indicate the order of traversal of the flat index in each case.

multi-index `(1, 2)` in "`C`" ordering, but corresponds to `(0, 2)` in "`F`" ordering (see Figure 3.3).

In general, we can convert between flat index and multi-index when given the element ordering convention "`C`" or "`F`" and the shape of the tensor. NumPy implements two functions to help the interconversion between the two indexing systems. `np.ravel_multi_index(multi_index, shape, order)` converts multi-indices to flat indices. Vice versa, `np.unravel_index(flat_indices, shape, order)` converts flat indices to multi-indices given tensor shape and order. For example, to convert from flat index to multi-index,

```
import numpy as np
shape = [2, 3, 3, 2]
flat_index = np.arange(0, np.prod(shape), 3)

r_multi_index = np.unravel_index(
    flat_index, shape, order="C"
)
print(np.stack([flat_index] + list(r_multi_index)).T)
# [[ 0  0  0  0  0]
#  [ 3  0  0  1  1]
#  [ 6  0  1  0  0]
#  [ 9  0  1  1  1]
#  [12  0  2  0  0]
#  [15  0  2  1  1]
#  [18  1  0  0  0]
```

```
# [21  1  0  1  1]
# [24  1  1  0  0]
# [27  1  1  1  1]
# [30  1  2  0  0]
# [33  1  2  1  1]]

c_multi_index = np.unravel_index(
    flat_index, shape, order="F"
)
print(np.stack([flat_index] + list(c_multi_index)).T)
# [[ 0  0  0  0  0]
#  [ 3  1  1  0  0]
#  [ 6  0  0  1  0]
#  [ 9  1  1  1  0]
#  [12  0  0  2  0]
#  [15  1  1  2  0]
#  [18  0  0  0  1]
#  [21  1  1  0  1]
#  [24  0  0  1  1]
#  [27  1  1  1  1]
#  [30  0  0  2  1]
#  [33  1  1  2  1]]
```

To further illustrate the relationship between the flat index and multi-index, we may derive the patterns of converting between these two indexing systems based on the above example. Let a tensor have shape $(n_0, n_1, \dots, n_d, \dots, n_{D-1})$, thus it has $r = \prod_i^{D-1} n_i$ elements, where D is the number of dimensions of the tensor. Then, to convert from flat index $[r_0, r_1, \dots, r_j, \dots]$, we can use the formula:

$$i_d^{(j)} = \left\lfloor \frac{r_j}{\eta_d} \right\rfloor \bmod n_d \tag{3.1}$$

where $\eta_{d<D-1} = \prod_{i=d+1}^{D-1} n_i$, $\eta_{D-1} = 1$ when using row-major ("C") ordering, and $\eta_{d>0} = \prod_{i=0}^{d-1} n_i$, $\eta_0 = 1$ when using column-major ("F") ordering; $\lfloor \cdot \rfloor$ is the floor division or integer division operation. We can implement our discovery as the following:

```
def ind2sub(flat_index, shape, order="C"):
    """Convert flat index to multi-index."""
    if order == "C":  # row-major
        eta = np.cumprod(shape[1:][::-1])
        eta = np.concatenate([[1], eta])[::-1]
    elif order == "F":  # column-major
        eta = np.cumprod(shape)[:-1]
        eta = np.concatenate([[1], eta])
```

```
9      else:
10         raise(ValueError(f"Unrecognized order: {order}"))
11 
12     # (num_indices, 1)
13     flat_index = np.asarray(flat_index)[:, None]
14     shape = np.asarray(shape)
15 
16     # Compute multi-index
17     return flat_index // eta % shape
```

- Lines 3–10 implement the array η depending on the order. Notice that the last element of `eta` is 1 when `order="C"` while the first element of it is 1 when `order="F"`.
- Lines 13 and 14 make sure the `flat_indices` and `shape` are both arrays in order to implement our formula in Equation 3.1.
- Line 17 implements the index conversion formula.

We can quickly verify our implementation against `np.unravel_index` as the following:

```
import numpy as np
shape = np.array([2, 3, 3, 2])
flat_index = np.arange(np.prod(shape))

# Use numpy unravel_index
r_multi_index = np.vstack(np.unravel_index(flat_index, shape, order="C")).T
c_multi_index = np.vstack(np.unravel_index(flat_index, shape, order="F")).T

# Call our own implementations
r_sub = ind2sub(flat_index, shape, order="C")
c_sub = ind2sub(flat_index, shape, order="F")

assert np.allclose(r_multi_index, r_sub)
assert np.allclose(c_multi_index, c_sub)
```

On the other hand, to convert the multi-index of a 4D tensor to a flat index using `np.ravel_multi_index`, e.g.,

```
import numpy as np
coordinates = [ # point coordinates
    (0, 1, 1, 0),
```

```
    (0, 2, 0, 1),
    (1, 0, 1, 0),
    (1, 1, 1, 1),
]
# one array for each dimension
multi_indices = np.array(coordinates).T.tolist()

shape = [2, 3, 3, 2]

# C-order
r_flat_index = np.ravel_multi_index(
    multi_indices, dims=shape, order="C"
)
print(np.concatenate([r_flat_index[:, None], coordinates], axis=1))
# [[ 8  0  1  1  0]
#  [13  0  2  0  1]
#  [20  1  0  1  0]
#  [27  1  1  1  1]]

# F-order
c_flat_index = np.ravel_multi_index(
    multi_indices, dims=shape, order="F"
)
print(np.concatenate([c_flat_index[:, None], coordinates], axis=1))
# [[ 8  0  1  1  0]
#  [22  0  2  0  1]
#  [ 7  1  0  1  0]
#  [27  1  1  1  1]]
```

We may discover the pattern of conversion like the following:

$$r_j = \sum_d i_d^{(j)} \eta_d \tag{3.2}$$

where $\eta_{d<D-1} = \prod_{i=d+1}^{D-1} n_i$, $\eta_{D-1} = 1$ when using row-major ("C") ordering, and $\eta_{d>0} = \prod_{i=0}^{d-1} n_i$, $\eta_0 = 1$ when using column-major ("F") ordering; this is identical to the previous definition in Equation 3.1. We can implement our findings as the following:

```
def sub2ind(multi_index, shape, order="C"):
    """Convert multi-index to flat index."""
    if order == "C":  # row-major
        eta = np.cumprod(shape[1:][::-1])
        eta = np.concatenate([[1], eta])[::-1]
    elif order == "F":  # column-major
```

```
7           eta = np.cumprod(shape)[:-1]
8           eta = np.concatenate([[1], eta])
9       else:
10          raise(ValueError(f"Unrecognized order: {order}"))
11
12      # (num_indices, d)
13      multi_index = np.asarray(multi_index)
14
15      return np.sum(multi_index * eta, axis=1)
```

- Lines 3–10 are identical to that of `ind2sub`, where we determine the array `eta`.
- Line 13 makes sure the input `multi_index` is an array.
- Line 15 implements the formula in Equation 3.2.

> **i Note**
>
> The function names `ind2sub` and `sub2ind` are from MATLAB.

It is important to understand that we cannot use flat indices directly on an un-flattened multidimensional tensor in NumPy, Tensorflow, or PyTorch. For example, if `X` is a 2D matrix, then `X[0]` does not retrieve the first element but rather the entire first row of the matrix. This is called ***slicing***, which we will look into more carefully in Section 3.2 in this chapter. To retrieve the first element according to the flat index, we need to either flatten the matrix, i.e. `X.flatten()[0]`, or use `np.take` which we will discuss in Section 3.4.1.

Indexing elements using a flat index only requires the linear values (from 0 to $N-1$, where N is the number of elements in the tensor), the shape of the tensor, and the order of traversal (e.g. C or F ordering). In contrast, indexing using a multi-index requires $D \times N$ values, where D is the dimension of the tensor. Thus, flat indices require less memory when enumerating the elements of the tensor while the dimensionality information can be disregarded. As we will see later in the chapter, operations such as gather and scatter on multi-dimensional tensors are usually performed on flat indices. However, multi-indices can encode the grouping of the elements. Slicing and groupwise reduction (e.g. sum over rows) are performed using multi-indices. Therefore, depending on the type of data manipulation needed, one may use a flat index at certain steps of the algorithm, but then switch to use a multi-index in another part. In Chapter 7, we will also see additional examples of how sparse matrices leverage both flat index and multi-index to store sparse data and how we can perform specialized operations on this data structure.

3.1.3 Boolean Indexing

Instead of using integers to select elements from the tensor, we can also use a tensor of boolean indicators whose shape is identical to the input tensor to select the desired elements. This way of indexing is helpful for us to select elements based on logical conditions. Let's suppose again we have a 2D matrix `X`, and we would like to select all the odd values from the tensor.

```
import numpy as np
X = np.array([
    [5, 2, 8, 3],
    [4, 9, 6, 7],
    [8, 11, 12, 10]
])

# Boolean condition of odd elements
# same shape as X
condition = X % 2 == 1

# Extract the odd elements
X[condition]
# array([5, 3, 9, 7, 11])
```

In Chapter 5, we will introduce the concept of masking. In fact, selecting elements from a tensor-based on a boolean condition can also be considered as a boolean masking operation (i.e. removing unselected elements).

On the other hand, we can also find the indices of the boolean conditions, using the `.where` operation. Consider the same 2D matrix as above, we can find the indices of the odd numbers in the matrix using

```
import numpy as np
X = np.array([
    [5, 2, 8, 3],
    [4, 9, 6, 7],
    [8, 11, 12, 10]
])

# indices of odd elements
row_index, col_index = np.where(X % 2 == 1)

# row_index: array([0, 0, 1, 1, 2])
# col_index: array([0, 3, 1, 3, 1])
```

Note that the `np.where` function returns two outputs, corresponding to the indices of the odd numbers along each dimension. In general, for a D-dimensional tensor, `np.where` will return D arrays of indices, corresponding to the indices of

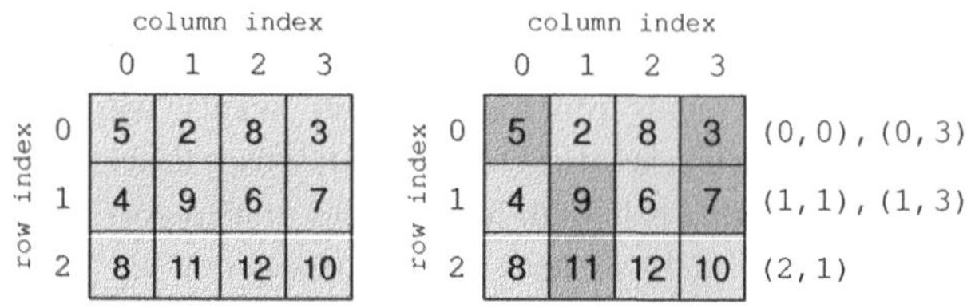

Figure 3.4 Illustration of extracting indices that satisfy boolean conditions.

the `True` conditions along each dimension. See Figure 3.4 for an illustration of the above example.

NumPy, Tensorflow (through `tf.experimental.numpy`), and PyTorch also implement a special `where` operation, `nonzero`, and its flat version `flatnonzero` (only in NumPy). As the names suggest, `nonzero` returns the indices (as a tuple) of the tensor along each dimension where the elements are not zeros (equivalently, `np.where(X != 0)`); `flatnonzero` returns the non-zero indices as if the tensor is being flattened (equivalently, `np.where(X.flatten() != 0)`). Since zeros are commonly used to fill missing data (e.g. padding, as we will see in Chapter 5), these convenient operations can sometimes be useful to determine the extent of valid data stored in the tensor.

3.2 Slicing

In Section 3.1, we have looked at ways to retrieve a single element using different indexing systems. However, sometimes, we would also like to select a segment or an interval of values from an array. This is known as **slicing**. In Python, we can "slice" an array by selecting elements between a start and an end index of a list `l` using the syntax `l[start:stop]`. For example, let `l = [1, 3, 5, 7, 9]`, and we would like to select all elements starting from index `1` to index `4` (excluded), using `l[1:4]`, which returns `[3, 5, 7]`. Similar to **slicing** is the concept of **striding**: instead of selecting all consecutive elements between two indices, **striding** selects elements between two indices by consistently skipping over a certain number of elements, with the syntax `l[start:stop:step]`. For example, suppose we would like to select every other element between index `1` and index `4` (excludes) (i.e. skipping 1 element when selecting between the indices, or selecting every second element). We can use `l[1:4:2]`, which returns `[3, 7]`. Additional commonly used slicing and striding syntaxes in Python are listed in Table 3.1.

In NumPy, TensorFlow, and PyTorch, slicing and striding operations can not only be applied to 1D vectors but can also generalize to each axis of a multi-dimensional tensor. Figures 3.5 and 3.6 illustrate several examples of slicing and striding operations in 2D matrix and 3D tensors, respectively.

Table 3.1 List of slicing and striding operations with example usages in 1D arrays.

Operations	Syntax	Example
Slicing between two indices	`start:stop`	`l[1:4] -> [3, 5, 7]`
Slicing using negative indices (translated to `start+n` and `stop+n`)	`start:stop`	`l[-4:-1] -> [3, 5, 7]`
Slicing from a start index to the last element	`start:`	`l[2:] -> [5, 7, 9]`
Slicing from the first element to a stop index	`:stop`	`l[:3] -> [1, 3, 5]`
Striding	`start:stop:step`	`l[1:4:2] -> [3, 7]`
Selecting all elements	`:`	`l[:] -> [1, 3, 5, 7, 9]`
Selecting all elements except the last element	`:-1`	`l[:-1] -> [1, 3, 5, 7]`
Selecting all elements except the last *N* element(s)	`:-N`	`l[:-2] -> [1, 3, 5]`
Reversing array	`::-1`	`l[::-1] -> [9, 7, 5, 3, 1]`
Striding from the first element to a stop index	`:stop:step`	`l[:4:2] -> [1, 5]`
Striding from a start index to the last element	`start::step`	`l[1::2] -> [3, 7]`
Striding from first to last element	`::step`	`l[::2] -> [1, 5, 9]`
Striding from last to first element	`::-step`	`l[::-2] -> [9, 5, 1]`

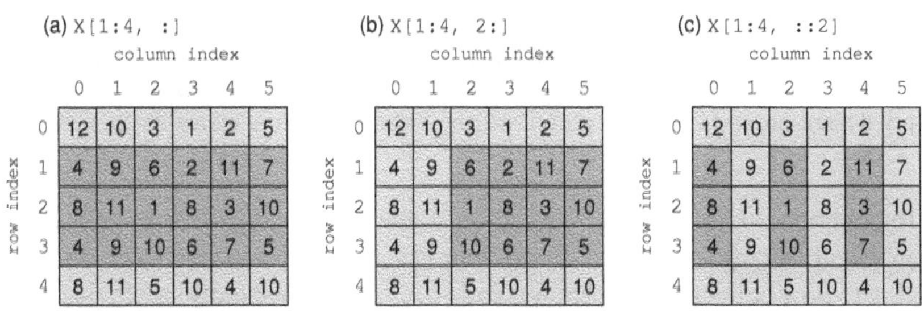

Figure 3.5 Examples of 2D matrix slicing and striding in NumPy, Tensorflow, or PyTorch tensors. (a) Slicing a subset of rows but keeping all the elements along the column: `X[1:4, :]`; (b) slicing a block of a matrix composed of a subset of rows and columns: `X[1:4, 2:]`; (c) slicing a subset of rows, and striding over every two columns of the matrix: `X[1:4, ::2]`.

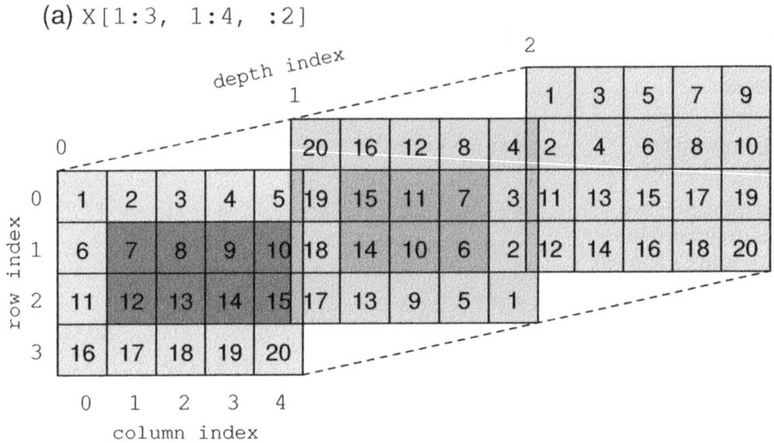

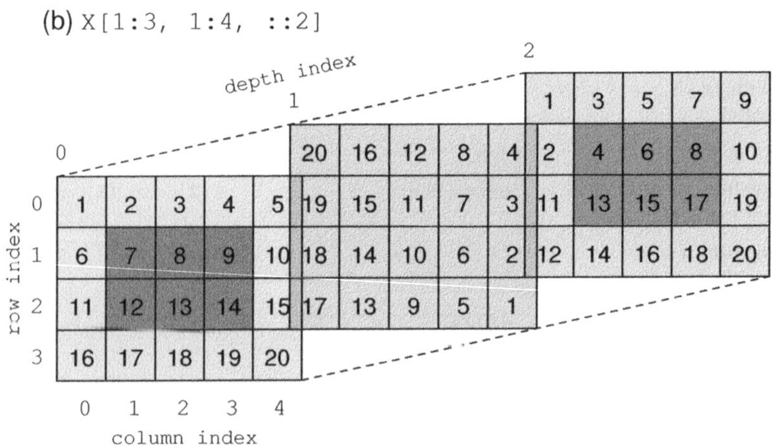

Figure 3.6 Examples of 3D tensor slicing and striding in NumPy, Tensorflow, or PyTorch tensors. (a) Slicing a subset of the tensor elements along the row, column, and depth dimensions: `X[1:3, 1:4, :2]`; (b) slicing first and third layers of the tensor along the depth dimension, and for each layer (2D matrix), select a subset of elements: `X[1:3, 1:4, ::2]`.

In addition to the syntaxes listed in Table 3.1, there are two additional convenient slicing operators, `None` (equivalently, `np.newaxis` in NumPy and `tf.newaxis` in Tensorflow) and `...` (or equivalently `Ellipsis`, a Python's builtin constant), that can be used only with NumPy, Tensorflow, and PyTorch tensor slicing, but not with Python `list`. Slicing a dimension with `None` or `np.newaxis` adds an extra singleton dimension at the axis of slicing.

The operation is equivalent to `np.expand_dims` in NumPy, `tf.expand_dims` in Tensorflow, or `torch.unsqueeze` in PyTorch, respectively. For example,

```
import numpy as np
X = np.random.rand(4, 5) # shape (4, 5)
# Add an extra 3rd dimension
# shape (4, 5, 1)
assert np.allclose(
    X[:, :, None].shape,
    np.expand_dims(X, axis=2).shape
)
# Insert an extra dimension in the first dimension
# shape (1, 4, 5)
assert np.allclose(
    X[None, :, :].shape,
    np.expand_dims(X, axis=0).shape
)
# Insert an extra dimension between the original
# 1st and 2nd dimensions: shape (4, 1, 5)
assert np.allclose(
    X[:, np.newaxis, :].shape,
    np.expand_dims(X, axis=1).shape
)
```

`...` (ellipsis) operator, on the other hand, allows us to simplify the expression when slicing all elements across multiple consecutive dimensions. For example, let `X` be a tensor of shape `[5, 2, 3, 8, 7]`, and we would like to select all elements along dimensions 2–5 but a subset along the first dimension. We can do `X[2:4, :, :, :, :]` or equivalently, `X[2:4, ...]`.

```
import numpy as np
X = np.random.rand(5, 2, 3, 8, 7)
# Selecting subset of a batch
assert np.allclose(
    X[2:4, :, :, :, :].shape,
    X[2:4, ...].shape
)
# Selecting the first slice of the last dimension
assert np.allclose(
    X[:, :, :, :, 0].shape,
    X[..., 0].shape
)
# Selecting a subset of the first dimension and
# the last slice from the last dimensions
assert np.allclose(
    X[1:3, :, :, :, -1].shape,
```

```
    X[1:3, ..., -1].shape
)
```

The use of ellipsis (`...`) for slicing can be convenient when implementing functions that accept variable dimensional tensors. For example, suppose we would like to implement a function that always extracts items in a sequence with one of the forms `(batch_size, seq_length)`, `(batch_size, seq_length, feature_size)`, or `(batch_size, seq_length, feature_size, num_instances)`. Notice that the sequence dimension is always at the `axis=1`, and the batch dimension is always at `axis=0`, then we can do

```
import numpy as np

def extract_last_event(X):
    return X[:, -1, ...]

# (batch_size, seq_length)
X1 = np.random.rand(16, 50)
assert np.allclose(
    extract_last_event(X1).shape,
    (16, )
)

# (batch_size, seq_length, feature_size)
X2 = np.random.rand(16, 50, 32)
assert np.allclose(
    extract_last_event(X2).shape,
    (16, 32)
)
# (batch_size, seq_length, feature_size, num_instances)
X3 = np.random.rand(16, 50, 32, 8)
assert np.allclose(
    extract_last_event(X3).shape,
    (16, 32, 8)
)
```

With the ellipsis `...`, the function can handle all three variants of the sequence inputs. Notice that in the case when the input has the form `(batch_size, seq_length)`, there are only two dimensions, while in our implementation, it looks like we are extracting from at least three dimensions. It turns out that the ellipsis operator `...` will ignore the slicing of the third dimension since the dimension does not exist.

Finally, there can only be one ellipsis ... operator when specifying a slice. Specifying more than one ellipsis will throw an error, for example:

```
X = np.arange(30).reshape(3, 2, 5)
X[..., 0, ...]
```

throws the error in NumPy:

```
IndexError: an index can only have a single ellipsis ('...')
```

This is because the slicing dimensions represented by the two ellipses are ambiguous. `X[..., 0, ...]` could potentially represent any of the following:

- `X[0, :, :]`
- `X[:, 0, :]`
- `X[:, :, 0]`

The program cannot determine which is the intended slicing operation.

Throughout the book, I will sometimes overload the ellipsis operator ... as the unspecified value of the argument when describing the usage of function, or used for omitting the enumeration of values in a list. However, these usages should never occur inside a code block (as all code blocks should be runnable as is). I will indicate the meaning of the ellipsis when the notation becomes confusing.

3.2.1 Reusing Slice Configuration

There may be times when we would like to apply identical slicing logic to more than 1 tensors, or perhaps we would like to pass the slicing configuration as a variable to a function. In these scenarios, we would need to construct the `slice` object. Python has a built-in method to build slices, `slice`, which is used as

- `slice(stop)`, which is equivalent to the slice as in `l[:stop]`;
- `slice(start, stop)`, which is equivalent to the slice as in `l[start:stop]`;
- `slice(start, stop, step)`, which is equivalent to the slice as in `l[start:stop:step]`.

Similarly, NumPy implements `np.s_` which can be used in an identical way but it uses the more commonly used colon : syntax than `slice`. Table 3.2 re-enumerates the examples of slicing operations as listed in Table 3.1 previously, but using `slice` and `np.s_` to build the slicing configurations.

Table 3.2 List of examples of building slicing and striding configurations using Python `slice` or NumPy's `np.s_`.

Operations	**Python `slice`**	**NumPy `np.s_`**
Slicing between two indices	`slice(start, stop)`	`np.s_[start:stop]`
Slicing using negative indices (translated to `start+n` and `stop+n`)	`slice(start, stop)`	`np.s_[start:stop]`
Slicing from a start index to the last element	`slice(start, None)`	`np.s_[start:]`
Slicing from the first element to a stop index	`slice(None, stop)`	`np.s_[:stop]`
Striding	`slice(start, stop, step)`	`np.s_[start:stop:step]`
Selecting all elements	`slice(None)`	`np.s_[:]`
Selecting all elements except the last element	`slice(None, -1)`	`np.s_[:-1]`
Selecting all elements except the last *N* element(s)	`slice(None,-N)`	`np.s_[:-N]`
Reversing array	`slice(None, None, -1)`	`np.s_[::-1]`
Striding from the first element to a stop index	`slice(None, stop, step)`	`np.s_[:stop:step]`
Striding from a start index to the last element	`slice(start, None, step)`	`np.s_[start::step]`
Striding from first to last element	`slice(None, None, step)`	`np.s_[::step]`
Striding from last to first element	`slice(None, None, -step)`	`np.s_[::-step]`

Note `np.s_` also supports convenient construction of slicing configurations of multiple dimensions, with the syntax `np.s_[start_0:stop_0:step_0, start_1:stop_1:step_1, ..., start_n:stop_n:step_n]`. This will return a tuple of `slice` objects, e.g. `(slice(start_0, stop_0, step_0), slice(start_1, stop_1, step_1),...,slice(start_n, stop_n, step_n))`. To clarify, I have overloaded the ellipsis `...` symbol here and it carries two meanings here:

- I intend the ellipsis "..." here as a notation to indicate there may be a variable number of dimensions when constructing the slices.

- However, putting the ellipsis operator `...` in the `np.s_` method is also valid here, which indicates that the slices across the omitted dimensions select all items within the dimension. This is just how ellipsis operator is being used previously. When `...` is used to construct multidimensional slices, the elided dimensions will be represented as Python's `Ellipsis` constant, e.g. `(slice(start_0, stop_0, step_0), slice(start_1, stop_1, step_2), Ellipsis, slice(start_3, stop_3, step_3))`.

Once we have built the `slice` configuration object, we can apply the slicing logic the same way as we did before, e.g.

```
import numpy as np
X = np.random.rand(5, 8, 6)
Y = np.random.rand(3, 9, 7)

# Build the slice object
s = np.s_[:4, 2:4, 5:]

# Shape may not necessarily be identical
print(X[s].shape) # shape (4, 2, 1)
print(Y[s].shape) # shape (3, 2, 2)
```

You may notice that the first dimension of tensor `Y` is of size 3 but we are slicing four items. Yet, the program did not throw an error, and returned three items in the first dimension, less than the four items asked by the slice. In general, Python's tensor slicing will take as many elements it can as possible; so if we specify the slice to take up to a `stop` index beyond the size of the tensor's dimension, it will return only whatever elements it can take in the slice. ***Taking a single element at a position beyond the length of the array will throw an error.*** To illustrate this behavior in another set of examples,

```
x = [1, 2, 3, 4, 5]

# stop beyond the size of the array
print(x[:7]) # [1, 2, 3, 4, 5]

# stop beyond the size of the array, stride 2
print(x[:7:2]) # [1, 3, 5]

# empty, no elements to take
# since the slice is beyond the length of the array
print(x[6:8]) # []

```

```
# The following, however, will throw an error
print(x[7]) # IndexError: list index out of range
```

3.3 Case Study: Get Consecutive Index

Suppose we have a list of integers `[-1, 1, 2, 3, 4, 7, 9, 10, 11, 12, 13, 15]`, we can recognize that there are two groups that are composed of consecutive integers, `[1, 2, 3, 4]` and `[9, 10, 11, 12, 13]`. We would like to find the start and end indices of these two blocks of integers. This can be helpful when we want to convert the list of integers into texts like `"-1, 1~4, 7, 9~13, 15"` which may be easier to read than the long list of integers. The following code implements our desired function:

```
1  import numpy as np
2
3  def get_consecutive_index(t, N=1, interval=True):
4      """
5      Given a sorted array of integers, find the start and the end
6      of consecutive blocks
7      e.g. t = [-1, 1,2,3,4, 7, 9,10,11,12,13, 15],
8      return [[1,4], [6,10]]
9
10     Inputs:
11     * t: the sorted array of integers
12     * N: filter for at least N consecutive. Default 1
13     * interval: if True, we are filtering by N consecutive
14             intervals instead of N consecutive numbers
15     """
16     # Determine if the (next element) = (1 + previous)
17     x = np.diff(t) == 1
18     # Determine which element is the start or end of the block
19     f = np.concatenate(([False], x)) != np.concatenate((x, [False]))
20     # Index of the start and end of the block
21     f = np.where(f)[0]
22     # reshape: the start and end must come in pairs
23     f = np.reshape(f, (-1, 2))
24     # filter for at least N consecutive
25     interval = 1 - int(interval)  # 0 if True, 1 if False
26     # Determine which interval has more than desired N consecutive
       ↪   numbers
27     f = f[(interval + np.diff(f, n=1, axis=1).T[0]) >= N, :]
28     return f
```

- Line 17 uses `np.diff` to determine if the current element is a consecutive integer of the previous element. If it is a consecutive integer, then its difference from the previous element should be 1, otherwise, it will be greater than 1. For instance, given the list `[-1, 1, 2, 3, 4, 7, 9, 10, 11, 12, 13, 15]`, the backward differences between the successive elements are `[2, 1, 1, 1, 3, 2, 1, 1, 1, 1, 2]` (note that this is 1 element shorter than the original input). Therefore, elements at indices `[2, 3, 4, 7, 8, 9, 10]` are elements that are 1 greater than the preceding elements.

- Line 19 determines which element is the start or end of the block. Given `x = [False, True, True, True, False, False, True, True, True, True, False]` in the previous step, we can observe that if the element is the start of the block, then the boolean value will transition from `False` to `True` (or if the first element is also the start of the block, then the first element in `x` should also be `True`). Similarly, if the element is the end of the block, then the boolean value will transition from `True` to `False` (or if the last element is the end of the block, then the last element should also be `True`). By padding `x` with a leading `False`, we offset `x` by 1 index backward; by appending `x` with a tailing `False`, we offset `x` by 1 index forward. Then, when we compare the values of the two offset boolean arrays, we are comparing:

```
[[(False), False], # -1 (start)
 [ False,   True], # 1
 [ True,    True], # 2
 [ True,    True], # 3
 [ True,   False], # 4
 [ False,  False], # 7
 [ False,   True], # 9
 [ True,    True], # 10
 [ True,    True], # 11
 [ True,    True], # 12
 [ True,   False], # 13
 [ False, (False)]] # 15 (end)
```

This construction would allow us to determine if we are transitioning from `False` to `True` or from `True` to `False`, or if there is no transition at all, between successive elements. We can observe that,

1. Values are both `False`, then the elements are not part of the consecutive block.
2. Values are both `True`, then the elements are in the middle of the block.
3. Values are `[False, True]`, the elements are the start of the block.
4. Values are `[True, False]`, the elements are the end of the block.

- Line 21 uses `np.where` to determine the indices of the start and the end block.

- Line 23 reshapes the 1D vector into a `(num_blocks, 2)` matrix. This is because there should always be a start and end of the block. Since the input array is already sorted, the end of the block should always occur after the start of the block, given a single block.
- Lines 25–27 perform a filtering for block size. There are two modes of this: either counting the number of elements or the number of intervals. The number of intervals can be computed by `end_index - start_index`, while the number of elements can be computed by `end_index - start_index + 1`. I use the flag `interval` to switch between the two cases: if `interval=True`, then I do not need to add the `1`; if `interval=False`, then I do need to add `1`; hence, the expression `int(~interval)`.

3.4 Take and Gather

Selecting elements and slicing from a tensor is also known as the **take** or **gather** operation. NumPy, Tensorflow, and PyTorch all implement these methods as concrete functions. Table 3.3 is a summary of a list of take and gather operations in these libraries. We will walk through each of the operations and study their usage patterns in this section.

3.4.1 Take

The `take` operation is a convenient function to select elements from a tensor using flat integer indices. To illustrate with a simple example,

Table 3.3 List of take and gather operations in NumPy, Tensorflow, and PyTorch.

Operation	NumPy	Tensorflow	PyTorch
Take elements/slices with flat index	`np.take`	`tf.experimental.numpy.take`	`torch.take`
Batchwise take	`np.take_along_axis`	`tf.experimental.numpy.take_along_axis`	`torch.take_along_dim`
Gather	`np.take`, `np.take_along_axis`	`tf.gather`	`torch.gather`
Take elements by multi-index	N/A	`tf.gather_nd`	N/A

```
import numpy as np
X = np.array([2, 3, 5, 7, 11, 13, 17, 19, 23, 29])
# 1D array slicing
flat_index = [2, 3, 5, 7]
elements = np.take(X, flat_index)
# array([5, 7, 13, 19])
```

This is equivalent to `X[index]` in the case of 1D vectors (see also Figure 3.7). For a multi-dimensional tensor, the `take` operation will first flatten the tensor into a 1D vector, and then select the elements based on the flat index. For instance,

```
import numpy as np
X = np.arange(1, 7)
X = np.reshape(X, (2, 3), order="F")
# array([[1, 3, 5],
#        [2, 4, 6]])

# Taking elements using flat indices
flat_index = [1, 3, 4]
elements = np.take(X, flat_index)
# array([3, 2, 4])
```

See also Figure 3.8 for a simple illustration. Note that the order of flattening is always "`C`" order instead of following the order of the input array. For instance, we can verify that

```
import numpy as np
X = np.arange(20).reshape(4, 5)
flat_index = [1, 2, 3, 4, 5]
assert np.allclose(
    np.take(np.asarray(X, order="C"), flat_index),
    np.take(np.asarray(X, order="F"), flat_index)
)
```

The shape of the `flat_index` also determines the shape of the output. If `flat_index` is a shape `(3, 2)` matrix, then the results of `take` is also a shape `(3, 2)` matrix (see also Figure 3.9 for an illustration). For example,

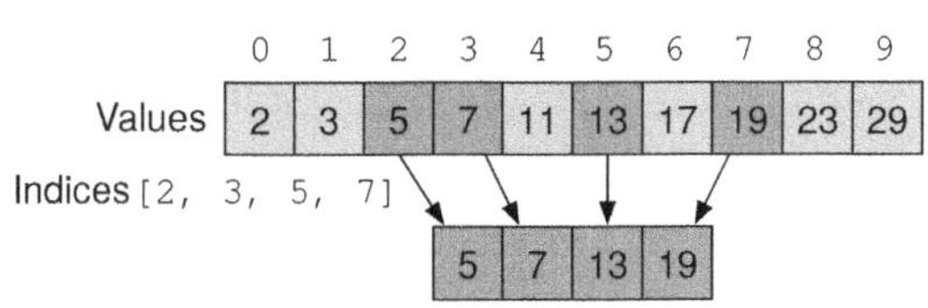

Figure 3.7 Example usage of `take` operation along a 1D vector.

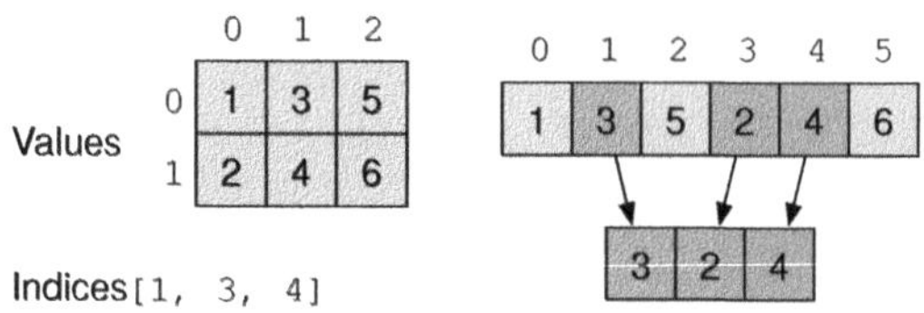

Figure 3.8 Example usage of flat `take` operation on a 2D matrix.

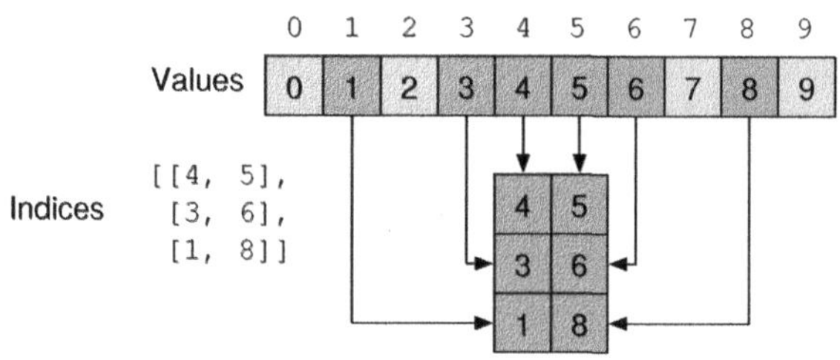

Figure 3.9 Example usage of multi-dimensional `take` operation on a vector.

```
import numpy as np
X = np.arange(10)
# Flat index with a desired output shape
flat_index = [[4, 5], [3, 6], [1, 8]] # (3, 2)
elements = np.take(X, flat_index)
# array([[4,  5],
#        [3,  6],
#        [1,  8]])
```

The `np.take` and `tf.experimental.numpy.take` operations allow an additional argument `axis` to specify along which axis the operation can extract a subset of items/slices. For examples,

```
import numpy as np
X = np.arange(30).reshape(5, 3, 2)
# Slicing along a single dimension
slice_index = [0, 3]
block = np.take(X, slice_index, axis=0)
# array([[[ 0,  1],
#         [ 2,  3],
#         [ 4,  5]],
#
#        [[18, 19],
#         [20, 21],
#         [22, 23]]])

# This operation is equivalent to slicing using indexing
assert np.allclose(
```

```
    block,
    X[slice_index, :, :]
)

# Slicing along a single dimension
# And arrange the slices across multiple dimensions
slice_index = [[0, 3], [1, 2]]
block = np.take(X, slice_index, axis=0) # (2, 2, 3, 2)
assert np.allclose(
    block,
    X[slice_index, :, :]
)
```

PyTorch's `torch.take` does not have the `axis` or `dim` argument, so more commonly simple slicing is used. Otherwise, one may also choose to use `torch.index_select(input, dim, index)`.

3.4.2 Take Along Axis

`take_along_axis` selects elements based on the integer index along a certain axis, for each item in the batch. The `axis` specifies the non-batch dimension while all other dimensions can be treated as batch dimensions (which parallelizes the operations); the `indices` specified are the indices along the dimension specified by the `axis`. Taking a simple example, suppose we have a batch of 1D sequences arranged as a 2D matrix whose shape is `(4, 5)`, and we would like to rearrange the order of the sequences one by one:

```
import numpy as np
X = np.array([
    [0, 1, 2, 3, 4],
    [1, 2, 3, 4, 5],
    [2, 3, 4, 5, 6],
    [3, 4, 5, 6, 7],
])
# Reordering index
reorder_index = np.array([
    [3, 1, 0, 4, 2],
    [2, 4, 3, 0, 1],
    [3, 0, 4, 1, 2],
    [0, 2, 1, 4, 3],
])
# Apply the reordering index
y = np.take_along_axis(X, reorder_index, axis=1)

```

```
# array([[3, 1, 0, 4, 2],
#        [3, 5, 4, 1, 2],
#        [5, 2, 6, 3, 4],
#        [3, 5, 4, 7, 6]])
```

Here, axis=0 is treated as the batch dimension, and axis=1 is treated as the indexing dimension. The integer values in indices=reorder_index do not exceed the size of the 2nd dimension (axis=1). In general, indices and input X must have the same number of dimensions (i.e. indices.ndim = X.ndim); the shape of indices specified needs to be the same or broadcastable (i.e. having singleton dimensions) to that of the input X. Let's consider a more advanced example usage of take_along_axis (note to allow the readers to track the reordered results manually, I have opted to spell out the exact matrix and reordering indices in the code below rather than using random generators):

```
import numpy as np
X1 = np.array([
    [0, 1, 2, 3, 4],
    [1, 2, 3, 4, 5],
    [2, 3, 4, 5, 6],
    [3, 4, 5, 6, 7],
])
X2 = np.array([
    [ 0, 10, 20, 30, 40],
    [10, 20, 30, 40, 50],
    [20, 30, 40, 50, 60],
    [30, 40, 50, 60, 70],
])
X3 = np.array([
    [-4, -3, -2, -1,  0],
    [-5, -4, -3, -2, -1],
    [-6, -5, -4, -3, -2],
    [-7, -6, -5, -4, -3],
])
# (4, 5, 3)
X = np.dstack([X1, X2, X3])

# Reordering index
reorder_index = np.array([
    [3, 1, 0, 4, 2],
    [2, 4, 3, 0, 1],
    [3, 0, 4, 1, 2],
    [0, 2, 1, 4, 3],
])[..., None] # last dimension broadcastable to X

```

```
# Apply the reordering index
y = np.take_along_axis(X, reorder_index, axis=1)

# Examine the results one matrix at a time
print(y[:, :, 0])
# array([[3, 1, 0, 4, 2],
#        [3, 5, 4, 1, 2],
#        [5, 2, 6, 3, 4],
#        [3, 5, 4, 7, 6]])
print(y[:, :, 1])
# array([[30, 10,  0, 40, 20],
#        [30, 50, 40, 10, 20],
#        [50, 20, 60, 30, 40],
#        [30, 50, 40, 70, 60]])
print(y[:, :, 2])
# array([[-1, -3, -4,  0, -2],
#        [-3, -1, -2, -5, -4],
#        [-3, -6, -2, -5, -4],
#        [-7, -5, -6, -3, -4]])
```

In this case, both `axis=0` and `axis=2` are treated as the batch dimensions. The input `X` can also be broadcasted to the indices at the singleton dimensions. For example,

```
import numpy as np
X = np.array([[0, 1, 2, 3, 4]])[:, :, None] # (1, 5, 1)
reorder_index = np.array([
    [3, 1, 0, 4, 2],
    [2, 4, 3, 0, 1],
    [3, 0, 4, 1, 2],
    [0, 2, 1, 4, 3],
])[..., None] # (4, 5, 1)
# Apply the reordering index
y = np.take_along_axis(X, reorder_index, axis=1)
assert np.allclose(
    y.shape,
    (4, 5, 1)
)
assert np.allclose(
    y,
    np.array([[3, 1, 0, 4, 2],
              [2, 4, 3, 0, 1],
              [3, 0, 4, 1, 2],
              [0, 2, 1, 4, 3]])[:, :, None]
)
```

Finally, let's look at the example where there is no broadcasting:

```
import numpy as np
X1 = np.array([
    [0, 1, 2, 3, 4],
    [1, 2, 3, 4, 5],
    [2, 3, 4, 5, 6],
    [3, 4, 5, 6, 7],
])
X2 = np.array([
    [ 0, 10, 20, 30, 40],
    [10, 20, 30, 40, 50],
    [20, 30, 40, 50, 60],
    [30, 40, 50, 60, 70],
])
X3 = np.array([
    [-4, -3, -2, -1,  0],
    [-5, -4, -3, -2, -1],
    [-6, -5, -4, -3, -2],
    [-7, -6, -5, -4, -3],
])
# (4, 5, 3)
X = np.dstack([X1, X2, X3])

# Reordering index
index_1 = np.array([
    [3, 1, 0, 4, 2],
    [2, 4, 3, 0, 1],
    [3, 0, 4, 1, 2],
    [0, 2, 1, 4, 3],
])
index_2 = np.array([
    [1, 0, 2, 3, 4],
    [3, 0, 4, 1, 2],
    [4, 0, 1, 2, 3],
    [0, 1, 2, 3, 4],
])
index_3 = np.array([
    [3, 0, 1, 4, 2],
    [2, 0, 1, 3, 4],
    [2, 1, 4, 3, 0],
    [0, 2, 4, 1, 3],
])
# (4, 5, 3)
reorder_index = np.dstack([index_1, index_2, index_3])

# Apply the reordering index
y = np.take_along_axis(X, reorder_index, axis=1)
# Examine several slices
```

```
print(y[0, :, 0])
# array([3, 1, 0, 4, 2])
print(y[1, :, 2])
# array([-3, -5, -4, -2, -1])
print(y[2, :, 1])
# array([60, 20, 30, 40, 50])
```

We can easily verify that

- `np.take(X[0, :, 0], reorder_index[0, :, 0]) == y[0, :, 0];`
- `np.take(X[1, :, 2], reorder_index[1, :, 2]) == y[1, :, 2];`
- `np.take(X[2, :, 1], reorder_index[2, :, 1]) == y[2, :, 1];`
- ...etc.

In general,

```
for ii in range(X.shape[0]):
    for jj in range(X.shape[2]):
        assert np.allclose(
            np.take(X[ii, :, jj], reorder_index[ii, :, jj]),
            y[ii, :, jj],
        )
```

Hence, `take_along_axis` is the vectorized version of `take` operation where it parallelizes the process of element selections from 1D vectors across batches (which may be arranged over multiple dimensions).

3.4.3 Gather

The `gather` operation is an alternative implementation of `take` and `take_along_axis` operations. Tensorflow and PyTorch implement the `gather` operation differently. `tf.gather` always selects the elements or slices along a certain dimension, specified by `axis`, which defaults to `None`. The value of `axis` is dependent on another argument, `batch_dims` (the number of batch dimensions starting from axis 0; this is not to specify which dimensions or axes are the batch dimensions), which defaults to `0`. It is necessary that `axis >= batch_dims` and `batch_dims <= index.ndim`. Despite the name, PyTorch's `torch.gather`, on the other hand, selects only single elements at a time given each index instead of selecting slices via broadcasting (i.e., `X` and `indices` do not broadcast). As a consequence, the results of `torch.gather` have the same shape as `indices`. Also, like `np.take_along_axis`, `torch.gather` requires the input `X` and `indices`

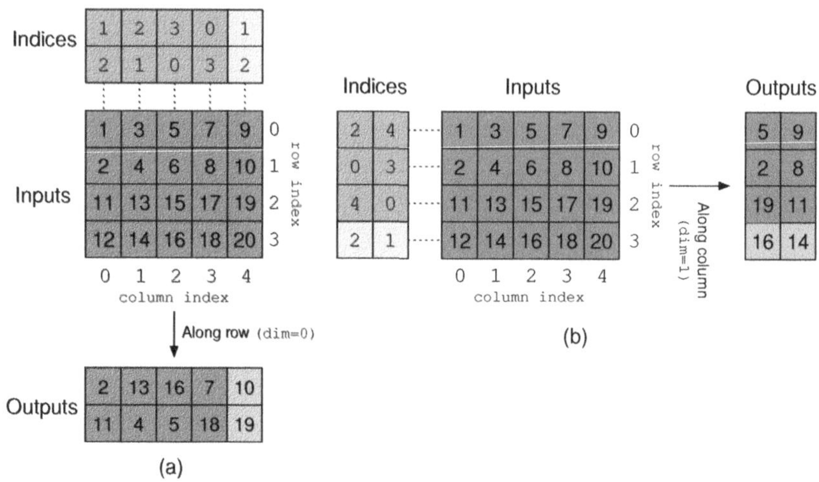

Figure 3.10 Illustrated examples of `torch.gather` operation on a 2D matrix. (a) Gathering across the rows (`torch.gather(X, dim=0, index=index)`). The `index` and inputs are both two dimensional matrices. The two matrices are aligned at the upper left corner (i.e. index `(0, 0)`). Columns of the `index` matrix correspond to the columns of the inputs, while the values within the `index` matrix correspond to the row indices. The number of columns in the `index` matrix can be less than (i.e. solid region) or equal to (i.e. solid and light region together) the number of columns in the inputs. The final outputs have the same shape as the `index` matrix; (b) gathering across the columns (`torch.gather(X, dim=1, index=index)`). The `index` and inputs are both two-dimensional matrices and are aligned at the upper left corner. Rows of `index` matrix correspond to the rows of the inputs, while the values within the `index` matrix correspond to the column indices. The number of rows in the `index` matrix again can be less than or equal to that of the number of rows in the inputs. The final output has the same shape as the `index` matrix.

to have the same dimensions. Figure 3.10 provides a couple of illustrated examples of how `torch.gather` works. We will discuss each operation's pattern of usage in the following with more details and examples.

For 1D tensors, `tf.gather(X, index)` and `torch.gather(X, dim=0, index=index)` are equivalent to `np.take(X, index)`. In this case, setting `axis=None` or `axis=0` has the same effect. For Tensorflow, `axis=None` pairing with `batch_dims=0` (both are the default) would result `axis=0`, where the elements are taken along the first (and the only) dimension of the 1D vector. PyTorch requires the user to specifically set `dim=0` (setting `dim=None` would throw an error). For NumPy, setting `axis=None` signals the function to flatten the 1D vector, which results in itself. Then, the values are taken along the same vector.

```
import numpy as np
import tensorflow as tf
import torch
X = np.arange(10)
index = [3, 5]
y_np = np.take(X, index)
y_tf = tf.gather(X, index)
y_torch = torch.gather(
    torch.tensor(X), 0, torch.tensor(index)
)
# all returns [3, 5]
```

For multi-dimensional tensors, `tf.gather(X, index)` is equivalent to `np.take(X, index, axis=0)`. Again, when `batch_dims=0` by default, `axis` is set to `0` (`axis=0`), thus `tf.gather(X, index)` is equivalent to `tf.gather(X, index, axis=0)`. For PyTorch's `torch.gather`, to return the entire slice, it is necessary to repeat the indices across the batching dimensions (thus in this case, it is recommended to use slicing directly):

```
import numpy as np
import tensorflow as tf
import torch
X = np.arange(20).reshape(4, 5)
index = np.array([1, 3])
y_np = np.take(X, index, axis=1)
y_tf = tf.gather(X, index, axis=1)
y_torch = torch.gather(
    torch.tensor(X),
    0,
    torch.tensor(np.repeat([index], X.shape[1], axis=0).T)
)
# all returns:
# [[ 5,  6,  7,  8,  9],
#  [15, 16, 17, 18, 19]]

# In contrast, if we do not repeat the index
y_torch_index = torch.gather(
    torch.tensor(X), 0, torch.tensor(index[:, None])
)
# tensor([[ 5],
#         [15]])
```

When specifying an integer value to the axis argument, tf.gather(X, index, axis=...) is equivalent to np.take(X, index, axis=...). index can also be multi-dimensional and will be handled the same way as np.take:

```
import numpy as np
import tensorflow as tf
X = np.arange(30).reshape(3, 5, 2)
slice_index = [[0, 3], [1, 1]]
block_np = np.take(X, slice_index, axis=1)
block_tf = tf.gather(X, slice_index, axis=1)
assert np.allclose(
    block_np, block_tf
)
```

tf.gather can also be used like np.take_along_axis (as well as torch.take_along_dims) to vectorize the gathering of elements over batches by specifying batch_dims to a value greater than 0. The difference though is that tf.gather only allows batching at the initial dimensions whereas np.take_along_axis allows the batch dimensions at arbitrary dimensions. For instance,

```
import numpy as np
import tensorflow as tf
import torch
X = np.random.randint(0, 10, size=(4, 3, 5))
reorder_index = np.random.randint(0, 5, size=(4, 3, 5))
y_np = np.take_along_axis(X, reorder_index, axis=2)
y_torch = torch.take_along_dim(
    torch.tensor(X), torch.tensor(reorder_index), axis=2
)
y_tf = tf.gather(X, reorder_index, axis=2, batch_dims=2)
```

When it comes to broadcasting, tf.gather will broadcast any dimensions not covered by the leading batch_dims batch dimensions and the axis dimension (hence, treating the rest of the dimensions as slices rather than single elements), while np.take_along_axis would require us to add additional singleton dimensions to signal broadcasting, i.e.

```
import numpy as np
import tensorflow as tf
import torch
```

```
X = np.random.randint(0, 10, size=(4, 3, 5, 6))
# 3rd dimension as the broadcasting dimension
reorder_index = np.random.randint(0, 5, size=(4, 3, 1, 7))
# Returns a tensor of shape (4, 3, 5, 7)
y_np = np.take_along_axis(
    X, reorder_index, axis=3
)
y_torch = torch.take_along_dim(
    torch.tensor(X),
    torch.tensor(reorder_index),
    axis=3
)
y_tf = tf.gather(
    # taking out the singleton dimension
    X, reorder_index[..., 0, :], axis=3, batch_dims=2
)
```

A heuristic on the expected output shape of `tf.gather` can therefore be derived (and also based on [Abadi et al., 2023a]) and implemented as

```
np.concatenate([
    inputs.shape[:axis],
    index.shape[batch_dims:],
    inputs.shape[axis+1:]
]).astype(int)
```

3.4.4 *N*-Dimensional Gather

Both NumPy and PyTorch use arrays of multi-indices to select multiple elements/slices. Tensorflow implements a special operation, `tf.gather_nd` to achieve the same purpose. The `tf.gather_nd` operation uses multi-index to gather elements from multiple dimensions instead of using flat index to gather from a single dimension like the `tf.gather` operation. As a simple example,

```
import numpy as np
import tensorflow as tf
import torch
X = np.array([
    [1, 3, 2, 4],
    [6, 8, 7, 9],
])
# 2D indices
index = np.array([
    [0, 1],
```

```
        [0, 3],
        [1, 0],
        [1, 2],
    ])

    # gather_nd
    y_tf = tf.gather_nd(X, index)
    # For numpy and torch, use indexing
    y_np = X[index.T[0], index.T[1]]
    y_torch = torch.tensor(X)[index.T[0], index.T[1]]
    # array([3, 4, 6, 7])
```

See Figure 3.11 for an illustration.

`tf.gather_nd` operation is unique to Tensorflow since Tensorflow only allows scalars and slices for indexing, and directly using an array of multi-indices to select multiple elements will throw an error:

```
import tensorflow as tf
X = tf.constant([
    [1, 3, 2, 4],
    [6, 8, 7, 9],
])
# 2D indices
index = tf.constant([
    [0, 1],
    [0, 3],
    [1, 0],
    [1, 2],
])
# This throws an error
X[index[:, 0], index[:, 1]]
```

will get the following error:

```
TypeError: Only integers, slices (':'), ellipsis ('...'),
↪  tf.newaxis ('None') and scalar tf.int32/tf.int64 tensors
↪  are valid indices, got <tf.Tensor: shape=(4,),
↪  dtype=int32, numpy=array([0, 0, 1, 1], dtype=int32)>
```

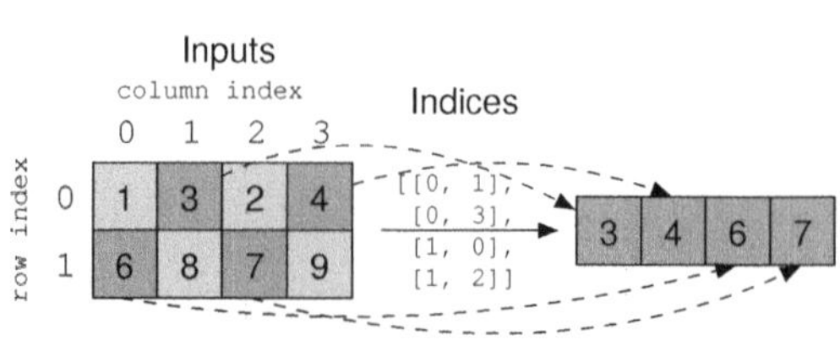

Figure 3.11 Illustration of multi-element multi-indexing using *N*-dimensional gather.

`tf.gather_nd` support slice gathering using multi-index. For instance,

```
import tensorflow as tf
X = tf.random.uniform((4, 5, 6))
# Multi-indexing the first 2 dimensions,
# using last dimension as the slice/block dimension
index = tf.constant([
    [0, 3],
    [1, 2],
    [3, 0],
    [2, 2],
]) #(4, 2)

# Gather 4 slices from the 3D tensor
y = tf.gather_nd(X, index) # (4, 6)
```

`index.shape[1]` number of leading dimensions in `X` is extracted using the multi-index, and the remaining lagging dimensions are left as the slice dimensions. If `index.shape[1] == X.ndim`, then 1D vectors are created by extracting single elements from the input `X`. Otherwise, a `(X.ndim - index.shape[1] + 1)`-dimensional tensor is created.

`tf.gather_nd` can also be used to gather slices arranged over a multi-dimensional batch.

```
import tensorflow as tf
X = tf.random.uniform((2, 3, 5, 4))
index = tf.experimental.numpy.random.randint(
    0, 4, (2, 3, 4, 2)
)
batch_dims = 2
y = tf.gather_nd(X, index, batch_dims=batch_dims)
```

In this example, we treat the first two dimensions of the `index` and `X` as the batch dimensions. Then, for each location within the batch, we take 4 of 2D points from each 5×4 matrix. See Figure 3.12 for an illustration.

Tensorflow's documentation on `tf.gather_nd` [Abadi et al., 2023b] provides even more examples. Readers are encouraged to study these examples carefully to understand the operation even further.

`tf.gather_nd` (as well as similar operations in NumPy and PyTorch) that gathers elements at indices rather than entire slices in a batch are quite useful for extracting heterogeneous regions of interests (ROIs) from high-dimensional data, such as medical imaging (e.g. fMRI brain structures) or video data (e.g. for moving objects), where sequences of 2D or 3D images are organized over time.

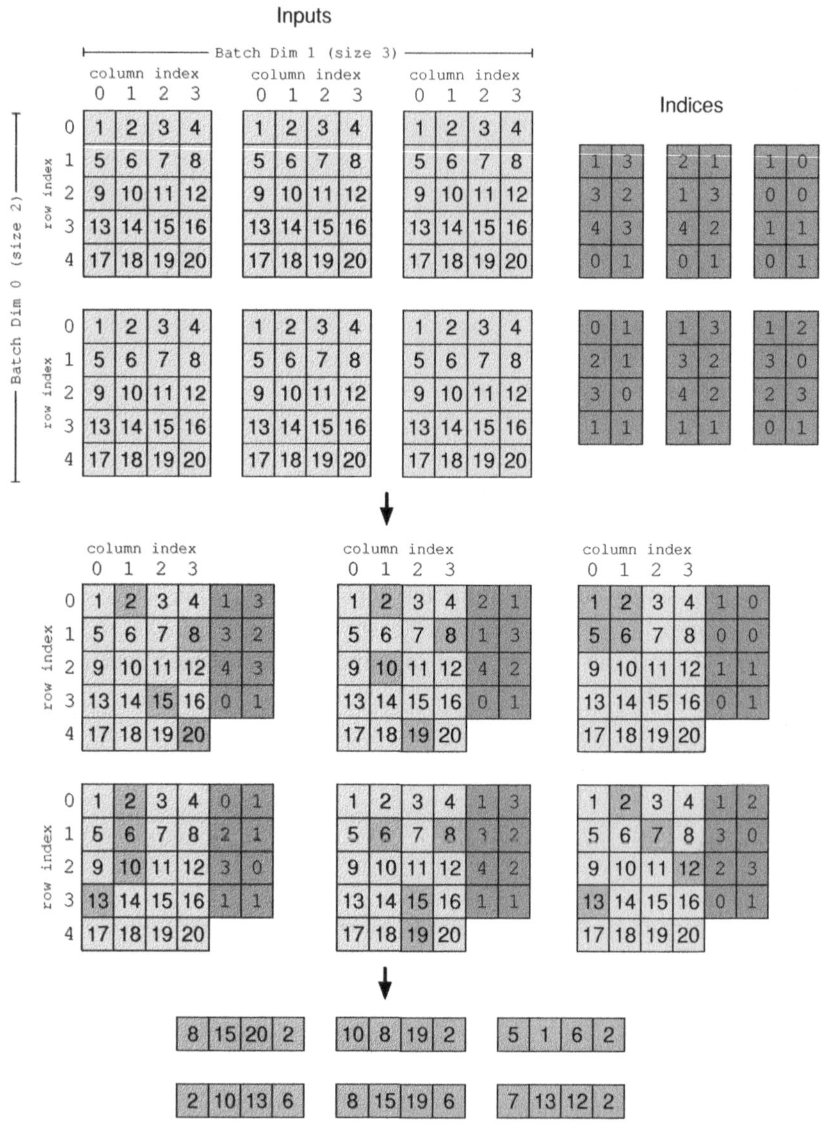

Figure 3.12 Illustration of `tf.gather_nd` with nonzero `batch_dims` argument. The initial two batch dimensions are unfolded into a 2 × 3 grid in this schematic representation.

3.5 Assign Values at Index

NumPy's `np.ndarray` and PyTorch's `torch.Tensor` are mutable, which means the values of the object can be updated. This behavior is similar to Python's `list` class. In contrast, Tensorflow's `tf.Tensor` is immutable [Abadi et al., 2023c], which is similar to Python's `tuple` or `set` class. This may be tied to Tensorflow's original operating principle which requires series of operations must be compilable into a computational graph before execution (i.e. graph mode), rather than executing as is, in place (i.e. eager mode, which is a more recent development since version 2.0). Both NumPy's `np.ndarray` and PyTorch's `torch.Tensor` can be modified in-place (i.e. without creating a new object) by specifying the indices of the element to change and the values to replace. For examples,

```
import numpy as np
import tensorflow as tf
import torch

# Python list
X_py = [1, 2, 3, 4]
X_py[3] = 100 # [1, 2, 3, 100]
# NumPy
X_np = np.array([1, 2, 3, 4])
X_np[3] = 100 # array([1, 2, 3, 100])
# PyTorch
X_torch = torch.tensor([1, 2, 3, 4])
X_torch[3] = 100 # tensor([1, 2, 3, 100])
# Tensorflow
X_tf = tf.constant([1, 2, 3, 4])
X_tf[3] = 100 # throws an error
```

The last line, which attempts to assign a value to a Tensorflow `tf.Tensor`, throws the following error:

```
TypeError: 'tensorflow.python.framework.ops.EagerTensor'
 ↪ object does not support item assignment
```

To illustrate that assigning a value at a given index is an in-place operation, let's look at the following example:

```
import numpy as np
X = np.array([1, 2, 3, 4])
# create a new variable that points to X
```

```
Y = X
# modify X
X[3] = 100
# print
print(X) # array([1, 2, 3, 100])
print(Y) # array([1, 2, 3, 100])
```

We will observe that both X and Y are modified, even though we only performed the assignment operation on X.

In addition to modifying values, we can also perform in-place value accumulations using Python's <operator>= syntaxes, as listed in Table 3.4:

Table 3.4 List of accumulation operators that can be used with NumPy and PyTorch.

Accumulation Operation	In-place Operator	Equivalent Expression
Addition	+=	x[index] += y $\Longleftrightarrow$ x[index] = x[index] + y
Subtraction	-=	x[index] -= y $\Longleftrightarrow$ x[index] = x[index] - y
Multiplication	*=	x[index] *= y $\Longleftrightarrow$ x[index] = x[index] * y
Division	/=	x[index] /= y $\Longleftrightarrow$ x[index] = x[index] / y
Integer division	//=	x[index] //= y $\Longleftrightarrow$ x[index] = x[index] // y
Modulus	%=	x[index] %= y $\Longleftrightarrow$ x[index] = x[index] % y
Exponentiation	**=	x[index] **= y $\Longleftrightarrow$ x[index] = x[index] ** y

As a quick example:

```
import numpy as np
X = np.array([1, 2, 3, 4])
X[3] += 5
print(X) # array([1, 2, 3, 9])
```

For multi-dimensional arrays, we may use multi-index to reference the element and assign the values at the index.

```
import numpy as np
X = np.arange(20).reshape(4, 5)
X[3, 2] -= 20 # 17 -> -3
print(X)
# array([[ 0,  1,  2,  3,  4],
```

```
#         [ 5,  6,  7,  8,  9],
#         [10, 11, 12, 13, 14],
#         [15, 16, -3, 18, 19]])
```

We can also replace the values of a slice with a single value.

```
import numpy as np
X = np.arange(20).reshape(4, 5)
# replacing the last column with -99
X[:, -1] = -99
print(X)
# array([[  0,   1,   2,   3, -99],
#         [  5,   6,   7,   8, -99],
#         [ 10,  11,  12,  13, -99],
#         [ 15,  16,  17,  18, -99]])
```

More generally, the values assigned can be broadcasted to the elements of the slice:

```
import numpy as np
X = np.arange(20).reshape(4, 5)
# replacing the last two columns with -88, -99, respectively
X[:, -2:] = np.array([-88, -99])
# array([[  0,   1,   2, -88, -99],
#         [  5,   6,   7, -88, -99],
#         [ 10,  11,  12, -88, -99],
#         [ 15,  16,  17, -88, -99]])
```

Lastly, we may also choose to assign different values at multiple indices we choose.

```
import numpy as np
X = np.arange(20).reshape(4, 5)
# Assign the diagonal entries with different values
rows = np.array([0, 1, 2, 3])
cols = np.array([0, 1, 2, 3])
values = np.array([-1, -2, -3, -4])
X[rows, cols] = values
# array([[-1,  1,  2,  3,  4],
#         [ 5, -2,  7,  8,  9],
#         [10, 11, -3, 13, 14],
#         [15, 16, 17, -4, 19]])
```

3.6 Put and Scatter

Put and scatter operations are the reverse operations of take and gather: instead of retrieving elements from a tensor at certain indices, these operations can replace and aggregate values from a source tensor to a destination tensor. Put and scatter operations are more general operations than in-place value assignment that we discussed in Section 3.5, as they can operate in batches more efficiently. Furthermore, put and scatter operations can be seen as a way to perform group-by and aggregate operations, which we will explore more thoroughly in Chapter 9. Table 3.5 provides a list of equivalent put and scatter operations among NumPy, Tensorflow, and PyTorch.

3.6.1 Put

`np.put` operation is the reverse operation of `np.take`. Similar to `np.take`, `np.put` uses flat indices to put values from a source vector to a destination tensor. The `np.put` operation is an in-place operation that modifies the input directly without returning a copy.

```
import numpy as np
X = np.zeros(10, dtype=int)
indices = np.array([2, 4, 6, 8])
values = np.array([1, 2, 3, 4])
# in-place operation, modifies X directly
np.put(X, indices, values)
# array([0, 0, 1, 0, 2, 0, 3, 0, 4, 0])
```

Like `np.take`, multi-dimensional targets are treated as if flattened (in the "`C`" order); `indices` are the flat index of the target tensor (see also Figure 3.13 for an illustration).

```
import numpy as np
X = np.zeros((3, 3), dtype=int)
indices = np.array([1, 2, 5, 8])
values = np.array([1, 2, 3, 4])
np.put(X, indices, values)
# array([[0, 1, 2],
#        [0, 0, 3],
#        [0, 0, 4]])
```

Multidimensional `indices` may not be utilized as expected, since they are treated as if the indices are flattened. The target's shape will be retained.

Table 5.5 List of put and scatter operations in NumPy, Tensorflow, and PyTorch.

Operation	NumPy	Tensorflow	PyTorch
Put using flat index	`np.put` (in-place)	N/A	`torch.put` (in-place: `torch.Tensor.put_`)
Put along axis	`np.put_along_axis` (in-place)	N/A	N/A
Put to shape using multi-index	N/A	`tf.scatter_nd`	N/A
Scatter with replacement	N/A	`tf.tensor_scatter_nd_update`	`torch.index_put(..., accumulate=False)`
Scatter with addition	`np.add.at`	`tf.tensor_scatter_nd_add`	`torch.index_put(..., accumulate=True)`
Scatter with subtraction	`np.subtract.at`	`tf.tensor_scatter_nd_sub`	N/A
Scatter with max	`np.maximum.at`	`tf.tensor_scatter_nd_max`	[a]`torch.index_reduce(..., reduce="amax")`, `torch.scatter_reduce(..., reduce="amax")`
Scatter with min	`np.minimum.at`	`tf.tensor_scatter_nd_min`	[a]`torch.index_reduce(..., reduce="amin")`, `torch.scatter_reduce(..., reduce="amin")`
Scatter with mean	N/A	N/A	[a]`torch.index_reduce(..., reduce="mean")`, `torch.scatter_reduce(..., reduce="mean")`
Scatter with product	`np.multiply.at`	N/A	[a]`torch.index_reduce(..., reduce="prod")`, `torch.scatter_reduce(..., reduce="prod")`

Note: a) The logic of these PyTorch scatter functions is not exactly the same as that of NumPy and Tensorflow. See discussions below.

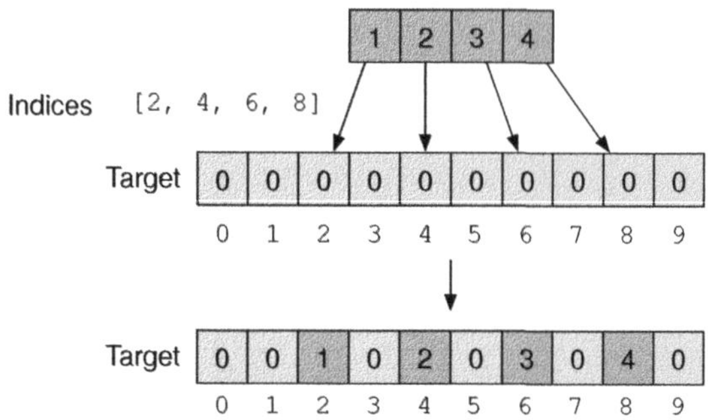

Figure 3.13 Illustration of np.put operation on a 1D vector target.

```
import numpy as np
X = np.zeros((3, 3), dtype=int)
indices = np.array([[1, 2], [5, 8]])
values = np.array([1, 2, 3, 4])
np.put(X, indices, values)
# array([[0, 1, 2],
#        [0, 0, 3],
#        [0, 0, 4]])
```

PyTorch's `torch.put` has identical behavior, but it will return a copy of the modified target tensor. PyTorch has an in-place version of the `put` method, implemented under `torch.Tensor.put_`. To modify the tensor in-place, one can do `X.put_(indices, values)`.

3.6.2 Put Along Axis

`np.put_along_axis` is the reverse operation of `np.take_along_axis`. This function is unique to NumPy. Like `np.take_along_axis`, the `axis` argument specified which dimension is the non-batching dimension, while the rest of the dimensions are the batch dimensions. The variable "`indices`" is indexing the items along the `axis`, whose value should not exceed `indices.shape[axis]`. The number of dimensions of `indices` must be the same as that of the target, i.e. `indices.ndim == X.ndim`. `values` also need to be broadcastable to `indices`. As illustrated in Figure 3.14, for instance,

```
import numpy as np
X = np.zeros((3, 4), dtype=int)
indices = np.array([
    [0, 2],
    [1, 3],
    [2, 1]
```

```
])
# (3, 1) broadcastable to (3, 2)
# this will reuse the value along the rows
values = np.array([
    [1],
    [2],
    [3],
])

np.put_along_axis(X, indices, values, axis=1)
# array([[1, 0, 1, 0],
#        [0, 2, 0, 2],
#        [0, 3, 3, 0]])
```

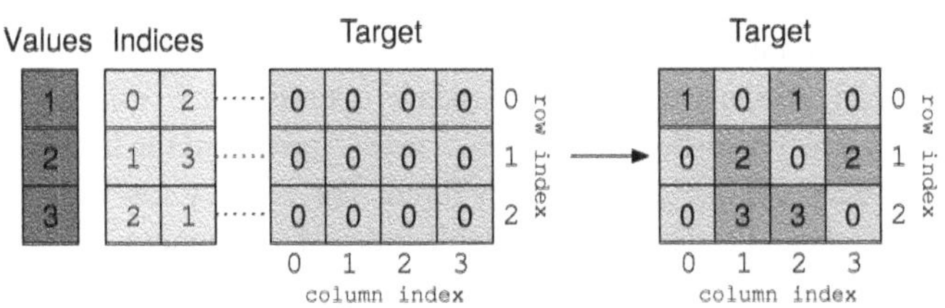

Figure 3.14 Illustrated example of `np.put_along_axis`.

3.6.3 Multi-index Scatter Replacement

Since Tensorflow's `tf.Tensor` are immutable objects, we need to rely on concrete functions to assign values at indices and return a copy of the modified tensor. Tensorflow does not implement an operation that uses flat index to assign values; i.e. there is not really a function that is the reverse of `tf.gather`. But Tensorflow does implement the *N*-dimensional scattering operation `tf.scatter_nd` which is the opposite operation of `tf.gather_nd` but with certain limitations (e.g. no support of scattering over a batch). In a subsequent case study below, we will try to implement a version of the batched scattering operation that uses flat indices along an axis. `tf.scatter_nd` is unique to Tensorflow, as it assumes the target tensor to be modified is always filled with 0s. Therefore, instead of asking for a specific tensor to replace values on, it asks for the shape of the target tensor. For example,

```
import tensorflow as tf
# [row, col]
indices = tf.constant([
    [0, 2],
```

```
    [1, 3],
    [2, 0],
    [3, 1],
    [4, 2],
])
values = tf.constant([1, 3, 5, 7, 9])
shape = (5, 4)
X_tf = tf.scatter_nd(indices, values, shape)
# <tf.Tensor: shape=(5, 4), dtype=int32, numpy=
# array([[ 0,  0,  1,  0],
#        [ 0,  0,  0,  3],
#        [ 5,  0,  0,  0],
#        [ 0,  7,  0,  0],
#        [ 0,  0,  9,  0]], dtype=int32)>
```

See also Figure 3.15 for an illustration.

Alternatively, we can also use the tensor update operation `tf.tensor_scatter_nd_update` in Tensorflow, or `torch.index_put` (and the in-place version, `torch.Tensor.index_put_`) in PyTorch. For NumPy (as well as PyTorch), the same effect can be achieved via simple indexed assignment,

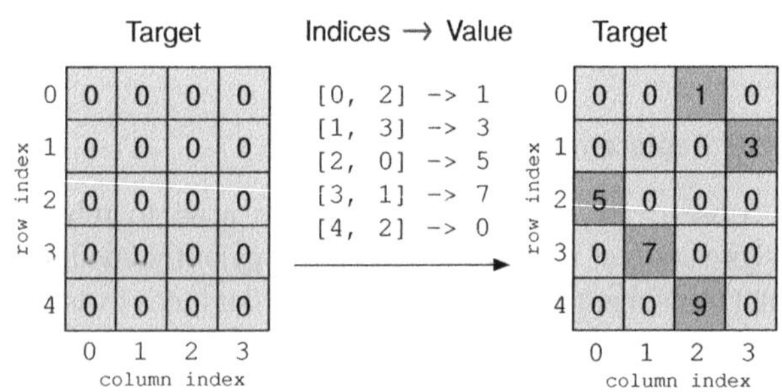

Figure 3.15 Illustration of scattering values over a 2D matrix using multi-index.

```
import numpy as np
import tensorflow as tf
import torch
# [row, col]
indices = tf.constant([
    [0, 2],
    [1, 3],
    [2, 0],
    [3, 1],
    [4, 2],
```

```
])
values = tf.constant([1, 3, 5, 7, 9])
shape = (5, 4)
X_tf = tf.zeros(shape, dtype=values.dtype)
X_tf = tf.tensor_scatter_nd_update(
    X_tf, indices, values
)

# equivalently in NumPy
X_np = np.zeros(shape, dtype=int)
X_np[tuple(indices.numpy().T)] = values
assert np.allclose(X_tf, X_np)

# 2 equivalent operations in PyTorch
# Converting to PyTorch dtypes
indices = torch.tensor(indices.numpy())
values = torch.tensor(values.numpy(), dtype=int)
# Method 1: Using indexed assignment like NumPy
X_torch = torch.zeros(shape, dtype=int)
X_torch[tuple(indices.T)] = values
assert np.allclose(X_tf, X_torch)

# Method 2: Using torch.index_put
X_torch = torch.zeros(shape, dtype=int)
X_torch = torch.index_put(X_torch, tuple(indices.T), values)

# Method 2 In-place: Using torch.Tensor.index_put_
X_torch = torch.zeros(shape, dtype=int)
# Also returns the modified tensor itself
X_torch.index_put_(tuple(indices.T), values)
```

`torch.index_put` has two modes, by setting the argument `accumulate=False` (default) or `accumulate=True`, respectively. As illustrated in the above example, PyTorch's `torch.index_put(..., accumulate=False)` is equivalent to Tensorflow's `tf.tensor_scatter_nd_update`. Just like `tf.tensor_scatter_nd_update`, the final replaced value of the target tensor is not deterministic on GPU. On the other hand, using `torch.index_put(..., accumulate=True)` in PyTorch is the same as using `tf.tensor_scatter_nd_add` in Tensorflow, or using `np.add.at` in NumPy. The behavior of `tf.scatter_nd(indices, values, shape)` is also the same as `tf.tensor_scatter_nd_add(tf.zeros(shape), indices, values)`. In these equivalent operations, duplicated indices will be summed together. Suppose the last index is duplicated with a different value to assign:

```
import numpy as np
import tensorflow as tf
import torch
indices = torch.tensor([
    [0, 2],
    [1, 3],
    [2, 0],
    [3, 1],
    [4, 2],
    [4, 2],
])
values = torch.tensor([1, 3, 5, 7, 9, 11])
shape = (5, 4)

# NumPy:
X_np = np.zeros(shape, dtype=int)
X_np[tuple(indices.T)] = values

# Tensorflow:
X_tf = tf.zeros(shape, dtype=tf.int32)
X_tf = tf.tensor_scatter_nd_update(X_tf, indices, values)

# PyTorch:
X_torch = torch.zeros(shape, dtype=int)
X_torch = torch.index_put(
    X_torch, tuple(indices.T), values, accumulate=False
)

assert np.allclose(X_np, X_tf)
assert np.allclose(X_tf, X_torch)

# On CPU, the last value of the duplicated index is used
# array([[ 0,  0,  1,  0],
#        [ 0,  0,  0,  3],
#        [ 5,  0,  0,  0],
#        [ 0,  7,  0,  0],
#        [ 0,  0,  11,  0]])

# Accumulate via summation
# NumPy:
X_np = np.zeros(shape, dtype=int)
np.add.at(X_np, tuple(indices.T), values) # in-place
# Alternatively,
# X_np[tuple(np.array(indices).T)] += np.array(values)

# Tensorflow:
X_tf = tf.zeros(shape, dtype=tf.int32)
X_tf = tf.tensor_scatter_nd_add(X_tf, indices, values)
X_tf_alt = tf.scatter_nd(indices, values, shape)
```

```

# PyTorch:
X_torch = torch.zeros(shape, dtype=int)
X_torch = torch.index_put(
    X_torch, tuple(indices.T), values, accumulate=True
)

assert np.allclose(X_np, X_tf)
assert np.allclose(X_tf, X_tf_alt)
assert np.allclose(X_tf, X_torch)

# array([[ 0,  0,  1,  0],
#         [ 0,  0,  0,  3],
#         [ 5,  0,  0,  0],
#         [ 0,  7,  0,  0],
#         [ 0,  0, 20,  0]])
```

NumPy, Tensorflow, and PyTorch also implement other scattering accumulation functions than simple summations, as illustrated in Table 3.5.

Though the name of the operation may imply the operation needs to scatter values over multiple dimensions, `tf.scatter_nd` can still be used to scatter values over a 1D vector:

```
import tensorflow as tf
indices = tf.constant([0, 2, 4])
values = tf.constant([1, 3, 5])
shape = (5, )
X = tf.scatter_nd(indices, values, shape)
```

However, this throws the following error:

```
InvalidArgumentError: {{function_node
 ↪  __wrapped__ScatterNd_device_/job:localhost/replica:0/task:0
 ↪  /device:CPU:0}} Dimensions [3,1) of input[shape=[5]] must
 ↪  match dimensions [0,1) of updates[shape=[3]]
 ↪  [Op:ScatterNd]
```

Instead, based on the documentation of `tf.scatter_nd` [Abadi et al., 2023d], we need to specify `indices` as a column vector of shape `(n, 1)`, i.e.

```
import tensorflow as tf
indices = tf.constant([0, 2, 4])[:, None]
```

```
values = tf.constant([1, 3, 5])
shape = (5, )
X = tf.scatter_nd(indices, values, shape)
# <tf.Tensor: shape=(5,), dtype=int32, numpy=array([1, 0, 3,
↪ 0, 5], dtype=int32)>
```

I found it helpful to understand this as if the target tensor is treated as a column vector as well, and the indices we are specifying are the indices of each row (i.e. first dimension), as illustrated in Figure 3.16. In general, the indices within the dimensions are arranged within a column, and indices of multiple dimensions are horizontally stacked together, which forms

```
[dim_0_index_0, dim_1_index_0, ..., dim_d_index_0,
 dim_0_index_1, dim_1_index_1, ..., dim_d_index_1,
 ...
 dim_0_index_n, dim_1_index_n, ..., dim_d_index_n]
```

PyTorch still requires the `indices` argument to be a tuple whose number of elements is the same as the number of dimensions, with each element being the 1D array of indices along the dimension:

```
import torch
indices = torch.tensor([0, 2, 4])
values = torch.tensor([1, 3, 5])
shape = (5, )
X_torch = torch.zeros(shape, dtype=int)
X_torch = torch.index_put(X_torch, (indices, ), values)
# tensor([1, 0, 3, 0, 5])
```

The scattering operations in NumPy, Tensorflow, and PyTorch also allow us to scatter multi-dimensional slices. See also Figure 3.17 for an illustration.

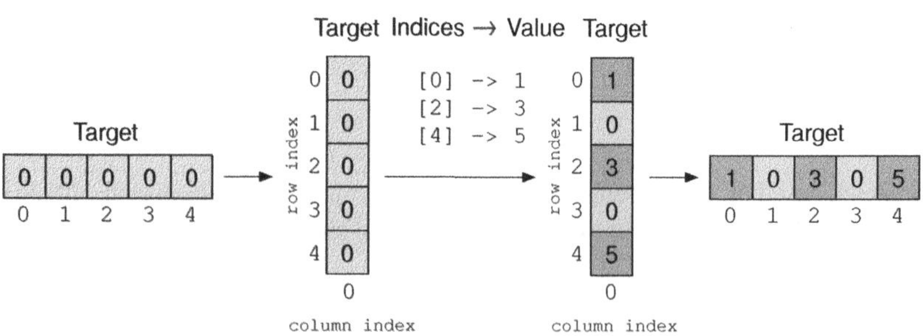

Figure 3.16 Illustration of `tf.scatter_nd` over a 1D target tensor.

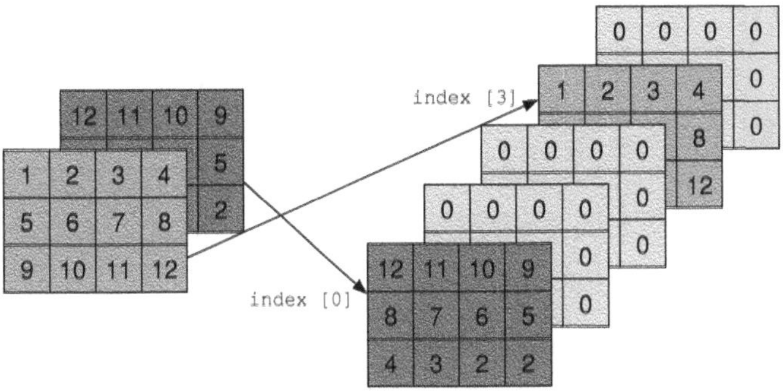

Figure 3.17 Illustration of scattering 2D slices over a 3D tensor.

```
import tensorflow as tf
values = tf.random.uniform((2, 3, 4))
indices = tf.constant([3, 0])[:, None]
X = tf.scatter_nd(indices, values, shape=(5, 3, 4))
```

In general, we can split the final target tensor shape into two parts: leading batch dimensions and lagging slice dimensions (Tensorflow's documentation also calls these outer dimensions and inner dimensions, respectively [Abadi et al., 2023d]). In the above example, the target tensor has a shape of `(5, 3, 4)`, with the batch dimension as `(5,)` and slice dimension as `(3, 4)`. We can observe that (i) the size of the first dimension (batch dimension) of `values` (used for updating the target tensor) is the same as the size of the first dimension of the `indices`; (ii) the size of the last two dimensions of the `values` are the same as the last two dimensions of the target tensor `X`; (iii) the size of the last dimension of the `indices` indicates how many dimensions the slices should have. By inductive reasoning, we can attempt an even more general example, where we want to rearrange the list of slices into multi-dimensional batches within the target tensor, as illustrated in Figure 3.18. The concept of scattering slices also applies to NumPy's `np.ufunc.at` and PyTorch's `torch.index_put`, with exactly the same behavior.

```
import tensorflow as tf
values = tf.random.uniform((2, 3, 4))
indices = tf.constant([[1, 0], [0, 1]]) # (2, 2)
X = tf.scatter_nd(indices, values, shape=(2, 2, 3, 4))
```

Tensorflow's `tf.scatter_nd` further generalizes the concept of batching, where the batch dimensions of the `values` and `indices` can also be arranged into multiple dimensions rather than staying as a flat list of slices or indices.

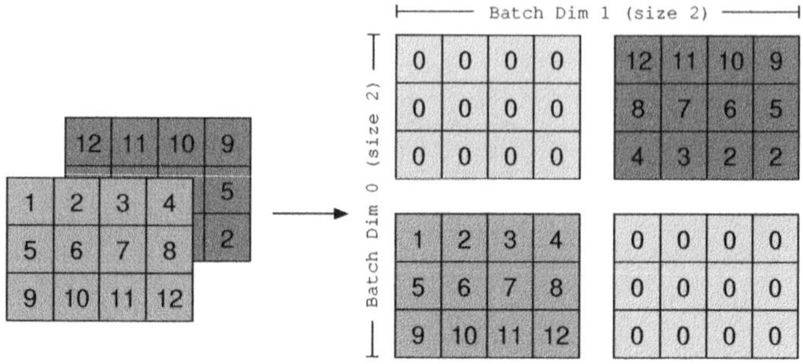

Figure 3.18 Illustration of scattering slices into multiple batch dimensions of the target tensor. The leading two batch dimensions are unfolded into a 2 × 2 grid in this schematic representation. The first slice is being put in position `(1, 0)` (second row, first column) of the 2 × 2 grid; the second slice is being put in position `(0, 1)` (first row, second column) of the 2 × 2 grid.

We only need to make sure that the leading batch dimensions for `values` and `indices` match exactly. Nonetheless, the `tf.scatter_nd` operation will still treat the multi-dimensional arrangement of batches of slices as if it were a flattened list of slices. This generalization is only for the sake of convenience when the input data batches are already arranged into multiple dimensions.

```
import tensorflow as tf
values = tf.random.uniform((4, 3, 4))
indices = tf.constant([
    [1, 0],
    [0, 1],
    [0, 0],
    [2, 2],
]) # (4, 2)
shape = (3, 3, 3, 4)
X = tf.scatter_nd(indices, values, shape=shape)

# fold indices and values into a 2 x 2 grid -> (2, 2, 2)
folded_indices = tf.reshape(indices, (2, 2, 2))
folded_values = tf.reshape(values, (2, 2, 3, 4))
X_folded = tf.scatter_nd(indices, values, shape=shape)

# same output results
assert np.allclose(X, X_folded)
```

3.6.4 Additional Scatter Operations from PyTorch

PyTorch implements two additional methods for accumulation, i.e. `torch.scatter_reduce` [Paszke et al., 2023c] and `torch.index_reduce`

[Paszke et al., 2023b]. Both methods are currently in beta at the time of writing. `torch.scatter_reduce` is a generalization of `torch.scatter` while `torch.index_reduce` complements `torch.index_add` to allow additional scatter-reduction methods including `prod`, `mean`, `max`, or `min` to be implemented. The usage pattern and the logic of the operation are not identical to that of `tf.tensor_scatter_nd_<reduce>` or `np.ufunc.at` for accumulation. In this section, I attempt to explain the differences between the two operations.

After comparing the equivalent Python `for-loop` equivalent code from the documentations for `torch.Tensor.scatter_reduce_` [Paszke et al., 2023c] (the in-place version of `torch.scatter_reduce`) and `torch.gather` [Paszke et al., 2023a], we can deduce that these two operations follow the same formulation. In general, `indices`, `values`, and target `X` should have the same number of dimensions, i.e. `indices.ndim == values.ndim == X.ndim`. It is also required that the size of `indices` is no greater than the size of input `X` or the scattering `values` at the corresponding axis dimensions, i.e. `indices.shape[dim] <= X.shape[dim]` and `indices.shape[dim] <= values.shape[dim]`. This means, we can specify an index that is shorter than the size of the inputs or the scattered `values`; the extra elements will be ignored during the scattering operation. Let us construct the following examples on how we can use `tf.scatter_reduce`, and see how similar it was to `torch.gather` (see also illustration in Figure 3.19):

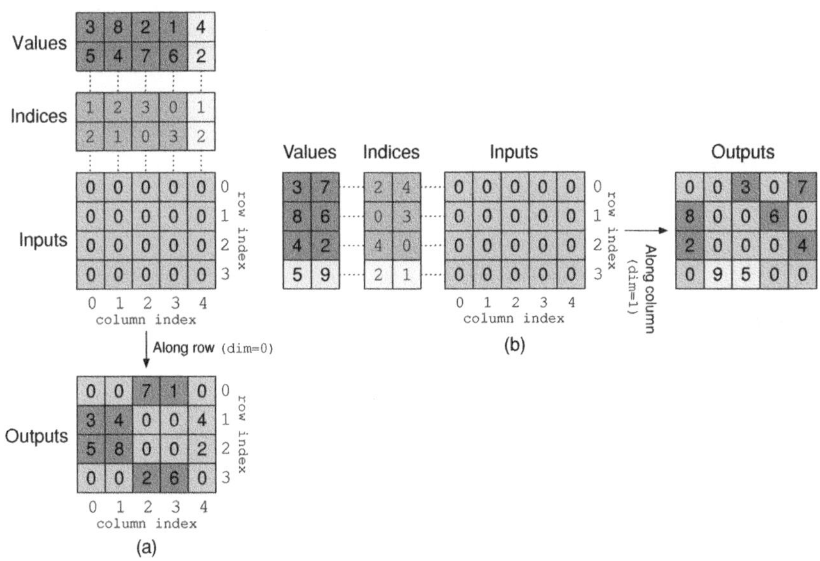

Figure 3.19 Illustration of `torch.scatter_reduce` on 2D matrices. (a) Scattering along `dim=0`; (b) Scattering along `dim=1`.

Example 3.1 Scattering along `dim=0`.

```
import torch
values = torch.tensor([
    [3, 8, 2, 1, 4],
    [5, 4, 7, 6, 2],
    # [1, 3, 2, 4, 6], this extra row will be ignored
], dtype=int)
indices = torch.tensor([
    [1, 2, 3, 0, 1],
    [2, 1, 0, 3, 2],
])

# Scatter along dim=0
X = torch.zeros((4, 5), dtype=int)
y = torch.scatter_reduce(
    input=X, dim=0, index=indices, src=values, reduce="sum"
)
# tensor([[0, 0, 7, 1, 0],
#         [3, 4, 0, 0, 4],
#         [5, 8, 0, 0, 2],
#         [0, 0, 2, 6, 0]])

# Not exactly matching the full columns:
# indices will be left aligned
y = torch.scatter_reduce(
    input=X,
    dim=0,
    index=indices[:, :-1],
    src=values[:, :-1],
    reduce="sum",
)
# tensor([[0, 0, 7, 1, 0],
#         [3, 4, 0, 0, 0],
#         [5, 8, 0, 0, 0],
#         [0, 0, 2, 6, 0]])
```

Example 3.2 Scattering along `dim=1`.

```
import torch
values = torch.tensor([
    [3, 7],
    [8, 6],
    [4, 2],
    [5, 9],
```

```
], dtype=int)
indices = torch.tensor([
    [2, 4],
    [0, 3],
    [4, 0],
    [2, 1],
])

# Scatter along dim=1
X = torch.zeros((4, 5), dtype=int)
y = torch.scatter_reduce(
    input=X, dim=1, index=indices, src=values, reduce="sum"
)
# tensor([[0, 0, 3, 0, 7],
#         [8, 0, 0, 6, 0],
#         [2, 0, 0, 0, 4],
#         [0, 9, 5, 0, 0]])

# Not exactly matching the full columns:
# indices will be left aligned
y = torch.scatter_reduce(
    input=X,
    dim=1,
    index=indices[:-1, :],
    src=values[:-1, :],
    reduce="sum",
)
# tensor([[0, 0, 3, 0, 7],
#         [8, 0, 0, 6, 0],
#         [2, 0, 0, 0, 4],
#         [0, 0, 0, 0, 0]])
```

On the other hand, `torch.index_reduce` resembles the slice scattering operation as we have seen in `tf.scatter_nd`. It also can be seen as the opposite operation of `torch.take_along_dim`, as it scatters the slices only along a certain dimension, using the 1D vector `indices` within that dimension. We can also compare it with NumPy's `np.put_along_axis`, though it has additional accumulation methods while lacking broadcasting capability since the `indices` must be a 1D vector. `torch.index_reduce` also requires the `values` and the target `X` having the same number of dimensions. We can construct the following example, which achieves the same effect of scattering two of 2D matrices into a 3D tensor using `torch.index_add` (for summation accumulation) as illustrated in Figure 3.17 previously:

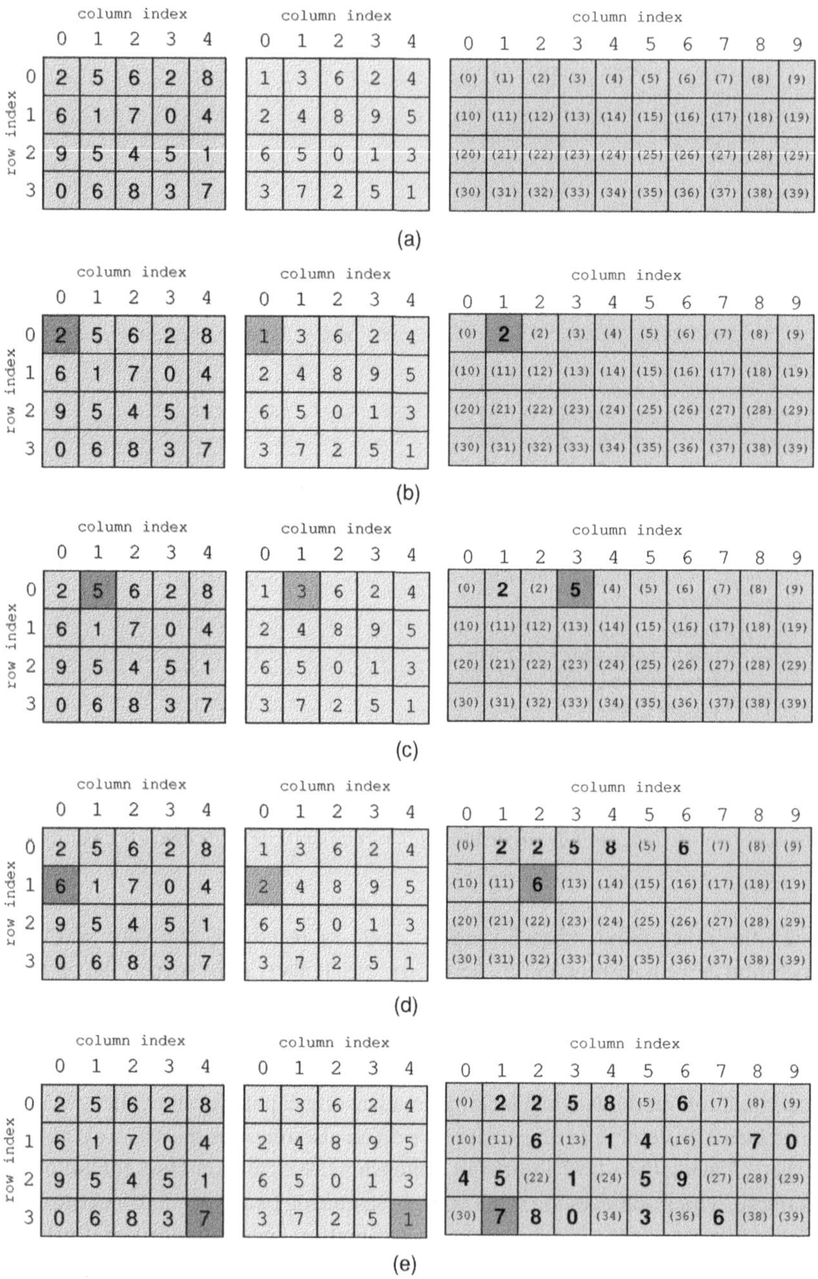
column index
row index

```
import torch
values = torch.rand((2, 3, 4), dtype=float)
indices = torch.tensor([3, 0])
X = torch.zeros((5, 3, 4), dtype=float)
X = torch.index_add(X, 0, indices, values)
```

3.7 Case Study: Batchwise Scatter Values

As we have seen in Section 3.6, PyTorch implements a version of the scattering operation `torch.scatter` that is the reverse operation of its gather operation `torch.gather`. Tensorflow does not implement `tf.scatter` which allows batched application of scattering by flat indices. However, as we will see in one of the case studies in Chapter 9, batchwise scattering is often needed to implement group by aggregate logic.

Suppose we have the following data (see also Figure 3.20a),

```
import tensorflow as tf
index = tf.constant([
    [1, 3, 6, 2, 4],
    [2, 4, 8, 9, 5],
    [6, 5, 0, 1, 3],
    [3, 7, 2, 5, 4],
], dtype=tf.int64)
inputs = tf.constant([
    [2, 5, 6, 2, 8],
    [6, 1, 7, 0, 4],
    [9, 5, 4, 5, 1],
    [0, 6, 8, 3, 7],
])
```

Figure 3.20 A step-by-step illustration of the batchwise scattering process using flat indices. We assume the rows are the batch dimension. The first column of matrices is the values whereas the second column of matrices stores the flat indices at the corresponding location. The third column of matrices is the targets, where the numbers in the parentheses are the flat indices. Assuming the tiles only with parenthesized indices are filled with 0 values. (a). Initial set up; (b). Results obtained from the first value (`row=0`, `col=0`); (c). Results obtained from the second value (`row=0`, `col=3`); (d). Results obtained from the first value of the second row (`row=1`, `col=2`); (e). Full results when all steps are completed.

We would like to scatter the inputs to a new target matrix based on the batch of indices specified by the index. This is easily done in PyTorch with

```
import torch
output = torch.scatter(
    torch.zeros((4, 10), dtype=int),
    1,
    torch.tensor(index.numpy(), dtype=int),
    torch.tensor(inputs.numpy(), dtype=int),
)
# tensor([[0, 2, 2, 5, 8, 0, 6, 0, 0, 0],
#         [0, 0, 6, 0, 1, 4, 0, 0, 7, 0],
#         [4, 5, 0, 1, 0, 5, 9, 0, 0, 0],
#         [0, 0, 8, 0, 7, 3, 0, 6, 0, 0]])
```

However, we can use Tensorflow's `tf.scatter_nd` method to achieve the same effect. More concretely, consider, step by step,

- For `index[0, 0] = 1`, we have a value at the corresponding index `input[0, 0] = 2`. We would like to put this value into a new tensor at the position such that `X[0, 1] = 2` (Figure 3.20b). Given the target tensor has shape `(4, 10)`, the multi-index `[0, 1]` can be converted into a flat index `1`. Therefore, this is the same as assigning a flat vector `X_flat[1] = 2`, and then reshaping the flat vector back into a 2D matrix.
- For the subsequent figures, Figure 3.20c–e, we are simply iterating over the elements of `index`, extracting the corresponding value from the `input`, and scatter it to the corresponding position on `X`.

Notice that the final output matrix has many zero entries. A matrix with many zero entries is called a sparse matrix, which we will introduce in more detail in Chapter 7.

With the above logic, we can implement the following code that specifically handles the 2D matrix case. Interested readers can further generalize this implementation to higher-dimensional tensors, or perhaps find alternative methods to achieve the same goal.

```
import tensorflow as tf

def scatter_batch_2d(index, inputs):
    # Determines the length of each row for the output
    row_length = tf.cast(tf.reduce_max(index), tf.int32) + 1
    row_length = tf.maximum(tf.shape(inputs)[1], row_length)
    # start and end index of each row
    # row 0: inc[0]:inc[1], row 1: inc[1]:inc[2], ...
```

```
9        inc = row_length * tf.range(tf.shape(inputs)[0]+1)
10       # Compute flat indices for each element
11       indices = tf.cast(index, inc.dtype) + tf.reshape(inc[:-1], (-1, 1))
12       # Flatten to column vector
13       indices = tf.reshape(indices, (-1, 1))
14       # Flatten updates
15       updates = tf.reshape(inputs, (-1,))
16       shape = (inc[-1], )
17       out = tf.scatter_nd(indices, updates, shape)
18       # reshape: (batch_size, row_length)
19       out = tf.reshape(out, (tf.shape(index)[0], -1))
20       return out
```

- Lines 5 and 6 compute the length of each row of the output. This is determined as the larger of the maximum index values from `index` and the number of columns of the `inputs`. In essence, we would like to return an output that has just enough number of columns to accommodate all the indices; we also want to return no fewer than the number of columns from the `input`. Note that this is in preparation for a case study in Chapter 9. Alternatively, we can specify an argument `output_shape` to let the user control the size of the output.
- Line 9 computes the start and end indices for each row as a 1D vector. In the above example, we will have `inc = [0, 10, 20, 30, 40]`, where `0:10` is the slice for the first row, `10:20` is the slice for the second row, etc.
- Line 11 offsets the row-specific flat indices by the start of the row index. For example, if an index 4 occurs on the second row of the original index matrix, then the corresponding flat index will be `4 + inc[1] = 4 + 10 = 14`.
- Line 13 flattens the indices with offset, as a column vector.
- Lines 15–17 scatter the 1D array of values to the target 1D array, given the indices.
- Line 19 reshapes the 1D tensor back to the desired 2D matrix, with shape `(batch_size, row_length)`.

Bibliography

Martín Abadi, Ashish Agarwal, Paul Barham, Eugene Brevdo, Zhifeng Chen, Craig Citro, Greg S. Corrado, Andy Davis, Jeffrey Dean, Matthieu Devin, Sanjay Ghemawat, Ian Goodfellow, Andrew Harp, Geoffrey Irving, Michael Isard, Yangqing Jia, Rafal Jozefowicz, Lukasz Kaiser, Manjunath Kudlur, Josh Levenberg, Dan Mane, Rajat Monga, Sherry Moore, Derek Murray, Chris Olah, Mike Schuster, Jonathon Shlens, Benoit Steiner, Ilya Sutskever, Kunal Talwar, Paul

Tucker, Vincent Vanhoucke, Vijay Vasudevan, Fernanda Viegas, Oriol Vinyals, Pete Warden, Martin Wattenberg, Martin Wicke, Yuan Yu, and Xiaoqiang Zheng. Tf.gather. https://www.tensorflow.org/api_docs/python/tf/gather, 2023a.

Martín Abadi, Ashish Agarwal, Paul Barham, Eugene Brevdo, Zhifeng Chen, Craig Citro, Greg S. Corrado, Andy Davis, Jeffrey Dean, Matthieu Devin, Sanjay Ghemawat, Ian Goodfellow, Andrew Harp, Geoffrey Irving, Michael Isard, Yangqing Jia, Rafal Jozefowicz, Lukasz Kaiser, Manjunath Kudlur, Josh Levenberg, Dan Mane, Rajat Monga, Sherry Moore, Derek Murray, Chris Olah, Mike Schuster, Jonathon Shlens, Benoit Steiner, Ilya Sutskever, Kunal Talwar, Paul Tucker, Vincent Vanhoucke, Vijay Vasudevan, Fernanda Viegas, Oriol Vinyals, Pete Warden, Martin Wattenberg, Martin Wicke, Yuan Yu, and Xiaoqiang Zheng. Tf.gathernd. https://www.tensorflow.org/api_docs/python/tf/gather_nd, 2023b.

Martín Abadi, Ashish Agarwal, Paul Barham, Eugene Brevdo, Zhifeng Chen, Craig Citro, Greg S. Corrado, Andy Davis, Jeffrey Dean, Matthieu Devin, Sanjay Ghemawat, Ian Goodfellow, Andrew Harp, Geoffrey Irving, Michael Isard, Yangqing Jia, Rafal Jozefowicz, Lukasz Kaiser, Manjunath Kudlur, Josh Levenberg, Dan Mane, Rajat Monga, Sherry Moore, Derek Murray, Chris Olah, Mike Schuster, Jonathon Shlens, Benoit Steiner, Ilya Sutskever, Kunal Talwar, Paul Tucker, Vincent Vanhoucke, Vijay Vasudevan, Fernanda Viegas, Oriol Vinyals, Pete Warden, Martin Wattenberg, Martin Wicke, Yuan Yu, and Xiaoqiang Zheng. Introduction to Tensors. https://www.tensorflow.org/guide/tensor, 2023c.

Martín Abadi, Ashish Agarwal, Paul Barham, Eugene Brevdo, Zhifeng Chen, Craig Citro, Greg S. Corrado, Andy Davis, Jeffrey Dean, Matthieu Devin, Sanjay Ghemawat, Ian Goodfellow, Andrew Harp, Geoffrey Irving, Michael Isard, Yangqing Jia, Rafal Jozefowicz, Lukasz Kaiser, Manjunath Kudlur, Josh Levenberg, Dan Mane, Rajat Monga, Sherry Moore, Derek Murray, Chris Olah, Mike Schuster, Jonathon Shlens, Benoit Steiner, Ilya Sutskever, Kunal Talwar, Paul Tucker, Vincent Vanhoucke, Vijay Vasudevan, Fernanda Viegas, Oriol Vinyals, Pete Warden, Martin Wattenberg, Martin Wicke, Yuan Yu, and Xiaoqiang Zheng. Tf.scatternd. https://www.tensorflow.org/api_docs/python/tf/scatter_nd, 2023d.

Adam Paszke, Sam Gross, Francisco Massa, Adam Lerer, James Bradbury, Gregory Chanan, Trevor Killeen, Zeming Lin, Natalia Gimelshein, Luca Antiga, Alban Desmaison, Andreas Köpf, Edward Yang, Zach DeVito, Martin Raison, Alykhan Tejani, Sasank Chilamkurthy, Benoit Steiner, Lu Fang, Junjie Bai, and Soumith Chintala. Torch.gather. https://pytorch.org/docs/stable/generated/torch.gather .html, 2023a.

Adam Paszke, Sam Gross, Francisco Massa, Adam Lerer, James Bradbury, Gregory Chanan, Trevor Killeen, Zeming Lin, Natalia Gimelshein, Luca Antiga, Alban Desmaison, Andreas Köpf, Edward Yang, Zach DeVito, Martin Raison, Alykhan Tejani, Sasank Chilamkurthy, Benoit Steiner, Lu Fang, Junjie Bai, and Soumith

Chintala. Torch.Tensor.indexreduce. https://pytorch.org/docs/stable/generated/torch.Tensor.index_reduce:.html#torch-tensor-index-reduce, 2023b.

Adam Paszke, Sam Gross, Francisco Massa, Adam Lerer, James Bradbury, Gregory Chanan, Trevor Killeen, Zeming Lin, Natalia Gimelshein, Luca Antiga, Alban Desmaison, Andreas Köpf, Edward Yang, Zach DeVito, Martin Raison, Alykhan Tejani, Sasank Chilamkurthy, Benoit Steiner, Lu Fang, Junjie Bai, and Soumith Chintala. Torch.Tensor.scatterreduce. https://pytorch.org/docs/stable/generated/torch.Tensor.scatter_reduce:.html#torch.Tensor.scatter_reduce:, 2023c.

4

Linear Algebra

In Chapter 2, we introduced element-wise operations on tensors or a pair of tensors. Specifically, when multiplying two tensors element-wise, we are performing the mathematical operation called Hadamard Product, which is expressed as $X \circ Y$. In this chapter, we will first examine operations associated with matrix and tensor multiplications, which are expressed as $X \cdot Y$ or simply XY mathematically. We will then look at several additional common linear algebra routines, such as finding the matrix inverse and eigenvalue decomposition. Performing multiplication between vector, matrix, and higher dimensional tensors, finding matrix inverse, and computing eigenvalue decomposition are common tasks when implementing machine learning models that are typically described using mathematical terms.

4.1 Tensor Multiplications

Let us start our discussion with an overview of the tensor multiplication operations. Table 4.1 provides a summary of commonly used tensor multiplication operations implemented by the NumPy, Tensorflow, and PyTorch libraries.

We will then dedicate the following two sections to discussing `matmul` (Section 4.2) and `tensordot` (Section 4.3) operations.

4.2 The `matmul` Operation

The `np.matmul` operations perform general matrix multiplications that are applied over multidimensional batches.

- When the input tensors are 2D matrices of shapes (n, k) and (k, m), then it is the conventional matrix multiplication resulting in a matrix of shape (n, m).

Vectorization: A Practical Guide to Efficient Implementations of Machine Learning Algorithms, First Edition. Edward DongBo Cui.

Table 4.1 Examples of tensor multiplication operations in NumPy, Tensorflow, and PyTorch.

Operation	NumPy	Tensorflow	PyTorch
Dot product[a)]	`np.dot`	`tf.experimental.numpy.dot`	`torch.dot`
Complex conjugate dot product[b)]	`np.vdot`	`tf.experimental.numpy.vdot`	`torch.vdot`
Inner product[c)]	`np.inner`	`tf.experimental.numpy.inner`	`torch.inner`
Outer (cross) product[d)]	`np.outer`	`tf.experimental.numpy.outer`	`torch.outer`
Matrix multiplication	`np.matmul`	`tf.matmul`	`torch.matmul`
Tensor dot product	`np.tensordot`	`tf.tensordot`	`torch.tensordot`

Notes:

a) NumPy's `np.dot` can support dot product between 1D vectors, 2D matrices (which is equivalent to `np.matmul`), or N-D tensor to tensor multiplications. In contrast, `torch.dot` only supports 1D vectors. Tensorflow implements the equivalent operation under the `tf.experimental.numpy` module.

b) `np.vdot` operation applies complex conjugate to the first input before taking the dot product, whereas `np.dot` does not perform complex conjugation. Mathematically, i.e. for two complex vectors $\boldsymbol{a}, \boldsymbol{b}$, $\text{dot}(\boldsymbol{a}, \boldsymbol{b}) = (x_a + iy_a) \cdot (x_b + iy_b)$, where i is the imaginary number. By contrast, $\text{vdot}(\boldsymbol{a}, \boldsymbol{b}) = \bar{a} \cdot b = \overline{(x_a + iy_a)} \cdot (x_b + iy_b) = (x_a - iy_a) \cdot (x_b + iy_b)$, where the overline $\overline{(\ldots)}$ indicates complex conjugates. Note that this definition of dot product between two complex vectors (i.e. `vdot`) is consistent with MATLAB's implementation, and perhaps many engineering applications such as Discrete Fourier Transform. However, the mathematical convention for the dot product between two complex vectors uses $\langle a, b \rangle = \sum_j a_j \bar{b}_j$ instead, where the second argument is complex conjugated. Thus, special attention needs to be paid when translating from mathematics to code when using this function, as the results arising from these two conventions are complex conjugate of each other. When used on multi-dimensional tensors, `np.vdot` will flatten the tensors into 1D vectors before executing the operation. Similar to `torch.dot`, `torch.vdot` only supports 1D vectors for complex conjugate dot products. Tensorflow implements the equivalent operation under the `tf.experimental.numpy` module.

c) `np.inner` operates on a pair of 1D vectors, or at least treating the last dimension of a tensor as 1D vectors. There used to be a function `numpy.core.umath_tests.inner1d` which performs batchwise inner product along rows of a pair of matrices, however, it was deprecated in newer versions of NumPy. Instead, one can use `np.sum(a * b, axis=1)`, or `einsum` operation (which we will introduce in Section 4.4) `np.einsum("ij, ij->i", a, b)`, to achieve the same effect. Note that this `einsum` operation is not the same as `np.inner(a, b)` as `np.inner` will result in matrix multiplication $a \cdot b^T$ rather than a batchwise operation. We can verify the distinction using the following example:

```
a = np.random.rand(5, 8)
b = np.random.rand(5, 8)
z1 = np.sum(a * b, axis=1) # shape (5, 8)
z2 = np.inner(a, b) # shape (5, 5)
np.allclose(
    z2,
    np.matmul(a, b.T)
)
```

d) `np.outer` will flatten any tensors with more than one dimension into vectors before performing the cross product. `torch.outer` only supports 1D vector inputs and will throw an error when receiving multi-dimensional inputs. Tensorflow implements the `outer` operation under the `tf.experimental.numpy` module.

- When the inputs are tensors with >2 dimensions, then the last two dimensions are used as the multiplication dimensions, while other dimensions are treated as batch dimensions and either must be identical or can be broadcasted to each other. For example, if we have tensors of shapes (a, b, n, k) and $(1, 1, k, m)$, then the dimension of the output tensor is (a, b, n, m).
- If the first input is a 1D vector, then it is treated as a row vector with singleton dimensions added before the vector length dimension as $(1, \dots, 1, n)$ and follows the broadcasting rules. However, after multiplication, the extra added dimensions are removed. For example, a vector of length $(k,)$ is multiplied with a tensor of shape (a, b, n, k), then the resulting tensor has shape (a, b, n).
- If the second input is a 1D vector, then it is treated as a column vector with singleton dimensions added after the vector length dimensions as $(m, 1, \dots, 1)$. However, after multiplication, the extra added dimensions are removed. For example, a tensor of shape (k, a, b, n) is multiplied with a vector of length $(k,)$, then the resulting tensor has shape (a, b, n).
- If both inputs are 1D vectors, then it is equivalent to a dot product.

Unlike NumPy and PyTorch, `tf.matmul` requires the two inputs to have at least two dimensions, so it will throw an error if any of the inputs are vectors. Additionally, Tensorflow's implementation of `tf.matmul` allows transpose (`is_transpose`) and conjugate (`is_adjoint`, which is the transpose and complex conjugate) preprocessing on the two inputs before performing the matrix multiplication step. Suppose we would like to multiply two matrices of shapes $(5, 3)$ and $(5, 3)$.

```
import tensorflow as tf

a = tf.complex(
    tf.random.uniform((5, 3)),
    tf.random.uniform((5, 3))
)
b = tf.complex(
    tf.random.uniform((5, 3)),
    tf.random.uniform((5, 3))
)
# This will throw an error because of shape mismatch
tf.matmul(a, b)
# This will pass, with shape (5, 5)
tf.matmul(a, b, transpose_b=True)
# This will pass, with shape (3, 3)
tf.matmul(a, b, adjoint_a=True)
```

Furthermore, `tf.tensordot` can also be used for matrix multiplication, using, for instance, `tf.tensordot(a, b, 1)` given two matrix inputs.

4.2.1 The @ Operator

The @ operator was introduced in Python 3.5 (PEP465 [Van Rossum and Drake, 2014]). It can help naturally express tensor multiplications, hence, enhancing the readability of the codes that implement certain mathematical operations. It also help simplify calls like `np.matmul(np.matmul(np.matmul(a, b), c), d)` to `a @ b @ c @ d`. This operator contrasts the `*` operator which indicates element-wise multiplication. Its usage rules are the same as `matmul` operations in each library. Python will assume which library's `matmul` operation to call based on the object type of the tensor representations. It is interesting to ask the question of which library's `matmul` operation will take priority when there is a mix of variables. We can find out with some experiments, as illustrated in the following code example:

```
import numpy as np
import tensorflow as tf
import torch

a = np.random.rand(7, 4, 3, 5)
b = np.random.rand(5)
c = np.random.rand(1, 1, 5, 3)
# this is working as expected, returns numpy.ndarray
a @ b
# This throws an error,
# Torch and NumPy is not compatible with @ operator
a @ torch.tensor(b)
# This will throw an error,
# since Tensorflow's matmul is called
a @ tf.constant(b)
# This will pass, returns a tf.Tensor
a @ tf.constant(c)
# This will pass, returns a tf.Tensor
# It does not matter if the first or
# second argument of matmul is tf.Tensor
tf.constant(a) @ c
# This will pass, returns a tf.Tensor
tf.constant(a) @ torch.tensor(c)
```

The conclusions are:

- PyTorch requires the inputs to be both `torch.Tensor` before executing the operation, so we cannot use NumPy's `np.ndarray` and Torch tensors in a mix.
- NumPy and Tensorflow's tensors can be used in a mix (at least in eager mode). But Tensorflow's method will take higher priority, and the results returned are cast into a `tf.Tensor`. Broadcasting also works between NumPy and Tensorflow.

- Torch and Tensorflow's tensors can also be used in a mix. The behavior is the same as if the Torch tensor is a NumPy tensor with respect to broadcasting. Tensorflow's method is taking higher priority and the result is a `tf.Tensor`.

4.3 The `tensordot` Operation

The `tensordot` (a.k.a tensor contraction) operation is a generalization of all of the above matrix product operations. It performs dot product (or equivalently, element-wise multiplication followed by summation reduction operation) along certain axes. It takes not only two inputs to be multiplied together but also a third argument `axes` to indicate how many axes (if integer argument) or along which specific pairs of axes (input a pair of list argument) the multiplication results are summed together. Several mathematical operations we introduced in Section 4.2 can be expressed using `tensordot`:

- Dot product between vectors, $a \cdot b$, using `tensordot(a, b, 1)`:

```
a = np.random.rand(5)
b = np.random.rand(5)
np.allclose(
    np.dot(a, b),
    np.tensordot(a, b, 1)
)
```

  Note that complex conjugate dot product `vdot` can be implemented by `tensordot` using `np.tensordot(a, b.conj(), 1)`
- Cross (outer) product between vectors, $a \times b$, using `tensordot(a, b, 0)`:

```
a = np.random.rand(5)
b = np.random.rand(3)
np.allclose(
    np.outer(a, b),
    np.tensordot(a, b, 0)
)
```

- 2D matrix multiplication ab:

```
a = np.random.rand(5, 3)
b = np.random.rand(3, 5)
np.allclose(
    np.matmul(a, b),
    np.tensordot(a, b, 1)
)
```

A typical usage of `tensordot` is to implement double matrix/tensor contraction (commonly denoted as $a : b = \mathrm{Tr}(a \cdot b^T)$), using `tf.tensordot(a, b, 2)`, given two matrix inputs:

```
a = np.random.rand(5, 3)
b = np.random.rand(5, 3)
np.allclose(
    np.tensordot(a, b, 2),
    np.trace(a @ b.T)
)
```

`tensordot` can be used for even more general tensor multiplications, especially when the `axes` argument is provided as a pair of lists of indices. Notice that `len(axes[0]) == len(axes[1])`. Values in the list of dimensions can also be negative. For example, we want to compute the matrix contraction between the combinations of a stack of three matrices and another stack of two matrices of shape (4, 2),

```
a = np.random.rand(2, 4, 2)
b = np.random.rand(3, 4, 2)
# this results a tensor of shape (2, 3)
np.tensordot(a, b, axes=[[1, 2], [1, 2]])
```

The above example is illustrated in Figure 4.1.

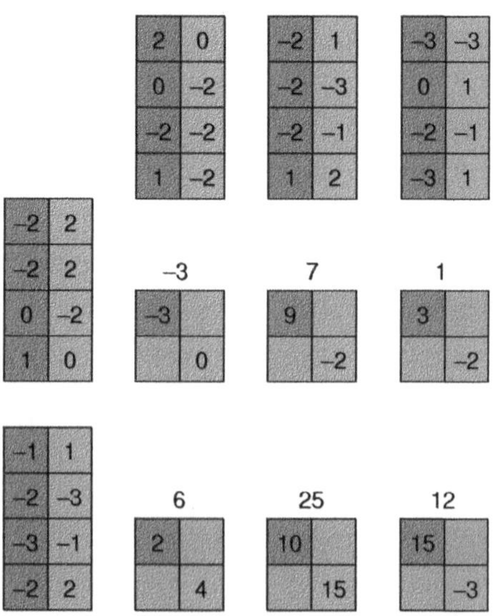

Figure 4.1 Illustration of an example tensordot operation between tensors of shape (2, 4, 2) and (3, 4, 2). Notice that the dot products between combinations of the first column vectors result in the upper left corner tiles of the 2 × 2 matrices, while dot products between combinations of the second column vectors result in the lower right tiles in the 2 × 2 matrices. The sum of the values along the diagonal of the 2 × 2 matrices is the values of the final matrix of shape (2, 3).

4.3.1 Heuristics of `tensordot` Operations

Notice from the above examples that the shape of the output tensor from `tensordot` operations varies when the `axes` argument changes. Different integer values of the `axes` argument seem to dictate different tensor multiplication behaviors. But we can derive a general rule of thumb by inductive reasoning.

When `axes=0`, given two vector inputs $(m,)$, and $(n,)$, we have seen that the resulting tensor has shape (m, n), as if an `outer` product is taken. In general, when given two tensors a with shape $(m, n, \ldots)$ and b with shape $(p, q, \ldots)$, the tensordot with `axes=0` has shape of $(m, n, \ldots, p, q, \ldots)$. This is as if every single element from tensor a is multiplying with every single element from tensor b to form the final tensor without summation to reduce any dimensions. Figure 4.2 illustrates this generalization. The following code snippet demonstrates an example of this heuristic:

```
a = np.random.rand(2, 3, 4, 5)
b = np.random.rand(6, 7, 8)
np.allclose(
    np.tensordot(a, b, 0).shape,
    [2, 3, 4, 5, 6, 7, 8]
)
```

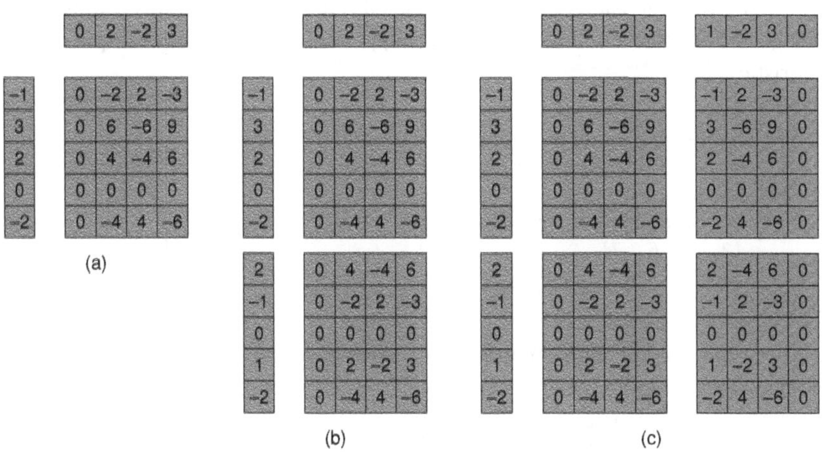

Figure 4.2 Illustration of `tensordot` operations when `axes=0`. (a) For a pair of 1D vectors of shapes $(5,)$ and $(4,)$, a combination of each element between the two vectors forms the outer product, having shape $(5, 4)$. (b) Adding one row dimension in the first input, the problem becomes the tensordot operation between a matrix of shape $(2, 5)$ and a vector of shape $(4,)$. Hence, the resulting tensor has shape $(2, 5, 4)$ (notice that singleton dimensions would be squeezed out). (c) Adding one additional column dimension in the second input forms a problem of tensordot between a matrix of shape $(2, 5)$ and one of $(4, 2)$. Then, the resulting tensor has shape $(2, 5, 4, 2)$.

When axes=1, we have seen that when the two inputs are 1D vectors, the dot product (which is a scalar) of the two tensors is returned. In general, given tensors a of shape $(m, n, \dots, i)$ and b of shape $(i, \dots, p, q, \dots)$, the tensor dot product with of the two tensors with axes=1 has shape $(m, n, \dots, p, q, \dots)$, where the last dimension of a and first dimension of b are reduced together. This is as if we are performing a series of dot products between combinations of two vector batches. Figure 4.3 illustrates examples of this generalization. The following code snippet demonstrates an example of this heuristic:

```
a = np.random.rand(2, 3, 4, 5)
b = np.random.rand(5, 6, 7)
np.allclose(
    np.tensordot(a, b, 1).shape,
    [2, 3, 4, 6, 7]
)
```

When axes=2, the two input tensors must have at least two dimensions. We have seen that the double tensor contraction operation can be performed when given two matrices. In general, given tensors a of shape $(m, n, \dots, i, j)$ and b of shape $(i, j, \dots, p, q, \dots)$, the tensor dot product between a and b with axes=2 has the shape $(m, n, \dots, p, q, \dots)$, where the last two dimensions of a and first two dimensions of b are contracted. Figure 4.4 illustrates examples of this generalization. The following code snippet demonstrates an example of this heuristic:

```
a = np.random.rand(2, 3, 4, 5)
b = np.random.rand(4, 5, 6, 7)
np.allclose(
    np.tensordot(a, b, 2).shape,
    [2, 3, 6, 7]
)
```

After exploring the above heuristics for when axes=0, axes=1 and axes=2, we may derive a more general rule for when axes is an integer argument: given the first input tensor a has shape $(m, n, \dots, s_1, s_2, \dots, s_k)$, if we would like to perform tensor dot product with axes=k, then it is necessary to make the second input tensor b has the shape $(s_1, s_2, \dots, s_k, \dots, p, q, \dots)$. Each tensor of shape $(s_1, s_2, \dots, s_k)$ from the batched tensor a is multiplied together with the tensor of the same shape from b element-wise and then summed together to form a single scalar. The resulting tensor will then have shape $(m, n, \dots, p, q, \dots)$, with the last k dimensions of the first input a and first k dimensions of the second input b contracted (i.e. reduced, or summed together).

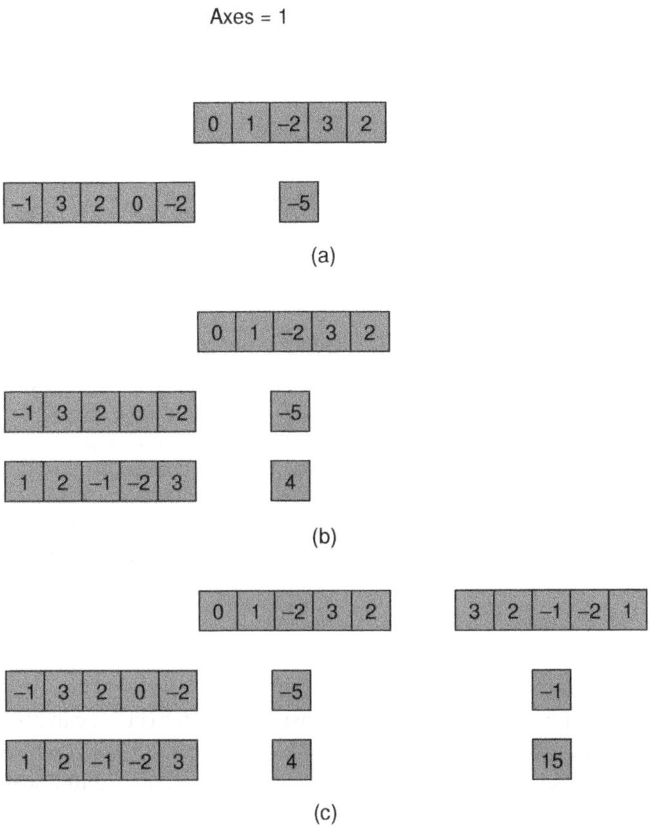

Figure 4.3 Illustration of `tensordot` operations when `axes=1`. (a) For a pair of 1D vectors each having a shape of (5,), the tensor dot product along `axes=1` is equivalent to simple vector inner dot product, which results in a scalar; (b) adding another vector along the first dimension of *a*, the operation becomes dot products between two vectors in *a* (shape (2, 5), each against *b* (shape (5,). This should yield a column vector of shape (2, 1). But, since the implicit singleton dimension is squeezed, the resulting vector has shape (2,). (c) Adding an additional column dimension on the second input *b*, this results in an operation that performs pair-wise dot product between each vector in *a* (shape (2, 5)) and *b* (shape (5, 2)). The resulting tensor has shape (2, 2), which is equivalent to matrix multiplication *ab*.

The `axes` argument can also be specified as a pair of lists of dimensions for contraction. In fact, in some specific cases, integer value arguments for `axes` can have a corresponding `list` version. For examples:

- `np.tensordot(a, b, 0)` is equivalent to `np.tensordot(a, b, [[], []])` where *a* and *b* can be any arbitrary tensor.

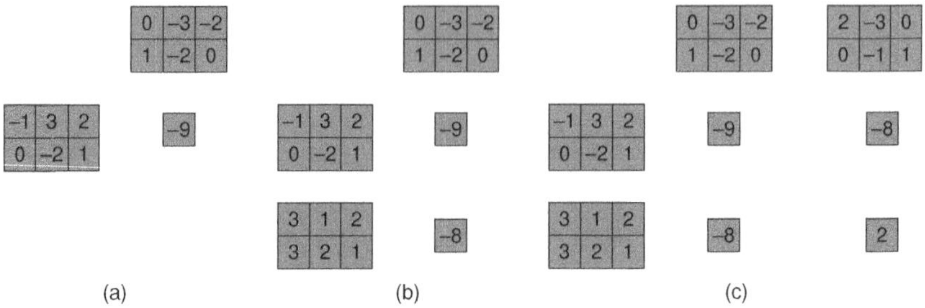

Figure 4.4 Illustration of `tensordot` operations when `axes=2`. (a) For a pair of 2D matrices each having a shape of (2, 3), the tensor dot product along `axes=2` is equivalent to tensor double contraction $a : b = \text{Tr}(ab^T)$, which yields a scalar; (b) after adding one leading dimension, making the first tensor a have shape (2, 2, 3), the tensor dot product problem becomes 2 matrix contraction operations, for each matrix within a against matrix b; (c) after adding one trailing dimension, making the second tensor b have shape (2, 3, 2), the tensor dot product becomes a set of matrix contraction operations for each combination of matrices between a and b. The matrices in a are taken along the axes 1 and 2, making `axis=0` the batch dimension. The matrices in b are taken along the axes 0 and 1, making `axis=2` the batch dimension. The resulting tensordot product has shape (2, 2).

- `np.tensordot(a, b, 1)` is equivalent to either `np.tensordot(a, b, [[1], [0]])` when a is a matrix and b is at least a 1D vector (but it can also be greater dimensions), or `np.tensordot(a, b, [[0], [0]])` when a and b are both vectors. More generally, `axes=[[-1], [-1]]` will cover both cases.
- `np.tensordot(a, b, 2)` is equivalent to `np.tensordot(a, b, [[0, 1], [0, 1]])` when a and b are both 2D matrices, or more generally, `np.tensordot(a, b, [[-2, -1], [0, 1]])` for arbitrary tensors with a number of dimensions greater than or equal to 2.

We can similarly derive a general rule here: let the first input tensor a have shape $(m, n, \dots, s_i, \dots, \dots)$ and the second input tensor b have shape $(p, q, \dots, s_{\beta_1}, \dots, r_j, \dots)$, and let $s_i == r_j$, for $i, j \in 1, 2, \dots, k$, where k is the number of dimensions to contract. We then specify the `axes` argument as `axes=[[..., i, ...], [..., j, ...]]` to indicate contractable dimensions. Then, the elements along the corresponding contractable dimensions are multiplied together element-wise and then summed together to obtain a single scalar. The operation is repeated (i.e. vectorized) over all combinations of dimensions that are not contracted, and the resulting tensor has shape $(m, n, \dots, p, q, \dots)$.

4.4 Einsum

Einstein Summation (or einsum) Convention is a generalization of various forms of vector, matrix, and tensor summations and multiplications. Suppose we want to perform matrix multiplication between two tensors X and Y, each element of the matrix should be expressed mathematically as

$$[\mathbf{XY}]_{i,j} = \vec{x}_{i,} \cdot \vec{y}_{,j} = \sum_k x_{i,k} \cdot y_{k,j}$$

and using a for-loop, we would compute the product like the following, e.g.:

```
import numpy as np
X = np.random.rand(5, 3)
Y = np.random.rand(3, 7)
Z = np.zeros((5, 7)) # expected shape
for ii in range(5):
    for jj in range(7):
        Z[ii, jj] = X[ii, :].dot(Y[:, jj])
```

i Note

Various sources I have found on the history of the Einstein Summation Convention pointed to Albert Einstein's original manuscript on General Relativity, "Die Grundlage der allgemeinen Relativitätstheorie," or in English, "The Foundation of the General Theory of Relativity" [Einstein, 1996]. But, based on what was described in Section 4.5, Einstein credited "Ricci and Levi-Civita" (i.e. Gregorio Ricci-Curbastro and Tullio Levi-Civita) for these notations.

In einsum notation, the matrix multiplication operation will track only the indices, `ik` and `kj`, respectively. Then, matrix multiplication is expressed as `ik,kj->ij`. Notice that the repeated index `k` is collapsed or summed together after multiplication along its dimension.

There are two modes of expressing the einsum equations. The above example is called the "explicit mode," in which the equation string has a right arrow `"->"` to indicate the ordering of the indices of the resulting tensor. If there are no indices to the right of the arrow, then a constant is assumed to be returned.

By contrast, in "implicit" mode, the equation string does not have a right arrow `->` specifying the resulting indices, and the results are computed with the following assumptions:

- Indices are sorted alphabetically. If indices in the equations are specified not in alphabetical order, then the returned tensor will have the indices sorted alphabetically. For example, `"ji"` (whose equivalent explicit mode is `"ji->ij"`) is equivalent to X^T.
- Repeated indices are summed together. For example, `"ij,j"` (whose equivalent explicit mode is `"ij,j->i"`) is equivalent to $\sum_j X_{i,j} \cdot y_j = z_i$, and to NumPy operation `np.sum(X * y, axis=1)`.

Note that certain operations can only be expressed in explicit mode. See various examples in Table 4.2.

In Sections 4.5 and 4.6, we will look at several examples of implementing mathematical formulas and algorithms using tensor multiplications.

Table 4.2 Examples of einsum equations and equivalent NumPy operations.

Operation	Equation String (Explicit)	Equation String (Implicit)	Equivalent NumPy Operations
Inner product[a]	`"i,i->"`	`"i,i"`	`np.inner`
Outer product[b]	`"i,j->ij"`	`"i,j"`	`np.outer`

Notes:

a) Dot product of two vectors.

```
a = np.random.rand(8)
b = np.random.rand(8)
np.allclose(
    np.inner(a, b),
    np.einsum("i,i", a, b)
)
```

b) Outer product of two vectors.

```
a = np.random.rand(8)
b = np.random.rand(8)
np.allclose(
    np.outer(a, b),
    np.einsum("i,j", a, b)
)
```

Table 4.2 (Continued)

Operation	Equation String (Explicit)	Equation String (Implicit)	Equivalent NumPy Operations
Extract diagonal[c)]	`"ii->i"`	NA	`np.diagonal, np.diag`
Trace[c)]	`"ii->"`	`"ii"`	`np.trace`
Transpose[d)]	`"ij->ji"`	`"ji"`	`np.tranpose`
Swapaxes[d)]	`"ijk->kji"`	`"kji"`	`np.swapaxes`

c) Diagonal extraction vs. Trace. Even though the two expressions are similar, the Trace operation should return the sum of the diagonal as a scalar, while diagonal extraction should return the diagonal entries as a vector.

```
X = np.random.rand(8, 8)
np.allclose(
     np.trace(X),
     np.einsum("ii->", X)
)

np.allclose(
     np.diag(X),
     np.einsum("ii->i", X)
)
```

d) Transpose and Swapaxes. These two are similar operations in einsum. The implicit mode takes advantage of the assumption that the axes of the returned tensor will be sorted alphabetically.

```
# Tranpose of matrix
X = np.random.rand(4, 5)
np.allclose(
     np.transpose(X),
     np.einsum("ij->ji", X)
)
np.allclose(
     np.transpose(X),
     np.einsum("ji", X)
)
# Swapaxes
Y = np.random.rand(4, 5, 6)
np.allclose(
     np.swapaxes(Y, 0, 2),
     np.einsum("ijk->kji", Y)
)
np.allclose(
     np.swapaxes(Y, 0, 2),
     np.einsum("kji", Y)
)
```

(continued)

Table 4.2 (Continued)

Operation	Equation String (Explicit)	Equation String (Implicit)	Equivalent NumPy Operations
Sum along row[e]	`"ij->j"`	NA	`np.sum(..., axis=0)`
Sum along column[e]	`"ij->i"`	NA	`np.sum(..., axis=1)`
Sum all[f]	`"ij->"`	NA	`np.sum(...)`
Identity[f]	`"ij->ij"`	`"ij"`	`np.array`

e) Sum along rows and columns.

```
X = np.random.rand(3, 4)
# Sum along rows
np.allclose(
    np.sum(X, axis=0),
    np.einsum("ij->j", X)
)
# Sum along columns
np.allclose(
    np.sum(X, axis=1),
    np.einsum("ij->i", X)
)
```

f) Sum All vs. Identity (an operation that gives output identical to input, not to be confused with a mathematical identity matrix). "Sum All" can only be expressed using the explicit mode, whereas "Identity" has an explicit mode `"ij->ij"` as well as the implicit mode `"ij"`. This is because when there is no value to the right of the arrow, the equation is always interpreted as summing into a constant; by contrast, when there is no right arrow (i.e. in implicit mode), summation only happens when indices are repeated.

```
X = np.random.rand(4, 5)
# Identity function
np.allclose(
    X,
    np.einsum("ij->ij", X)
)
np.allclose(
    X,
    np.einsum("ij", X)
)
# Sum all elements
np.allclose(
    X.sum(),
    np.einsum("ij->", X)
)
```

Table 4.2 (Continued)

Operation	Equation String (Explicit)	Equation String (Implicit)	Equivalent NumPy Operations
Hadamard product[g)]	`"ij,ij->ij"`	NA	`np.multiply`
Matrix multiplication[h)]	`"...ij,...jk->...ik"`	`"...ij,...jk"`	`np.matmul`
Dot product[i)]	`"...j,ijk->...ik"`	`"...j,ijk"`	`np.dot`

g) The Hadamard multiplication example illustrates the case of element-wise multiplication between a pair of 2D matrices. In general, to construct a Hadamard multiplication using einsum, we can use the format of `"ijk...,ijk...->ijk..."`, in which the corresponding dimensions will be multiplied together. It's also convenient to construct the Hadamard multiplication when one of the tensor's dimensions needs to be swapped or transposed. For example, if we want to perform element-wise multiplication between a tensor of shape `(2, 5, 3)` and another one of shape `(2, 3, 5)`:

```
X = np.random.rand(2, 5, 3)
Y = np.random.rand(2, 3, 5)
np.allclose(
    X * Y.swapaxes(1, 2),
    np.einsum("ijk,ikj->ijk", X, Y)
)
```

h) Multidimensional matrix multiplication requires dimensions other than the last two dimensions broadcastable. The ellipses ... represent the broadcasting dimensions. For example,

```
a = np.random.rand(1, 6, 3)
b = np.random.rand(8, 3, 7)
# shape (8, 6, 7)
np.allclose(
    np.einsum("...ij,...jk", a, b),
    np.matmul(a, b)
)
```

i) For multidimensional dot products, the last axis of the first input is multiplied and summed together with the second to last axis of the second input. For the first input, we can omit the indices for all the dimensions other than the last dimension, and replace them with ellipsis .... For the second input, we need to specify all the indices. For example

```
a = np.random.rand(2, 5, 3, 6)
b = np.random.rand(7, 6, 4)
# shape (2, 5, 3, 7, 4)
np.allclose(
    np.einsum("...j,ijk->...ik", a, b),
    np.dot(a, b)
)
```

(continued)

Table 4.2 (Continued)

Operation	Equation String (Explicit)	Equation String (Implicit)	Equivalent NumPy Operations
Tensor dot product[j)]	Case specific[j)]	NA	`np.tensordot`

j) Tensor dot product allows the sum of product reduction operation to be performed across multiple dimensions more flexibly than simple multidimensional dot product. In the einsum equation string, we need to specify the indices we want to product sum together using the same letters. For example,

```
a = np.random.rand(1, 2, 3, 4, 5)  # (a, i, b, j, c)
b = np.random.rand(6, 4, 8, 2, 10) # (d, j, e, i, f)
# shape (1, 3, 5, 6, 8, 10)
# axes=(1, 3) vs. axes=(3, 1)
# shape=(2, 4) vs. shape(4, 2)
np.allclose(
    np.einsum("aibjc,djeif->abcdef", a, b),
    np.tensordot(a, b, axes=([1, 3], [3, 1]))
)
```

4.5 Case Study: Pair-wise Pearson's Cross-Correlation

In Chapter 2, we implemented a way to compute Pearson's correlation between corresponding columns of a pair of matrices. However, during exploratory data analysis, we sometimes need to examine the $n \times n$ cross-correlation matrix between each feature. This would require us to perform a pair-wise Pearson's cross-correlation calculation. Recall the formula of Pearson's cross-correlation is

$$r_{ij} = \sum_k \left(\frac{\tilde{x}_k^{(i)}}{\sqrt{\sum_p (\tilde{x}_p^{(i)})^2}} \cdot \frac{\tilde{y}_k^{(j)}}{\sqrt{\sum_q (\tilde{y}_q^{(j)})^2}} \right)$$

where i, j are indices of the columns of the feature matrices X and Y, respectively; k, p, q are the indices of the elements of the vectors. The cross-correlation matrix of shape (N, M) between two inputs X and Y, each of the shape of (K, N) and (K, M) is implemented as the following:

```
import numpy as np

def faster_cross_corr(X, Y):
    """
```

```
5        faster way to compute Pearson's correlation
6        between each combination of pairs of two matrices
7        returns a cross-correlation matrix
8        [[x1:y1, x1:y2, x1:y3,...., x1:ym],
9         [x2:y1, x2:y2, x2:y3,...., x1:ym],
10        [...]
11        [xn:y1, xn:y2, xn:y3,...., xn:ym]
12       """
13       X = X - np.mean(X, axis=0, keepdims=True) # zero mean
14       Y = Y - np.mean(Y, axis=0, keepdims=True) # zero mean
15       X = X / np.sqrt(np.sum(X**2, axis=0, keepdims=True)) # L2 normalization
16       Y = Y / np.sqrt(np.sum(Y**2, axis=0, keepdims=True)) # L2 normalization
17       R = X.T.dot(Y)
18
19       return R
```

Notice that the code was almost identical to what we implemented in Chapter 2, except in Line 15. Instead of doing element-wise multiplication and reducing along `axis=1`, we perform a matrix multiplication using the `dot` operator.

> **i Note**
>
> I originally implemented this algorithm in MATLAB, so I kept the convention that each column corresponds to a feature. I also often apply this function to my pandas dataframe to compute feature cross-correlations, where each feature is also in a column of `pd.Series` within the data frame. However, it may not be the most efficient in NumPy since its vectors and matrices are stored using the row-major convention. Nonetheless, at least historically and for the majority of use cases, this custom cross-correlation function that I implemented here appears to be more efficient than `pd.DataFrame.corr`. We leave this as a homework exercise to compare the performance of `faster_cross_corr` with `pd.DataFrame.corr` and `np.corrcoef`.

4.6 Case Study: Hausdorff Distance

Back in 2012–2013, I was a research scientist doing schizophrenia studies using brain functional magnetic resonance imaging (fMRI) scans. A common problem with fMRI is that the patients undergoing a brain scan may move their heads, thus weakening the functional signal collected during the scan. As an analogy, this is as if someone takes a picture of a bird flying away with a slow-speed shutter (and fMRI's "shutter speed" is in the magnitude of seconds); the image of the bird would be very blurry. So, in one of our projects, we were trying to find a way to correct the effect of head movement during the scan [Cui et al., 2018]. But, how can we track the head position in a dark environment inside the MRI machine? A smart engineer from our team thought of a simple way: we printed a pair of dark squares on a piece of white paper and fixed these fiducial markers on a pair of swimming goggles. Then, the patient undergoing the scan wore these goggles while an infrared camera took pictures of the fiducial markers. After collecting the data, we initially thought of using simple image processing steps like thresholding and segmentation to find the centroids of the pair of fiducial markers, and the mean position of the two centroids over time would form a time series of head positions relative to the starting position. However, due to lighting and patient head orientations, the image quality of the data collected was not always the best (see Figure 4.5, e.g. where images taken could be skewed and affected by lighting artifacts). Still, if a human being looked at the images, they could pinpoint the mean position of the fiducial marker very well. Nowadays, image segmentation and object tracking problems are mostly solved by deep-learning-based solutions. But, back in 2013, this was a pretty difficult problem to solve at scale. Probably due to my skill level at the time, despite trying many tools in MATLAB's image processing library and Python's OpenCV and Scipy, none of them helped us consistently track the positions of the markers (which we verified through visual inspections after processing the images through the algorithm).

Therefore, I was tasked to determine a robust, alternate algorithm that would allow us to automate the process of accurately determining the mean position of the markers. My thought at the time was that we only needed to extract part of the image that would match the template just well enough, so that the lighting artifact or any noise in the image would be filtered out during the template-matching process. The template-matching process also needed to be robust enough so that

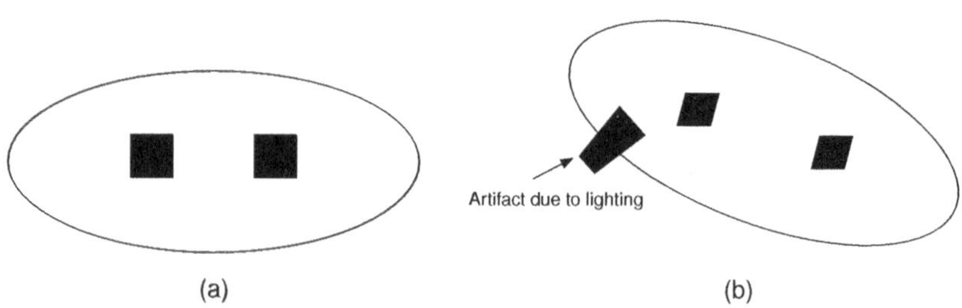

Figure 4.5 Fiducial markers used to track head positions. (a) The original template of the marker; (b) in reality (drawn as a schematic for illustration purposes), the image taken from the marker on the goggle can be skewed and affected by light artifacts.

it would not be affected by the skew of the image. To find the pair of squares, I would first segment the image into multiple patches and then iterate over all combinations of pairs of patches to compute some distance metric against the template. Some combinations would be between one of the squares in the fiducial marker and some lighting artifacts, but only one of them would be a pair of squares matching exactly to the template. This pair should have the smallest distance to the template image. The core template-matching distance metric I found was the Hausdorff Distance [Dubuisson and Jain, 1994].

The original Hausdorff Distance is expressed by the following formula:

$$\max(h(A, B), h(B, A))$$

where $h(A, B) = \max_i(d(A_i, B)) = \max_i \left(\min_j \|A_i - B_j\|\right)$, and $\|\cdot\|$ is the Euclidean distance. To implement this formula through vectorization, the key here is to compute the Euclidean distance term $\|A_i - B_j\|, \forall i, j$. This creates a matrix of shape (n, m), where n, m are the number of pixels or points in image A and B, respectively. Recall that for a pair of 2D points, the Euclidean distance is defined as

$$\begin{aligned}
D_{ij} &= \|A_i - B_j\| \\
&= \sqrt{(x_i - x_j)^2 + (y_i - y_j)^2} \\
&= \sqrt{(x_i - x_j)^2 + (y_i - y_j)^2} \\
&= \sqrt{x_i^2 + x_j^2 - 2x_ix_j + y_i^2 + y_j^2 - 2y_iy_j} \\
&= \sqrt{(x_i^2 + y_i^2) + (x_j^2 + y_j^2) - 2(x_ix_j + y_iy_j)} \\
&= \sqrt{A_i \cdot A_i^T + B_j \cdot B_j^T - 2A_i \cdot B_j^T}
\end{aligned}$$

Under the square root, the first term is an inner product between row vector A_i and itself; similarly, the second term is an inner product between row vector B_i and itself. Alternatively, the two terms can be seen as the diagonal entries of the matrix products AA^T and BB^T, respectively. The third term is the combination of the dot product between each point in matrix A and each point in matrix B. We can therefore implement the Hausdorff distance in the following as

```
import numpy as np

def HausdorffDist(A, B):
    """
    Compute the Hausdorff distance between two point clouds.
    Let A and B be subsets of metric space (Z, dZ),
    The Hausdorff distance between A and B, denoted by dH(A,
↪   B),
```

```
8       is defined by:
9       dH(A, B) = max(h(A, B), h(B, A)),
10      where h(A, B) = max(min(d(a, b))
11      and d(a, b) is a L2 norm
12      dist_H = hausdorff(A, B)
13      A: First point sets (M1 x N, with M1 observations in N dimension)
14      B: Second point sets (M2 x N, with M2 observations in N dimension)
15      ** A and B may have different number of rows, but must have the same
16      number of columns.
17      """
18      # Find pairwise distance: (A - B)^2 = A^2 + B ^2 - 2AB
19      A_sq = np.einsum('ij,ij->i', A, A)[:, None]
20      B_sq = np.einsum('ij,ij->i', B, B)
21      AB = np.dot(A, B.T)
22      D_mat = A_sq + B_sq - 2 * AB
23      D_mat = np.sqrt(D_mat)
24      # Find DH
25      d_AB = np.min(D_mat, axis=0)
26      d_BA = np.min(D_mat, axis=1)
27      dH = np.maximum(
28          np.max(d_AB),
29          np.max(d_BA)
30     )
31      return(dH)
```

1. In Line 19, we can compute the A^2 term as batchwise dot product for each row of A against itself using the einsum notation `"ij,ij->i"`. Notice we want this to be a column vector by expanding the second dimension, making it shape $(n, 1)$ to facilitate later broadcasting.
2. In Line 20, we similarly compute B^2. This is a row vector and will be treated as shape $(1, m)$ implicitly during broadcasting in the later step.
3. Line 21 implements the third term as a pair-wise dot product between combinations of rows of A and B. This forms a matrix of shape (n, m).
4. Line 22 implements the pair-wise distance by combining all three terms together via broadcasting. This forms a matrix of shape (n, m).
5. Line 23 takes the square root of the pair-wise distance matrix, resulting in the final Euclidean distance matrix D.
6. Lines 25 and 26 implement the $d(a, B)$ and $d(b, A)$ terms. Notice that we implement the asymmetric function by taking the minimum value along different dimensions of the matrix D.
7. Lines 27–30 implement the final Hausdorff distance formula.

4.7 Common Linear Algebraic Routines

Table 4.3 provides a list of commonly used linear algebra routines implemented by NumPy, Tensorflow, and PyTorch.

4.8 Case Study: Fitting Single Exponential Curves

In my previous studies, I have often found the need to fit exponential curves in the form of $y = a\exp(bx) + c$ to describe my data or to forecast a time series trend. I initially used `scipy`'s curve fitting implementation `scipy.optimize.curve_fit`, but found that the results are sensitive to the initial guesses when the provided data points are noisy. Then, I later discovered the mathematical treatise "Regressions et Equations Integrales" (translated to English as "Regression and Integral Equations") by Jean Jacquelin [Jacquelin, 2014]. I cannot read French, but the math formula translated well enough such that I was able to follow them step by step. (At the time of the writing, however, I discovered that Joseph R. Fox-Rabinovitz translated Jean Jacquelin's treatise into English; interested readers can find this work at https://scikit-guess.readthedocs.io/en/latest/appendices/references.html#regressions-et-equations-integrales, along with a version of implementation for the methods described in the original paper.) In my description below, I will follow the formulation of the exponential curve as $y = a\exp(bx) + c$ instead of $y = a + b\exp(cx)$ in the original paper.

> **i Note**
>
> Many people would argue that more sophisticated machine learning and deep learning algorithms are simply glorified curve fitting. Still, curve fitting may have served as the first introduction to machine learning to many people, including myself.

Suppose we have two vectors, x and y, storing the xy-coordinates of the data points. We initialize $S_0 = 0$ and build an array of S iteratively, as

$$S_k = S_{k-1} + \frac{1}{2}(y_k + y_{k-1})(x_k - x_{k-1}) \tag{4.1}$$

Then, to solve b, or the coefficient inside the exponential, we construct the equation

$$\begin{bmatrix} \cdot \\ b \end{bmatrix} = N^{-1}M = \begin{bmatrix} \sum_k (x_k - x_0)^2 & \sum_k (x_k - x_0)S_k \\ \sum_k (x_k - x_0)S_k & \sum_k S_k^2 \end{bmatrix}^{-1} \begin{bmatrix} \sum_k (y_k - y_0)(x_k - x_0) \\ \sum_k (y_k - y_0)S_k \end{bmatrix} \tag{4.2}$$

Table 4.3 List of linear algebraic routines in NumPy, Tensorflow, and PyTorch.

Operation	NumPy	Tensorflow	PyTorch
Matrix inverse[a)]	`np.linalg.inv`	`tf.linalg.inv`	`torch.linalg.inv`
Moore–Penrose pseudo-inverse[b)]	`np.linalg.pinv`	`tf.linalg.pinv`	`torch.linalg.pinv`
Eigenvalue decomposition[c)]	`np.linalg.eig`	`tf.linalg.eig`	`torch.linalg.eig`
Singular value decomposition[d)]	`np.linalg.svd`	`tf.linalg.svd`	`torch.linalg.svd`

Notes:

a) This method finds the square matrix inverse (if the matrix is invertible). Mathematically, let the matrix be $X \in \mathbb{R}^{m\times m}$ (real valued) or $X \in \mathbb{C}^{m\times m}$ (complex valued). Then, the inverse of the square matrix X^{-1} is defined as the $XX^{-1} = X^{-1}X = I$, where I is the identity matrix. The method can be applied in batches, i.e. we can have the input tensor `X` be of shape `(..., m, m)` and the `.linalg.inv` method will find the inverse using the matrices formed by the last two dimensions, resulting inverse tensors of the same shape as `X`.

b) The Moore–Penrose pseudo-inverse generalizes to non-square matrices. Mathematically, let $X \in \mathbb{R}^{m\times n}$ or $X \in \mathbb{C}^{m\times n}$. Then, the pseudo-inverse is defined as $X^{+} = (X^{*}X)^{-1}X^{*}$, where X^{*} is the Hermitian of X (complex conjugate, then transpose; for real matrices, this would simply be X^T) [Barata and Hussein, 2012]. Like the `.linalg.inv` method, `.linalg.pinv` can also operate on batches of matrices by solving along the last two dimensions of the input tensor.

c) Suppose we have a square matrix $X \in \mathbb{R}^{m\times m}$. If v is a unit vector and λ is a scalar value, which satisfies the equation $Xv = \lambda v$, then v is the eigenvector and λ is the eigenvalue for the matrix X. The `.linalg.eig` method returns the `eigenvalues` (not ordered) and `eigenvectors` whose each column is a unit eigenvector. The original X can be recovered via $X = \lambda v v^{-1}$, or `(eigenvalues * eigenvectors) @.linalg.inv(eigenvectors)`. Like the inverse methods, `.linalg.eig` can also operate on batches of matrices by solving along the last two dimensions of the input tensor.

d) Singular value decomposition (SVD) is commonly used for classic machine learning algorithms like principal component analysis (PCA). It is defined as $X = U\Sigma V^T$ (for complex matrices, V^T would become the Hermitian V^{*}), where $U \in \mathbb{R}^{m\times m}$ is the left singular factors, $\Sigma \in \mathbb{R}^{m\times n}$ is the singular value matrix whose diagonal values are the singular values, $V \in \mathbb{R}^{n\times n}$ is the right singular factor matrix. Matrices U and V both have the property $UU^T = U^TU = I_m$ and $VV^T = V^TV = I_n$. Therefore, $XX^T = (U\Sigma V^T)(V\Sigma^T U^T) = U\Sigma(V^TV)\Sigma^T U^T = U\Sigma^2 U^T$, and $X^TX = (V\Sigma^T U^T)(U\Sigma V^T) = V\Sigma^T(U^TU)\Sigma V^T = V\Sigma^2 V^T$ [Zhang, 2015]. Notice that this becomes the eigenvalue problem. We can therefore solve for U, V, Σ by performing eigenvalue decomposition on XX^T and X^TX. The `.linalg.svd` method returns `U`, `s`, and `V_hermitian` (corresponding to U, diagonal entries of Σ arranged in descending order, and V^T). We can recover the original matrix via `U @ (s * np.eye(U.shape[0], V_hermitian.shape[0])) @ V_hermitian`. Like all other linear algebra routines, `.linalg.svd` operates on the last two axes of the input tensor. As an example to illustrate:

```
import numpy as np
# Create a cloud of points inside of 3D ellipsoid
n = 2000 # number of points
a, b, c =10, 6, 1 # axes lengths
# randomly generate
x, y, z = [np.random.uniform(-d, d, n) for d in (a, b, c)]
inside =  ((x/a)**2 + (y/b)**2 + (z/c)**2) <= 1.0

# Use SVD to decompose the matrix
X = np.vstack([x[inside], y[inside], z[inside]]).T
X = X - X.mean(axis=0)
# s returns the standard deviation, sorted descendingly
U, s, Vh = np.linalg.svd(X)

# Recover
X_recovered = U @ (s * np.eye(U.shape[0], Vh.shape[0])) @ Vh

assert np.allclose(X, X_recovered)
```

Table 4.3 (Continued)

Operation	NumPy	Tensorflow	PyTorch
QR decomposition[e)]	`np.linalg.qr`	`tf.linalg.qr`	`torch.linalg.qr`
Linear least square[f)]	`np.linalg.lstsq`	`tf.linalg.lstsq`	`torch.linalg.lstsq`

e) QR decomposition is defined as $X = QR$, where $X \in \mathbb{R}^{m\times n}$, $Q \in \mathbb{R}^{m\times n}$ is a matrix whose columns are orthonormal to each other (unit vectors, whose dot products between the columns are zeros), and $R \in \mathbb{R}^{n\times n}$ is a upper triangular matrix. The `.linalg.qr` method returns `Q` and `R` and can recover the original matrix by `Q @ R`. Like all other linear algebra routines, `.linalg.qr` operates on the last two axes of the input tensor.

f) `.linalg.lstsq` solves the least squares problem for linear equations in the form $X\Theta = Y$, where $X \in \mathbb{R}^{m\times k}$ is the feature vectors (the independent variable) that contains k features with m observations, $\Theta \in \mathbb{R}^{k\times n}$ is the slope term of the linear equation, and $Y \in \mathbb{R}^{m\times n}$ is the observation outcome (the dependent variable), for n dimensions. It minimizes the square loss between the left- and right-hand side of the equation, i.e. $|X\Theta - Y|_2 = \sum_i^m \sum_j^n (x_i \cdot \theta_j - y_{ij})^2$. As an example, suppose we would want to estimate the linear equation $y = 2x + 5$ from a set of data points with random noise:

```python
import numpy as np
x = np.random.randn(100)
Y = 2 * x + 5 + np.random.randn(100)
# Two features, one feature for the input data x,
# and another feature as constant, independent of the data.
X = np.vstack([x, np.ones_like(x)]).T
theta, res, rank, s = np.linalg.lstsq(X, Y)
print(theta)
# array([2.04210113, 5.07653151])
```

Note that the anti-diagonal terms of the first inverse term are the same. The first element of the column vector on the left-hand side $(\cdot)$ is ignored in our computation. In the original paper, this term is the product ac. However, we will use another formula to compute these two terms

$$\begin{bmatrix} a \\ c \end{bmatrix} = N^{-1}M = \begin{bmatrix} n & \sum_k \theta_k \\ \sum_k \theta_k & \sum_k \theta_k^2 \end{bmatrix}^{-1} \begin{bmatrix} \sum_k y_k \\ \sum_k y_k \theta_k \end{bmatrix} \tag{4.3}$$

where $\theta_k = \exp(bx_k)$ and n is the number of data points. Again, the anti-diagonal terms of the first inverse term are the same. The following code implements this algorithm:

```python
import numpy as np
def fit_exp_with_offset(x, y):
    """
    Fitting y = a * exp(b * x) + c
    using a non-iterative method based on
    Regressions et Equations Integrales by Jean Jacquelin
```

```
    x and y are vectors of coordinates of the data points.
    They need to be sorted in ascending order,
    e.g. (0, 0), (0, 1), ... (1, 0), ..., (3, 4), (3, 5), ...
    """
    # Start algorithm
    S = np.zeros_like(x)
    S[1:] = 0.5 * (y[:-1] + y[1:]) * np.diff(x)
    S = np.cumsum(S)
    # for k in range(1, len(S)):
    #     S[k] = S[k-1] + 1/2 * (y[k] + y[k-1]) * (x[k] -
    ↪  x[k-1])

    M = np.empty((2, 2))
    N = np.empty((2, 1))

    # Getting b
    M[0, 0] = np.sum((x - x[0])**2)
    M[0, 1] = np.sum((x - x[0]) * S)
    M[1, 0] = M[0, 1]
    M[1, 1] = np.sum(S**2)

    N[0, 0] = np.sum((y - y[0]) * (x - x[0]))
    N[1, 0] = np.sum((y - y[0]) * S)

    B = np.matmul(np.linalg.inv(M),  N)
    b = B[1, 0]

    # Getting a and c
    theta = np.exp(b * x)
    M[0, 0] = len(x)
    M[0, 1] = np.sum(theta)
    M[1, 0] = M[0, 1]
    M[1, 1] = np.sum(theta**2)

    N[0, 0] = np.sum(y)
    N[1, 0] = np.sum(y * theta)

    AC = np.matmul(np.linalg.inv(M), N)
    c = AC[0, 0]
    a = AC[1, 0]

    return a, b, c
```

- Lines 13–15 build the vector S, following the formula described by Equation 4.1 but with vectorization. This is equivalent to the for-loop method that I commented below, in Lines 16 and 17.
- Lines 19 and 20 initialize two empty matrices to store the intermediate matrices used to find the three parameters.
- Lines 23–26 implements the M term in Equation 4.2.
- Lines 28 and 29 implements the N term in Equation 4.2.
- Line 31 implements the multiplication of the two terms in Equation 4.2.
- Line 32 extracts b as the second element of the column vector produced by the above matrix multiplication step.
- Line 35 computes the intermediate variable θ.
- Lines 36–39 implements the M term in Equation 4.3.
- Lines 41–42 implements the N term in Equation 4.3.
- Line 44 implements the multiplication of the two terms in Equation 4.3.
- Lines 45 and 46 extract a and c from the column vector produced by the above matrix multiplication step.
- Line 48 returns the three parameters a, b, c.

Let us test this implementation. Suppose we would like to fit a single exponential curve on a set of noisy points generated by the equation $y = 0.5\exp(-2x) + 3$. The following code illustrates how this function can be used. The result of the curve fitting is shown in Figure 4.6.

```
import matplotlib.pyplot as plt
x = np.random.rand(500) * 10
x = np.sort(x) # sorting x, required by the algorithm
y = 0.5 * np.exp(-2 * x) + 3 + np.random.randn(500) * 0.025
# fit the curve
a, b, c = fit_exp_with_offset(x, y)
# recreate the curve
x_pred = np.linspace(0, 10, 100)
y_pred = a * np.exp(b * x_pred) + c

# plot the data points and the fit
plt.scatter(x, y, label="data")
plt.plot(x_pred, y_pred, color="r", label="fit")
plt.xlabel("x")
plt.ylabel("y")
plt.legend()
```

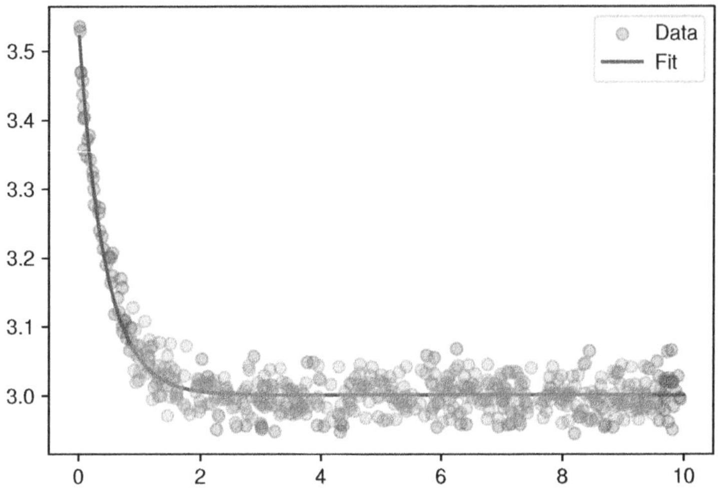

Figure 4.6 Fit single exponential to noisy data points.

Bibliography

J. C. A. Barata and M. S. Hussein. The Moore–Penrose pseudoinverse. A tutorial review of the theory. *Brazilian Journal of Physics*, 42(1–2):146–165, April 2012. ISSN 0103-9733, 1678-4448. doi: 10.1007/s13538-011-0052-z.

Edward D. Cui, Richard J. Maddock, Michael H. Buonocore, Meric Ozturk, and Jong H. Yoon. Quantifying the effects of head movement on magnetic resonance spectroscopy estimates of gamma-aminobutyric acid, July 2018.

M.-P. Dubuisson and A. K. Jain. A modified Hausdorff distance for object matching. *Proceedings of 12th International Conference on Pattern Recognition*, volume 1, pages 566–568, 1994. doi: 10.1109/ICPR.1994.576361.

Albert Einstein. The foundation of the general theory of relativity. In *The Berlin Years: Writings, 1914–1917*, Princeton University Press, volume 6 of *The Collected Papers of Albert Einstein*. July 1996.

Jean Jacquelin. *Regressions et Equations Integrales*. January 2014.

Guido Van Rossum and Fred L. Drake. PEP 465 –A dedicated infix operator for matrix multiplication. https://peps.python.org/pep-0465/, February 2014.

Zhihua Zhang. The singular value decomposition, applications and beyond, October 2015.

5

Masking and Padding

Our discussions on vectorized operations have thus far assumed that we are dealing with data that can be consistently represented by a multidimensional dense tensor given a certain shape. For example, a stack of images can be represented by a dense tensor of shape `(batch_size, width, height, channel)`. Tabular data that contains several numerical columns can be represented as a matrix of shape `(batch_size, num_columns)`. However, in various real-world machine learning applications, we also frequently need to deal with data that comes in varying sizes. For example, a collection of audio recordings may have variable durations. So to make our operations vectorizable across our collection of data, we need to seek ways to reconfigure our data so that it can be represented by a tensor with a single consistent shape. This can be achieved by **padding** the sequences of audio recordings to the same predefined length. On the other hand, the audio recording may have silent periods, which we do not want the model to pay attention to during the learning process. We can therefore use **masking** to tell the model to concentrate the learning on the parts that contain audio we would like to model. Therefore, masking and padding are important techniques to vectorize our machine learning modeling operations when dealing with variable-sized data.

5.1 Masking

Masking is a data filtering technique that allows the model to focus on relevant parts of the data that we determined heuristically while allowing our vectorization functions to operate on tensors with a consistent data shape. We will first look at several special types of masks implemented by our vectorization libraries. Then, we will look at four techniques to perform the masking operation.

Vectorization: A Practical Guide to Efficient Implementations of Machine Learning Algorithms, First Edition. Edward DongBo Cui.

5.1.1 Triangular and Diagonal Masks

NumPy, Tensorflow, and PyTorch implement several special types of masking operations. Table 5.1 shows a list of triangular and diagonal masking and mask construction operations in NumPy, TensorFlow, and PyTorch.

5.1.2 Changing Elements Using the `where` Operation

Other than using pre-existing masking operations to replace values, more often, we need to replace/mask values based on conditions. For example, given a grey-scale image of size 256×256, we would like to apply a threshold to the image, such that pixel values above this threshold are marked as the foreground (bright), and pixel values below this threshold are marked as the background (dark), so that the processed image can be used for subsequent tasks such as image segmentation. We can do this by using `np.where(condition, x, y)`, where `condition` is our boolean or binary mask.

```
import numpy as np
import matplotlib.pyplot as plt
import skimage # pip install scikit-image==0.21.0
image = skimage.data.coins()
# image thresholding to remove background
image_thresh = np.where(image > 150, image, 0)
# plot the images
fig, axs = plt.subplots(1, 2)
axs[0].imshow(image, cmap="gray")
axs[0].set_title("Original")
axs[1].imshow(image_thresh, cmap="gray")
axs[1].set_title("Thresholded")
```

The code generates the plots as illustrated in Figure 5.3.

The `where(condition, x, y)` syntax of the `where` operations implemented in NumPy, Tensorflow, and PyTorch is helpful for such "thresholding"-like operations, where it conditionally replaces the value in `x` using values from `y`, where `x`, `y`, and `condition` need to be broadcastable to each other. In the specific case of masking, we would typically replace the elements at the index corresponding to the `True` values in `conditions` with some masking values such as 0s.

5.1.3 Use Multiplication to Apply Masks

The masks created by the operations introduced in Section 5.1 are often binary tensors (0s and 1s) with element dtypes being numerical types such as floats and

Table 5.1 List of masking and mask indices construction operations.

Operation	NumPy	Tensorflow	PyTorch
Identity matrix[a)]	`np.eye`	`tf.eye`	`torch.eye`
Diagonal matrix[a)]	`np.diag`	`tf.linalg.diag`	`torch.diag`
One hot/binary[b)]	N/A	`tf.one_hot`	`torch.nn.functional.one_hot`
Lower triangular matrix[c)]	`np.tril`	`tf.experimental.numpy.tril`	`torch.tril`
Upper triangular matrix[c)]	`np.triu`	`tf.experimental.numpy.triu`	`torch.triu`
Lower triangular mask[c)]	`np.tri`	`tf.experimental.numpy.tri`	N/A
Diagonal band[d)]	N/A	`tf.linalg.band_part`	N/A

Notes:

a) `tf.eye` and `torch.eye` can only return the identity matrix along the main diagonal, while `np.eye` can also return off-diagonal matrices by setting k to a non-zero value. `tf.linalg.diag` and `torch.diag` can create off-diagonal matrices whose diagonal values do not have to be 1.

b) One hot encoding is typically used to encode categorical features. A single categorical value is mapped to an index $(0, 1, \dots, N-1)$, then a vector of size N composed of 0s and a single 1 is created, where the position of the single 1 is at the index of the categorical value. When the feature is a set of categorical features (i.e. a multivalued categorical feature), it is typical to construct a binary vector where multiple 0s and 1s are present in the vector. There are several ways to construct this binary or multi-hot vector. One way is to convert each categorical value in the feature into a single one-hot vector, then reduce-sum them together. However, this will not be memory efficient as it creates extra 0 entries in the intermediate step. In Section 5.1.6, we will use the `put` operation introduced in Chapter 3 to create the binary/multi-hot mask. Furthermore, even though NumPy does not implement a one-hot encoding function, Scikit-learn's `sklearn.preprocessing.OneHotEncoder` is more commonly used for encoding categorical values. On the other hand, one can also use `np.eye(num_values)[index]` to create a batch of one-hot encoded categorical features, where `num_values` is the size of the categorical variable, and `index` is the indices of the categorical value.

c) Both `tril` and `triu` can be used for the purpose of constructing the lower and upper triangular mask, or performing the masking operation on the existing matrix. See Figure 5.1 for illustrations of several examples. Beyond square matrices, both operations can also handle non-square matrices as well; the main diagonal (`k=0`) of non-square matrices still starts at the upper left corner of the matrix. When handling multidimensional tensors, it treats the last two dimensions as the matrix dimensions to extract/create triangular matrices from, while treating all the other leading dimensions as batch dimensions. To construct a lower triangular mask on the input matrix X, we can do `np.tril(np.ones_like(X), k=...)`; to construct an upper triangular mask on the input matrix, we can do `np.triu(np.ones_like(X), k=...)`. Notice that `tril` and `triu` are complementary to each other, i.e. `np.tril(X, k=-1) + np.triu(X, k=0) == np.tril(X, k=0) + np.triu(X, k=1) == np.tril(X, k=1) + np.triu(X, k=2) == ... == X`. The `tri` function is a dedicated function to create a 2D lower triangular binary mask. It either takes the shape of the mask `(N, M)` or reuses the shape from the reference matrix via the `like` argument (not supported by Tensorflow).

d) This can create diagonal bands. At the time of writing, only Tensorflow implements this diagonal band operation. See Figure 5.2 for an illustration of its usage. Notice that this is not an operation that fully replaces `triu` and `tril` (though some operations such as `tril(..., k=0)` can also be expressed using `band_part(..., -1, 0))`, since the results always need to include the main diagonal entries. Although NumPy does not implement this generalized operation, the same can be achieved via Scipy's `scipy.sparse.dia_matrix` (an even more flexible operation) which we will introduce in Chapter 7. Since such masks are usually static in data type and shapes, in practice, we can use Scipy to create the diagonal band mask when initializing the model and apply it during training and inference on PyTorch models as well.

(continued)

Table 5.1 (Continued)

Operation	NumPy	Tensorflow	PyTorch
Fill diagonal[e)]	`np.fill_diagonal`	`tf.linalg.set_diag`	`torch.Tensor.fill_diagonal_`
Indices of main diagonal[f)]	`np.diag_indices`, `np.diag_indices_from`	`tf.experimental.numpy.diag_indices`	N/A
Indices of tril[f)]	`np.tril_indices`, `np.tril_indices_from`	N/A	`torch.tril_indices`
Indices of triu[f)]	`np.triu_indices`, `np.triu_indices_from`	N/A	`torch.triu_indices`

e) This operation fills the diagonal of the matrix. In NumPy, this is an in-place operation (modifies the existing object; does not return anything, so use it as `np.fill_diagonal(a, ...)` instead of `a = np.fill_diagonal(a, ...)`). The NumPy implementation also only fills the main diagonal. In contrast, the Tensorflow implementation can fill off-diagonal entries and always returns a copy of the filled tensor. The PyTorch implementation is an in-place method within the `torch.Tensor` class itself. So this method needs to be called on an existing tensor object.

f) These operations return the multidimensional indices of `tril`, `triu` and main diagonal entries given the input matrix shape. The `*_from` variants of these functions reuse the shape from a reference input matrix. See the following code examples:

```
import numpy as np
X = np.arange(25).reshape(5, 5)
# Main diagonal indices
# (array([0, 1, 2, 3, 4]), array([0, 1, 2, 3, 4]))
np.diag_indices(5, ndim=2)
np.diag_indices_from(X)

# Higher dimensional diagonal indices
# (array([0, 1, 2]), array([0, 1, 2]), array([0, 1, 2]))
np.diag_indices(3, ndim=3)

# Tril indices
# (array([1, 2, 2, 3, 3, 3, 4, 4, 4, 4]),
# array([0, 0, 1, 0, 1, 2, 0, 1, 2, 3]))
np.tril_indices(5, k=-1)
np.tril_indices_from(X, k=-1)
```

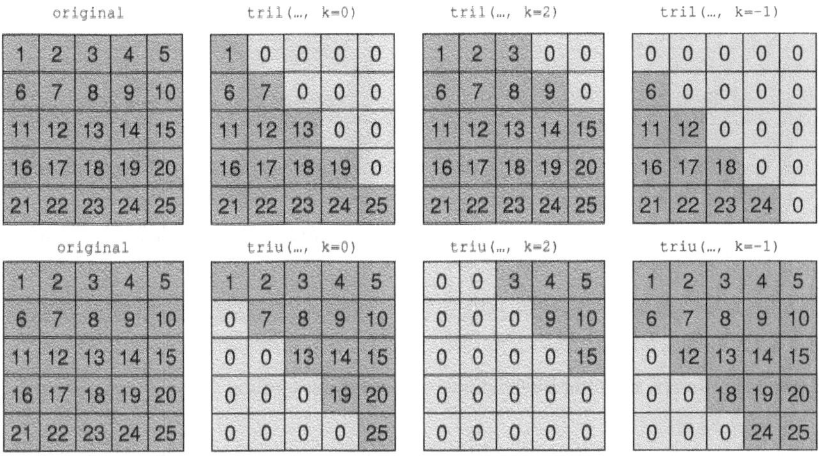

Figure 5.1 Example uses of `tril` and `triu` operations.

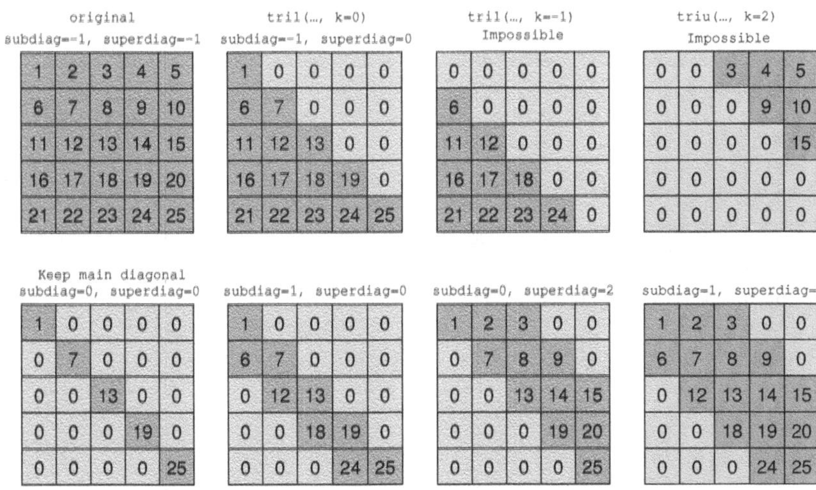

Figure 5.2 Example usage of `tf.linalg.band_part`. It is impossible to create generic `tril` or `triu` matrices since this operation will always keep the main diagonal as part of its output.

integers. Instead of converting the numerical mask into a boolean tensor and using the `where` operation to do the replacement, we may use element-wise multiplication to apply the mask, provided that we would like to have the elements to be masked to have values of 0s. For example, suppose we have a mask whose diagonal

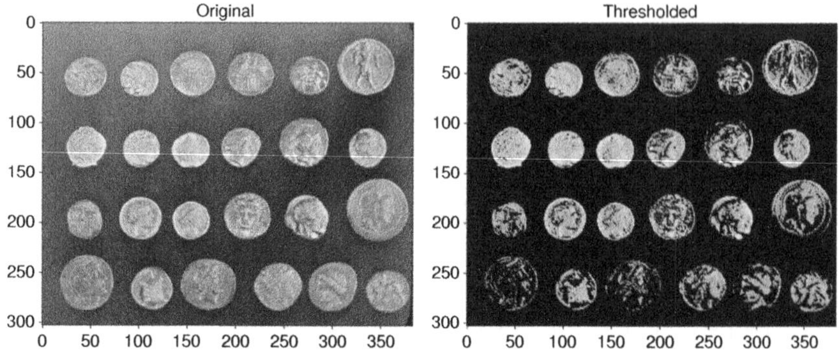

Figure 5.3 Before and after thresholding to remove background.

and anti-diagonal entries are 1s while all the other entries are zeros. We would like to apply this mask to a matrix of the same shape:

```
import numpy as np
X = np.arange(25).reshape(5, 5) + 1
# array([[ 1,  2,  3,  4,  5],
#        [ 6,  7,  8,  9, 10],
#        [11, 12, 13, 14, 15],
#        [16, 17, 18, 19, 20],
#        [21, 22, 23, 24, 25]])

# Create the cross-like mask
mask = np.zeros_like(X)
np.fill_diagonal(mask, 1)
mask = mask[:, ::-1] # now the anti-diagonal is filled
np.fill_diagonal(mask, 1)

# array([[1, 0, 0, 0, 1],
#        [0, 1, 0, 1, 0],
#        [0, 0, 1, 0, 0],
#        [0, 1, 0, 1, 0],
#        [1, 0, 0, 0, 1]])

# Apply the mask with multiplication
# equivalent to np.where(mask > 0, X, 0)
X = X * mask

# array([[ 1,  0,  0,  0,  5],
#        [ 0,  7,  0,  9,  0],
#        [ 0,  0, 13,  0,  0],
#        [ 0, 17,  0, 19,  0],
#        [21,  0,  0,  0, 25]])
```

5.1.4 Use Arithmetic Operations as Boolean Operations to Apply and Combine Masks

As introduced in Chapter 2, there are four major boolean operations, AND, OR, NOT, and XOR. When a mask tensor is expressed as booleans, we could take advantage of boolean operators `&`, `|`, `~`, and `^`, or dedicated operations such as Tensorflow's `logical_and`, `logical_or`, `logical_not`, and `logical_xor`. However, masks can be created as numerical tensors instead of boolean tensors, so type casting is needed. If both our masks and data are already in numerical dtypes, it may be simpler and easier to combine and apply the masks using only numerical operations.

To illustrate how boolean operations can be expressed as arithmetic operations, suppose we have two masks `mask_a = [1, 0, 0, 1, 0]` and `mask_b = [0, 0, 0, 1, 1]`. If we convert them to boolean masks, then `mask_a_bool = [True, False, False, True, False]` and `mask_b_bool = [False, False, False, True, True]`.

The NOT operation on `mask_a_bool` results `[False, True, True, False, True]`, or converting to binary, `[0, 1, 1, 0, 1]`. Notice that this is simply `1 - mask_a`. Therefore `1 - mask` can be used for the NOT operation.

The AND operation between `mask_a_bool` and `mask_b_bool` results in `B = [False, False, False, True, False]`. The AND operation yields `True` if and only if both elements from the two arrays at the corresponding index are `True`, or numerically, `1`. We can multiply the two numerical masks together, yielding `S = [0, 0, 0, 1, 0]`. Notice that this corresponds well to the boolean mask `B` given `0` corresponds to `False` and `1` corresponds to `True`.

The OR operation between `mask_a_bool` and `mask_b_bool` results in `B = [True, False, False, True, True]`. The OR operation yields `True` when either element from the two arrays at the corresponding index is `True`, or numerically corresponding to `1`. We can add the two numerical masks together, yielding `S = [1, 0, 0, 2, 1]`. This corresponds well to the combined boolean mask, where `0` corresponds to `False` and any value greater than `0` corresponds to `True`. To retain the binary (`0` and `1` only) nature of the combined mask, we can clip the value of the combined numerical mask to have a max value of `1`, with `np.minimum(S, 1) = [1, 0, 0, 1, 1]`.

Lastly, the XOR operation between `mask_a_bool` and `mask_b_bool` results in `B = [True, False, False, False, True]`. The XOR operation yields `True` when and only when the values of the two elements are not the same (`True` vs. `False` or `False` vs. `True`) and yields `False` otherwise (`True` vs. `True` or `False` vs. `False`). Using the AND, OR, and NOT operators, XOR is equivalent to `a ^ b = (a | b) & ~(a & b)`. So the more difficult way is to use the above arithmetic versions of the AND, OR, and NOT boolean operations to reconstruct XOR. However, a simpler way is to use subtraction and absolute value, i.e. `np.abs(a - b) = np.abs([1, 0, 0, 0, -1]) = [1, 0, 0, 0, 1]`.

The following code snippet implements the arithmetic version of the boolean operations.

```
import numpy as np

def combine_mask(a, b=None, operation="not"):
    if operation != "not":
        assert b is not None, f"b is required for {operation}"
    if operation == "not": # b is ignored
        return 1 - a
    elif operation == "and":
        return a * b
    elif operation == "or":
        return np.minimum(a + b, 1)
    elif operation == "xor":
        return np.abs(a - b)
    else:
        raise(ValueError(f"Unrecognized operation {operation}."))
```

5.1.5 Select Elements Based on Masking

The masking methods we have introduced so far all replace the values within the tensors while retaining the shape and dimensions of the input tensor. However, there are cases where we would like to discard the masked elements and therefore reduce the size of the input. NumPy's `np.compress` (or equivalently Tensorflow's `tf.boolean_mask` and PyTorch's `torch.masked_select`) is used for such purpose. In the simplest case, boolean masking is the same as using boolean indices to select elements. For example,

```
import numpy as np

X = np.array([5, 3, 2, 6, 9])
mask = np.mod(X, 2) == 0 # even number
assert np.allclose(
    X[mask], # array([2, 6])
    np.compress(mask, X) # array([2, 6])
)
```

For a more advanced example, suppose we have a 2D square matrix of size `(N, N)` and we would like to remove all the diagonal entries so that it becomes

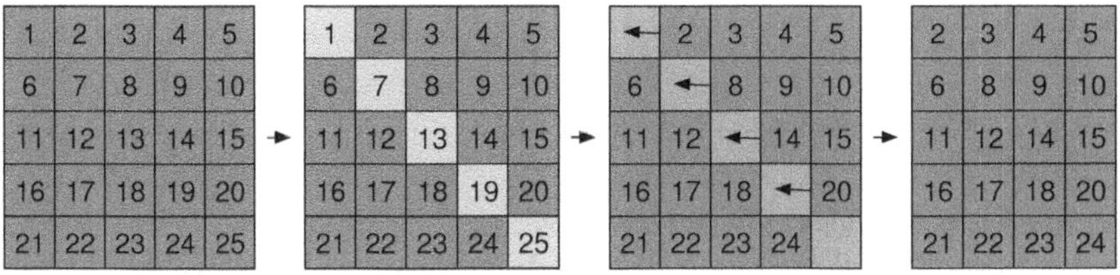

Figure 5.4 Boolean tensor masking and compressing.

a rectangular matrix of size `(N, N-1)`. We can do this using the following procedure in PyTorch (see also Figure 5.4):

```
import torch
X = torch.arange(25).reshape(5, 5) + 1
mask = torch.ones_like(X, dtype=bool)
mask.fill_diagonal_(False)
X_masked = torch.masked_select(X, mask).reshape(5, 4)
```

The equivalent operation can be implemented in Tensorflow but not using NumPy's `np.compress` since the `mask` is required to be 1D and operates along a certain dimension of the input tensor `X`.

5.1.6 Case Study: Top-k Masking

Suppose we have a batch of arrays (e.g. a 2D matrix). For each array, we would like to retain only the top three values, while setting all the other entries to be zeros (see Figure 5.5a). We can do this by creating a binary mask and applying the binary mask to the original matrix. The binary mask should have 1s at indices where the value of the original matrix is among the top three values, and 0s otherwise (see Figure 5.5b).

The code snippet below implements the top-k masking operation using NumPy.

```
import numpy as np
def top_k_masking(y, top_k, masked_value=0):
    """Mask the input and retain only top_k values."""
    # Get top k indices
    index = np.argsort(y)[..., -top_k:][..., ::-1]
    # Initialize the zeros mask
    mask = np.zeros_like(y)
    # scatter the indices (in-place)
    np.put_along_axis(mask, index, 1, axis=-1)
    # Adjust values such that the masked elements
```

```
11        # are replaced with masked_value
12        y_masked = y * mask + masked_value * (1 - mask)
13        return y_masked
```

- Line 5 gets the index of the top-k values
- Line 7 initializes the mask which will indicate the values to keep as 1s and values to discard as 0s.
- Line 9 does an in-place replacement to put 1s at indices where the values will be kept.
- Line 12 applies the masking. `mask` is binary (1s and 0s). The element-wise multiplication `y * mask` will set all masked/discarded values to be zeros first. Then, the `masked_value * (1 - mask)` will then set the masked entries to the `masked_value` (i.e. when the value is to be masked, `value:= 0 + masked_value = masked_value`; otherwise, `value:= value + 0 = value`).

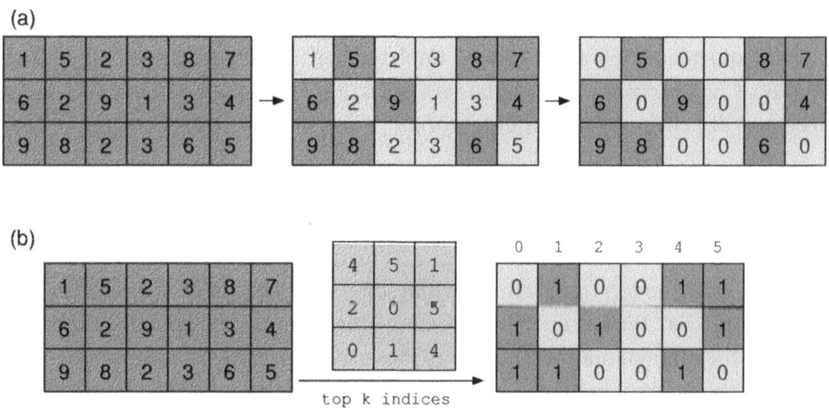

Figure 5.5 Masking top K values. (a) Given an example batch of arrays, we would like to keep the top three values within each array and set the rest of the elements in the array to be 0s. (b) To create the mask, we first sort the values by each array and identify the indices of the top three values. Then we create a binary mask whose shape is identical to the original matrix, and values are set to 1s at the corresponding indices of the top three values.

```
y = np.array([
    [1, 5, 2, 3, 8, 7],
    [2, 6, 3, 1, 9, 4]
])

```

```
top_k_masking(y, top_k=3, masked_value=0)
# array([[0, 5, 0, 0, 8, 7],
#        [0, 6, 0, 0, 9, 4]])
```

5.2 Padding

Padding adds elements to our data tensor to make sure the data are consistently represented in a desired shape. Like masking, the added data is intended to be ignored by the model while the data passes through our vectorized operations.

NumPy [Harris et al., 2023] and Tensorflow [Abadi et al., 2023] both implement three types of padding modes, `constant`, and `reflect`, and `symmetric` (with differences in how the modes are named or spelled); PyTorch also implements `constant` and `reflect` (with limitations), but not `symmetric`; it also implements two additional modes, `replicate` (which simply repeats the border values, equivalent to NumPy's `edge` mode) and `circular` (equivalent to NumPy's `warp` mode). NumPy implements several additional modes, such as padding with `mean`, `median`, `maximum`, or `minimum` values along each dimension.

In `constant` padding mode (NumPy: `"constant"`, Tensorflow: `"CONSTANT"`, PyTorch: `"constant"`), constant values are added at the beginning or the end of the tensor along certain dimensions. In NumPy and Tensorflow, we can specify the number of values to pad and pad either before or after the tensor along the dimensions, using the format `[(start_axis_0, end_axis_0), (start_axis_1, end_axis_1), ..., (start_axis_n, end_axis_n)]`. However, PyTorch's `torch.nn.functional.pad` has a different implementation and requires the padding to be specified in the reverse order and as a flattened array, i.e. `[start_axis_n, end_axis_n, ..., start_axis_0, end_axis_0]`. For example, given a 3×4 matrix, we would like to add three more zeros along its rows at the beginning and two more zeros along its columns at the end, we then do

```
import numpy as np
import tensorflow as tf
import torch.nn.functional as F
# shape(3, 4)
X = np.array([
    [7, 1, 4, 4],
    [5, 7, 6, 4],
    [7, 3, 1, 2],
])
# padding zeros -> shape (6, 6)
y_np = np.pad(X, [(3, 0), (0, 2)], mode="constant", constant_values=0)
```

```
y_tf = tf.pad(X, [(3, 0), (0, 2)], mode="CONSTANT", constant_values=0)
y_torch = F.pad(
    torch.tensor(X), [0, 2, 3, 0], mode="constant", value=0
)

# array([[0, 0, 0, 0, 0, 0],
#        [0, 0, 0, 0, 0, 0],
#        [0, 0, 0, 0, 0, 0],
#        [7, 1, 4, 4, 0, 0],
#        [5, 7, 6, 4, 0, 0],
#        [7, 3, 1, 2, 0, 0]])
```

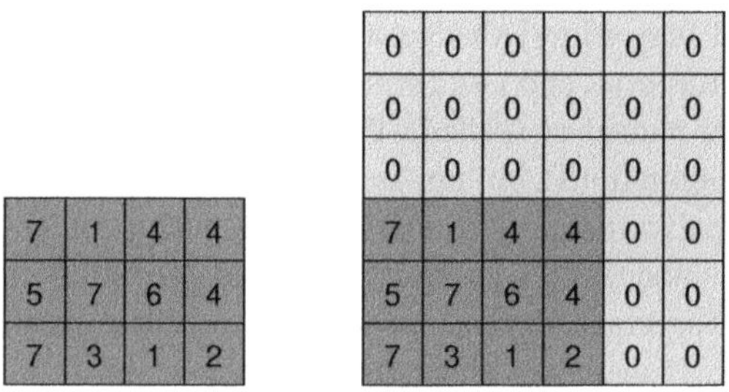

Figure 5.6 Example illustration of `constant` padding.

See also Figure 5.6 for an illustration.

In NumPy (i.e. this is not allowed in Tensorflow or PyTorch), the `constant_values` argument can also be a pair of values (`(start_value, end_value)` for all dimensions) or a list of tuples of values (`[(start_value_0, end_value_0), (start_value_1, end_value_1), ...]`), as illustrated in the following examples:

```
import numpy as np
# shape (3, 4)
X = np.array([
    [7, 1, 4, 4],
    [5, 7, 6, 4],
    [7, 3, 1, 2],
])

# padding starts with 0 and ends with -1 for each dimension
y = np.pad(X, [(0, 1), (2, 0)], mode="constant", constant_values=(0, -1))
# array([[ 0,  0,  7,  1,  4,  4],
#        [ 0,  0,  5,  7,  6,  4],
#        [ 0,  0,  7,  3,  1,  2],
#        [ 0,  0, -1, -1, -1, -1]])

# padding row start with 0, row end with -1
# padding column start with -2, column end with -3
y = np.pad(
    X, [(1, 1), (1, 1)], constant_values=[(0, -1), (-2, -3)]
```

```
)
# array([[-2,  0,  0,  0,  0, -3],
#        [-2,  7,  7,  9,  5, -3],
#        [-2,  9,  4,  7,  5, -3],
#        [-2,  1,  4,  1,  3, -3],
#        [-2, -1, -1, -1, -1, -3]])
```

We may see that the ordering of padding proceeds as the order of the `constant_values` list of tuples: it first pads the row start and end, then the column start and end. Higher dimensional tensor paddings follow the same axes ordering. See Figure 5.7 for an illustration.

In `reflect` padding mode (NumPy: `reflect`, Tensorflow `REFLECT`, PyTorch: `reflect`), the paddings are added as if a mirror is being put at the border values of the tensor. That is, at padding index $-1, -2, \ldots$ (start of the dimension), values at index $1, 2, \ldots$ are used; at padding index $N, N+1, \ldots$ (end of the dimension), values at $N-1, N-2, \ldots$ are used. Like `constant` mode, padding is added step-by-step by following the order of the axes of the tensor. `reflect` padding is useful before performing Fourier transform, for example, in the case of audio signal processing. See Figure 5.8 and the following code snippet for an illustration.

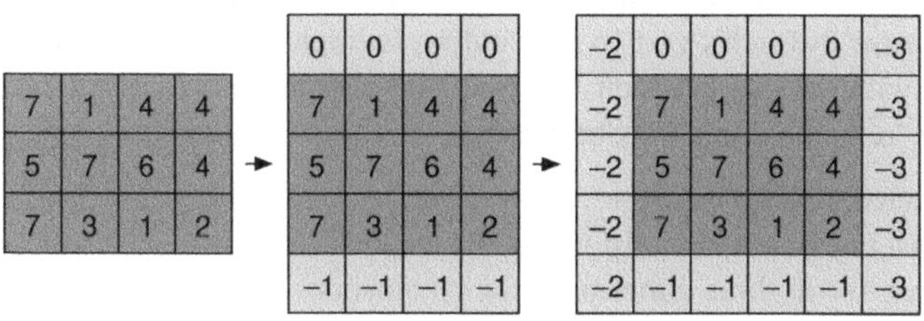

Figure 5.7 Illustration of `constant` padding order.

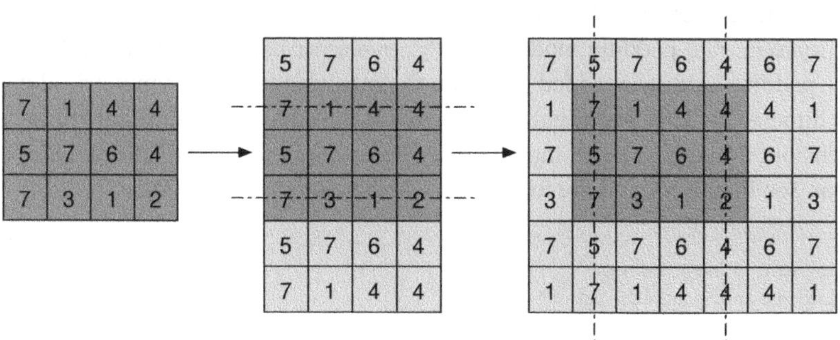

Figure 5.8 Example illustration of `reflect` padding. Dashed lines indicate axes of reflection.

```
import numpy as np
import tensorflow as tf
import torch.nn.functional as F
X = np.array([
    [7, 1, 4, 4],
    [5, 7, 6, 4],
    [7, 3, 1, 2],
])
y_np = np.pad(X, [(1, 2), (1, 2)], mode="reflect")
y_tf = tf.pad(X, [(1, 2), (1, 2)], mode="REFLECT")
y_torch = F.pad(
    torch.tensor(X[None, :, :], dtype=float),
    [1, 2, 1, 2],
    mode="reflect"
)[0, :, :]

# array([[7, 5, 7, 6, 4, 6, 7],
#        [1, 7, 1, 4, 4, 4, 1],
#        [7, 5, 7, 6, 4, 6, 7],
#        [3, 7, 3, 1, 2, 1, 3],
#        [7, 5, 7, 6, 4, 6, 7],
#        [1, 7, 1, 4, 4, 4, 1]])
```

Notice that in PyTorch's implementation, we need to add an additional dimension. This is because PyTorch's `reflect` padding is implemented in a way to specifically deal with tensors between two and five dimensions [Paszke et al., 2023], where the "reflection" can only be created:

- on the last dimension of 2D tensors;
- on either the last one or two dimensions for 3D tensors;
- on either the last two or three dimensions for 4D tensors;
- on the last three dimension for 5D tensors.

Also, only float values can be used for `reflect` padding operations in `torch.nn.functional.pad`.

In `symmetric` padding mode (NumPy: `symmetric`, Tensorflow: `SYMMETRIC`), padding values are added such that the resulting tensor is symmetric along each dimension. This is as if putting a mirror at the edge of the tensor along each dimension; unlike `reflect` mode, the border values will also be reflected in the mirror. See Figure 5.9 and the following code snippet for an illustration.

```
import numpy as np
import tensorflow as tf
X = np.array([
    [7, 1, 4, 4],
```

```
        [5, 7, 6, 4],
        [7, 3, 1, 2],
])
y_np = np.pad(X, [(1, 2), (1, 2)], mode="symmetric")
y_tf = tf.pad(X, [(1, 2), (1, 2)], mode="SYMMETRIC")
# no 'symmetric' mode in the torch.nn.functional.pad

# array([[7, 7, 1, 4, 4, 4, 4],
#         [7, 7, 1, 4, 4, 4, 4],
#         [5, 5, 7, 6, 4, 4, 6],
#         [7, 7, 3, 1, 2, 2, 1],
#         [7, 7, 3, 1, 2, 2, 1],
#         [5, 5, 7, 6, 4, 4, 6]])>
```

Both NumPy and PyTorch implement `edge` or `replicate` mode, respectively. In this mode, values at the borders along each dimension are repeated. PyTorch's `replicate` mode has similar limitations as its `reflect` mode. See Figure 5.10 and the following code snippet for illustrations.

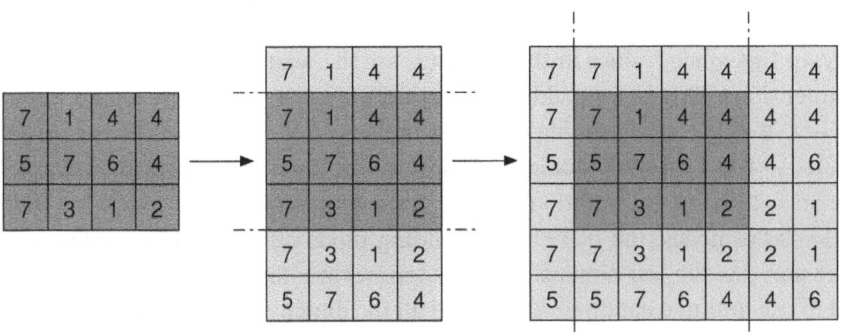

Figure 5.9 Example illustration of `symmetric` padding. Dashed lines indicate the axes of symmetry.

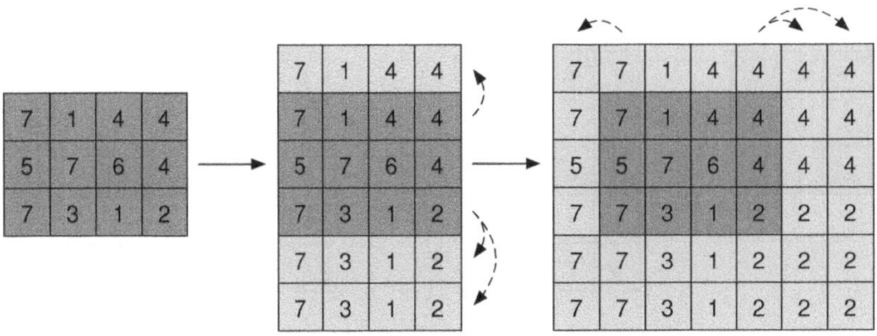

Figure 5.10 Example illustration of `edge` or `replicate` padding.

```
import numpy as np
import torch
X = np.array([
    [7, 1, 4, 4],
    [5, 7, 6, 4],
    [7, 3, 1, 2],
])
y_np = np.pad(X, [(1, 2), (1, 2)], mode="edge")
y_torch = F.pad(
    torch.tensor(X[None, :, :], dtype=float),
    [1, 2, 1, 2],
    mode="replicate"
)[0, :, :]

# array([[7, 7, 1, 4, 4, 4, 4],
#         [7, 7, 1, 4, 4, 4, 4],
#         [5, 5, 7, 6, 4, 4, 4],
#         [7, 7, 3, 1, 2, 2, 2],
#         [7, 7, 3, 1, 2, 2, 2],
#         [7, 7, 3, 1, 2, 2, 2]])
```

Both NumPy and PyTorch also implement `wrap` or `circular` mode, respectively. In this mode, values in the paddings are the wrapping of the tensor along its dimensions. That is, for padding values at indices $-1, -2, \ldots$, use tensor values at $N-1, N-2, \ldots$; for padding values at indices $N, N+1, \ldots$, use tensor values at $0, 1, \ldots$. For PyTorch's implementation at the time of writing this chapter (`torch==2.0.1`), in addition to the same limitation as the `reflect` and `replicate` modes, the `circular` padding mode only supports padding one dimension (i.e. the last dimension of the tensor) at a time. Therefore, we would need to use multiple padding steps to pad all the dimensions. See Figure 5.11 and the following code snippet for illustrations.

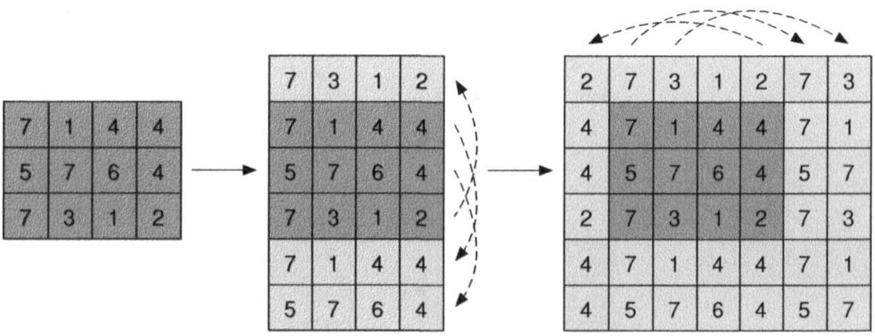

Figure 5.11 Example illustration of `wrap` or `circular` padding mode.

```
import numpy as np
import torch
X = np.array([
    [7, 1, 4, 4],
    [5, 7, 6, 4],
    [7, 3, 1, 2],
])
y_np = np.pad(X, [(1, 2), (1, 2)], mode="wrap")
# padding rows
y_torch = F.pad(
    torch.tensor(X, dtype=float).t()[None, :, :], [1, 2], mode="circular"
)
# padding columns
y_torch = F.pad(torch.swapaxes(y_torch, 1, 2), [1, 2], mode="circular")

# array([[2, 7, 3, 1, 2, 7, 3],
#        [4, 7, 1, 4, 4, 7, 1],
#        [4, 5, 7, 6, 4, 5, 7],
#        [2, 7, 3, 1, 2, 7, 3],
#        [4, 7, 1, 4, 4, 7, 1],
#        [4, 5, 7, 6, 4, 5, 7]])
```

5.2.1 Case Study: Padding in Convolutional Neural Networks

Understanding padding is quite useful when implementing convolutional neural networks . There are two typical padding modes in a convolutional layer: `valid` and `same`. `valid` simply means there is no padding added, and convolution results are smaller than the input. `same` means adding some zero padding to the heights and widths of the convolutional dimensions such that the result of the convolution has the same heights and widths as the input. See Figure 5.12 for an example that compares the differences between the two.

See also the below code example to verify the shapes of `same` vs. `valid` padding in convolutional operations.

```
import tensorflow as tf
X = tf.constant([
    [5, 3, 2, 7],
    [4, 1, 3, 7],
    [1, 4, 4, 1],
    [7, 6, 4, 2],
], dtype=tf.float32)[None, :, :, None]

Filters = tf.constant([
    [0, 2],
    [1, 1]
], dtype=tf.float32)[:, :, None, None]
```

```

y_valid = tf.nn.conv2d(
    X,
    Filters,
    1,
    "VALID",
    data_format='NHWC',
    dilations=None,
    name=None
)
assert y_valid.shape[1] == X.shape[1] - 1
assert y_valid.shape[2] == X.shape[2] - 1

y_same = tf.nn.conv2d(
    X,
    Filters,
    1,
    "SAME",
    data_format='NHWC',
    dilations=None,
    name=None
)
assert y_same.shape[1] == X.shape[1]
assert y_same.shape[2] == X.shape[2]
```

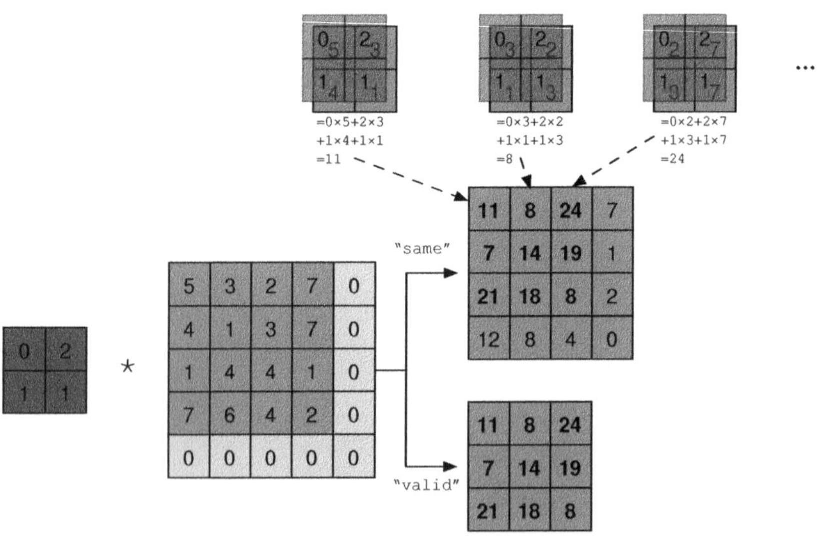

Figure 5.12 Example of `same` vs. `valid` padding in convolutional operations.

5.2.2 Case Study: Truncate or Pad Sequence to Desired Length

Suppose we have two batches of sequences, *X* and *y*. *X* has shape `(batch_size, seq_len_1)` and *y* has shape `(batch_size, seq_len_2)`. We do not know beforehand which sequence is longer, however, we would like to have *y* with sequence lengths identical to that of *X*. The following code uses the idea of padding and slicing to align the lengths of *X* and *y*.

```
1  import numpy as np
2  def align_lengths(X, y):
3      """Align y to the length (axis=1) of X."""
4      # Get the desired length
5      X_len = X.shape[1]
6      # Compute pad_size.
7      # No padding (pad_size=0) if y length is enough
8      pad_size = max(X_len - y.shape[1], 0)
9      # Pad if not enough
10     y = np.pad(y, [(0, 0), (0, pad_size)])
11     # Truncate if y is too long
12     y = y[:, :X_len]
13
14     return y
```

- Line 5 gets the desired length that we would like *y* to have.
- Line 8 computes the desired padding size. If *X* is longer, then we need to pad the difference between *X* and *y*; but if *X* is shorter, we should not pad (`pad_size=0`).
- Line 10 does the padding based on the `pad_size` we calculated. This will handle cases where *y* is shorter than *X* but has no effect when *y* is the same length or longer than *X*.
- Line 12 truncates any extra elements in *y*. If *y* is indeed longer than *X*, then after truncation, *y* should be the same size as *X*. If *y* is the same length as *X* or is shorter than *X*, padding from the previous step will first bring *y* to the same length as *X*. Then slicing up to the length *X* will have no effect.

We can quickly test out our implementation:

```
X1 = np.random.rand(16, 50)
X2 = np.random.rand(16, 30)
y = np.random.rand(16, 37)

y1 = align_lengths(X1, y)
assert np.allclose(X1.shape, y1.shape)
print(y1.shape)
```

```
y2 = align_lengths(X2, y)
assert np.allclose(X2.shape, y2.shape)
print(y2.shape)
```

A practical use case of the above algorithm is to align a generated sequence to some desired length. For example, a ground-truth speech audio sample has two words, lasts one seconds and contains 16,000 data points (i.e. 16 kHz sampling rate). However, the generated sample due to series of upsampling by factors of [16, 16, 32] from the two words creates a sequence of $2 \times 16 \times 16 \times 32 = 16{,}384$ data points. In order to learn from the audio sample by calculating a loss between the prediction and the ground-truth, we can use the above algorithm to align the two sequences (e.g. truncate the generated sequence to 16,000 data points). It is worth noting that many practical applications of machine learning that deal with sequence data, the predicted sequence y should in general be of similar length as the template or input sequence X. If the predicted sequence is significantly longer or shorter than expectation, it is important to understand if upsampling steps are implemented correctly, or hyper-parameters for upsampling are chosen appropriately.

5.3 Advanced Case Studies

For the rest of the chapter, let us explore several more advanced use cases of masking and padding in machine learning/deep learning models.

5.3.1 Scaled-Dot Product Attention

Transformer models introduced by the paper "Attention Is All You Need" [Vaswani et al., 2017] are probably the most successful type of sequence model by far. It has been used widely in modeling sequence data such as texts, audio, and images, for tasks including natural language understanding [Devlin et al., 2019], speech recognition [Gulati et al., 2020], singing voice synthesis [Liu et al., 2021], and image recognition [Dosovitskiy et al., 2021]. A key component of transformer models is the Scaled Dot-Product Attention mechanism, which teaches the model which part of the sequence to focus the learning on. Scaled Dot-Product Attention is defined as

$$A(Q, K, V) = \text{softmax}\left(\frac{QK^T}{\sqrt{d_K}}\right)V \tag{5.1}$$

where

- Q, K, V are the query, key, and value sequence.
- d_K is the "depth" or embedding dimensions of the keys.
- softmax(·) is the softmax activation function, defined as $\sigma(x_i) = \frac{\exp(x_i)}{\sum_j \exp(x_j)}$.

In Equation 5.1, query and key first perform a dot product QK^T before scaled by the square root of the depth of the `key` $\sqrt{d_K}$. Then, the computed attention score is transformed with the softmax function to convert into probabilities and applied to the `value` sequence via tensor multiplication.

In addition, there are two major flavors of this attention mechanism:

- **Self-attention**: `key`, `query`, and `value` are all the same sequence inputs. The attention weights are computed within the sequence itself, between pairs of its own elements.
- **Cross-attention**: `key`, `query`, and `value` are not the same sequence. Here, elements in the key and query sequences need to have identical embedding or so-called depth dimensions, while `key` and `value` need to have identical sequence lengths. Attention weights are computed between two distinct sequences represented by `key` and `query` and applied to a third sequence represented by `value`.

There are two major flavors of self-attention (see Figure 5.13):

- **Bidirectional self-attention**: A single token within the sequence can attend to not only the tokens coming before it (and itself) but also the tokens coming after it. This type of self-attention is implemented as part of the Encoder layers of the Transformer model, where all the contexts surrounding the current token help inform the model.
- **Causal self-attention**: A single token within the sequence can only attend to the tokens coming before it (and itself). This type of self-attention is implemented as part of the Decoder layers of the Transformer model, where only the historical token sequences are used to predict the tokens coming next.

Figure 5.14 demonstrates the attention mask of bi-directional and causal self-attentions. Notice that the causal self-attention mask forms a lower triangular matrix (including the diagonal).

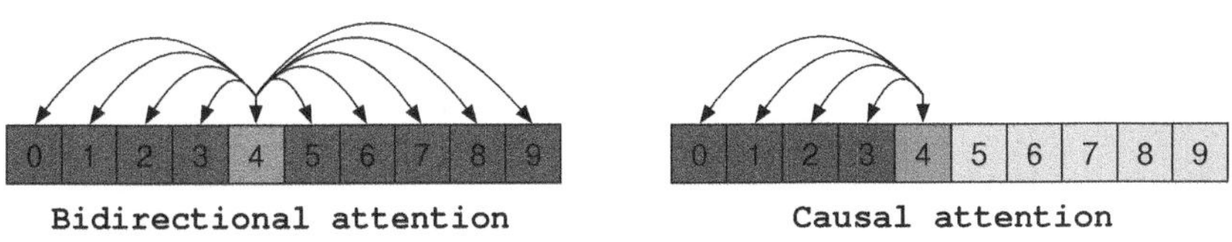

Figure 5.13 Bidirectional vs. causal self-attentions.

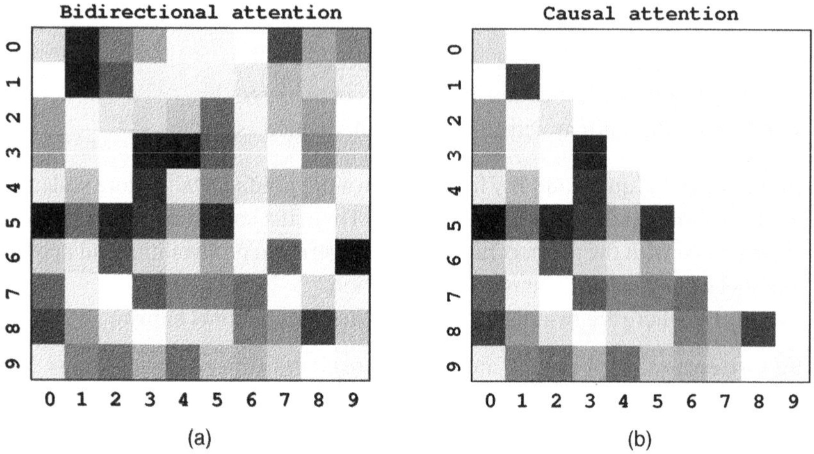

Figure 5.14 Attention mask. (a) Bidirectional attention mask, which allows the past tokens to attend future tokens and vice versa; (b) causal attention mask, which only allows future tokens attend to past tokens but past tokens will not be able to attend future tokens.

The following code snippet implements the Scaled Dot-Product Attention using Tensorflow and Keras.

```
import tensorflow as tf
from keras.activations import softmax

def scaled_dot_product_attention(
    query, key, value, mask=None, causal=True, droprate=0.0
):
    """
    Use the dense layer output from query, key, value,
    respectively, as input for attention

    Parameters:
    -----------
    * query: (..., l_q,        depth_src)
    * key:   (..., l_k == l_v, depth_src)
    * value: (..., l_v == l_k, depth_out)
    * mask: Float tensor of 0s and 1s, with
            shape broadcastable to (..., l_q, l_k).
    * causal: bool. Whether or not to apply causal attention masking.
    * droprate: dropout rate.

    - depth_*: depth x num_heads = embed_dim, which
               can be different, but are usually the same.
    - l_*: are lengths of input sequences which can be different,
           but are usually the same

```

```
26      Returns:
27      ----------
28      * attention: attention score (..., l_q, depth_out)
29      * weights: weights of the value when summing the value (..., l_q, l_k)
30      """
31      # Cosine/dot product similarity
32      # Q K.T, (..., l_q, l_k)
33      logit_score = tf.matmul(query, key, transpose_b=True)
34
35      # scale by dimension of key
36      d_k = tf.cast(tf.shape(key)[-1], tf.float32)
37      scaled_logit_score = logit_score / tf.sqrt(d_k)  # (..., l_q, l_k)
38
39      # Masking the region by a large negative number (est. -inf)
40      if mask is not None:
41          scaled_logit_score += mask * -1e9  # (..., l_q, l_k)
42      if causal:
43          l_q = tf.shape(query)[-2]
44          l_k = tf.shape(key)[-2]
45          # creates an upper triangular mask to retain lower triangular
            ↪ matrix
46          causal_mask = 1 - tf.linalg.band_part(
47              tf.ones((l_q, l_k), tf.float32), -1, 0
48          )
49          scaled_logit_score += causal_mask * -1e9
50
51      # weights: softmax normalization
52      weights = softmax(scaled_logit_score, axis=-1)
53      # Add dropout
54      weights = tf.nn.dropout(weights, droprate)
55
56      # attention (..., l_q, depth_out)
57      attention = tf.matmul(weights, value)
58
59      return attention, weights
```

- Lines 13 and 14: Key and value must have the same length, i.e. `l_k == l_v`.
- Line 33 implements the dot product part of the attention, between Key and Query QK^T. The `logit_score` has shape `(batch_size, [num_heads,] l_q, l_k)`, where optionally, `num_heads` is the number of heads in the case of multi-headed attention.
- Line 36 gets the embedding size of the key d_K from the input `key`.
- Line 37 computes the scaled dot product $\frac{QK^T}{\sqrt{d_K}}$.
- Lines 40 and 41: This handles a case where an additional binary mask is applied on top of the dot product attention. We apply the masking by adding the masked region with a large negative number. This is because, after applying the softmax activation the logit score of large negative numbers will become near 0. For example, `softmax([1, -3, 2, -1e9, 5])` = `1.7142441e-02, 3.1397477e-04, 4.6597987e-02,`

0.0000000e+00, 9.3594557e-01]. We choose to add the large negative number on the logit scores instead of performing softmax first, and then applying the binary mask because we want to make sure that the weights of the attention mask after applying softmax still sum to 1. Note that the mask is assumed to have or can be broadcasted to shape (batch_size, [num_heades,] l_q, l_k).

- Lines 42–49 implement the causal masking. Here, we use tf.linalg.band_part to create an upper triangular mask to retain the diagonal and lower triangular part of the matrix. Note that this implementation can handle cases where the value length does not equal to query length. The mask has the shape (l_q, l_k).
- In Line 49, we again apply the masking by adding a large negative number to the logit scores to the masked region. We take advantage of the broadcasting rule here as we know the scaled_logit_score has shape (batch_size, [num_heads,], l_q, l_k), and we apply identical upper triangular masking to all batches and all heads of the attention scores.
- Line 52 applies softmax on the last dimension of the batch of attention mask, such that the attention weights between key and query sums to 1.
- Line 54 adds a dropout layer. During inference time, we need to set droprate=0 to make sure the outputs are deterministic.
- Line 57 creates the attention-weighted sequence by multiplying the weights and value together. Here weights has shape (batch_size, [num_heads,] l_q, l_k) and value has shape (batch_size, [num_heads,], l_v, depth_out), knowing also that l_k == l_v. Recall that tf.matmul operates on the last two dimensions of the tensors. Then, attention has shape (batch_size, [num_heads,], l_q, depth_out).

5.3.2 Variable-Length Range via Masking

This case study creates a helper function that we will be using in Section 5.3.3 where we implement the Length Regulator module of FastSpeech2 [Ren et al., 2020].

Suppose we have a long sequence like the following [1, 1, 2, 2, 2, 2, 2, 3, 3, 3, 3] where each value indexes which group the current value belongs to. We can observe that there are 2 of 1s, 5 of 2s, and 4 of 3s. We would like to be able to enumerate the elements within each group by giving each number a within-group index [0, 1, 0, 1, 2, 3, 4, 0, 1, 2, 3] as a horizontal concatenation of 3 range arrays [0, 1], [0, 1, 2, 3, 4], and [0, 1, 2, 3], each having a different length. Tensorflow's RaggedTensor class

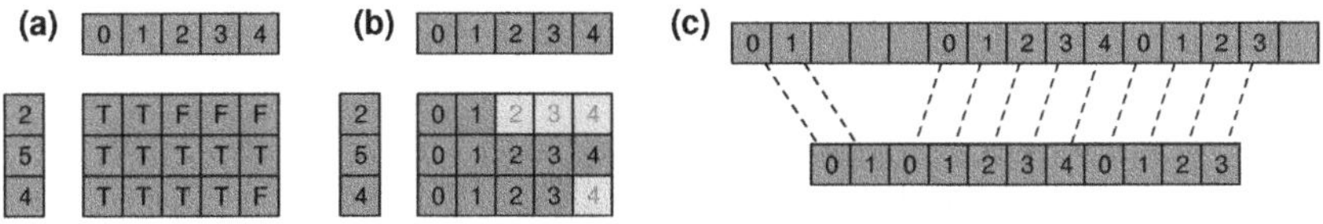

Figure 5.15 Building a ragged range vector. (a) Create the boolean mask by broadcasting the row vector of the range array and the column vector of array lengths. If the value in the range array is smaller than the value in the length vector at the corresponding row, then mark it as True `T`; otherwise, mark it as False `F`; (b) mask the stack of dense range arrays. The greyed-out values will be replaced with −1 in our implementation; (c) flatten the matrix and compress the vector by discarding the masked values.

implements the initializer, `tf.ragged.range`, which returns a batch of vectors of variable lengths. For example, `tf.ragged.range([2, 5, 4])` returns

```
[
    [0, 1],
    [0, 1, 2, 3, 4],
    [0, 1, 2, 3]
]
```

We can then flatten this ragged tensor, using `tf.ragged.range([2, 5, 4]).flat_values`, and obtain the within-group index we wanted before.

But what if we do not have such a convenient initializer in the library we are using? We can construct something similar to this using the idea of indexing and masking. The general strategy of our implementation is the following (see also Figure 5.15):

- We will first construct a single range array (i.e. $[0, 1, 2, \ldots, n]$ via `np.arange`) up to the maximum desired array length based on the input.
- Then, we mask out elements of the range array where its value is beyond the desired range and replace it with a padding value (e.g. −1). The mask can be constructed by broadcasting the range array row vector with the array length column vector together.
- Lastly, to return the desired 1D vector, we first flatten the matrix and then remove any padding values.

Here is an implementation using NumPy:

```
from typing import List
import numpy as np

def ragged_range(n: List[int], pad=-1, flatten=True):
```

```
5       """Implement ragged matrix using a dense matrix with
         ↪ paddings."""
6       # Find the maximum length
7       n = np.array(n)
8       max_n = np.max(n)
9       # Construct and repeat arange
10      r = np.arange(max_n)[None, :]
11      # Masking with broadcasting
12      mask = r >= n[:, None]
13      r = np.where(mask, pad, r)
14      # Flatten
15      if flatten:
16          # Flatten
17          r = np.ravel(r)
18          # Remove padding values
19          r = np.compress(r != pad, r)
20
21      return r
```

- Line 7 ensures that the list input `n` that specifies the desired lengths of the range arrays is an `np.ndarray` which will later be used for broadcasting.
- Line 8 determines the maximum size of the range array.
- Line 10 creates the initial range array, up to the maximum array length.
- Line 12 creates the mask to retain values less than the size of the array specified by `n`. This is done by broadcasting the column vector of array lengths and the row vector of the range array initialized in Line 10. The resulting mask is a matrix of size `(len(n), max_n)`. Note that we are sacrificing memory spaces in exchange for subsequent opportunities for vectorization. In some worst case scenarios, one range array may be much longer than the rest of the range arrays. In this case, the majority of the elements are padding values, or the matrix is sparse (a concept that we will discuss in Chapter 7).
- Line 13 uses `np.where` to replace values according to the `mask`. If the `mask` value is `True`, then we replace the value in `r` with the `pad` value, e.g. –1; otherwise, we retain the value inside `r`.
- Lines 15–19 implement the logic when we need to return a flat array. Otherwise, we will simply return the matrix with each row being a ranged array with paddings.
- Line 17 flattens the 2D matrix to a 1D vector.
- Line 19 removes the padding values.

In order to prepare for the case study in Section 5.3.3, I have also translated the above function from NumPy to Tensorflow. Notice that we need extra effort to make sure the data types are consistent before performing each operation.

```python
from typing import List
import tensorflow as tf

def tf_ragged_range(n: tf.Tensor, pad=-1, flatten=True, dtype=tf.int64):
    """Implement ragged matrix using a dense matrix with paddings."""
    # Find the maximum length
    n = tf.cast(n, dtype)
    pad = tf.cast(pad, dtype)
    max_n = tf.reduce_max(n)
    # Construct and repeat arange
    r = tf.expand_dims(tf.range(max_n, dtype=dtype), axis=0)
    # Masking with broadcasting
    mask = r >= tf.expand_dims(n, axis=1)
    r = tf.where(mask, pad, r)
    # Flatten
    if flatten:
        # Flatten
        r = tf.reshape(r, (-1, ))
        # Remove padding values
        r = tf.boolean_mask(r, r!= pad)

    return r
```

In Chapter 8, we will further explore ragged or jagged tensors in more detail.

5.3.3 Length Regulator Module of FastSpeech2

In a typical text-to-speech (TTS) application, a sequence of bytes or strings is converted into a time series of waveforms. This is the sequence-to-sequence problem in machine learning, where the sequential representation of a certain information modality is converted into another sequence, sometimes in another modality. Classic approaches to solve the TTS problem were using auto-regressive modeling (e.g. Hidden-Markov Model [Zen et al., 2007] and WaveNet [van den Oord et al., 2016]), where prompt waveforms and contexts are used to generate subsequent new speech waveforms. On the other hand, non-autoregressive approaches like FastSpeech/2 [Ren et al., 2019, 2020] use sequence-to-sequence (seq2seq) learning, where the input texts are converted into audio signals using only one pass.

One problem with the seq2seq learning approach between two different information modalities is that the input and output sequences may represent information at different densities and lengths. For example, a sentence like "I love machine learning" uses an array of 4-string tokens to form the full text, but it may require ~2 seconds of 16 kHz single-channel audio, or an array of 32,000 float32 numbers, to encode the speech signal. Depending on the words in the text, a sentence containing only four words may correspond to audios made up of either several thousands or tens of thousands of float32 values ("I love machine learning" vs. "How to pronounce Supercalifragilisticexpialidocious"); and depending on the speaker, the same sentence "I love machine learning" may be spoken under a second, or stretched over tens of seconds (e.g. "I looooooooooove machine learniiiiiiiiiiiii-ing"). This is the so-called many-to-many mapping problem.

FastSpeech2 proposed the Length Regulator module to solve the many-to-many mapping problem. During training time, tokens of the text are passed through the embedding layer which encodes the words in the sentence as a sequence of embedding vectors, forming the tensor `[batch_size, seq_len, emb_size]`, where `seq_len` here can be a variable sequence size and is the length of the longest sentence in the batch. The corresponding speech audio signals (1D vectors) are converted into 2D mel-spectrograms via Fourier Transform, forming a tensor of `[batch_size, time_length, num_channels]`, where `time_length` here can be variable and is the length of the longest audio in the batch, measured as the number of time steps or frames of the temporal domain of the mel-spectrogram. It is easy to observe that `seq_len != time_length`. However, we have the following observations in the TTS data:

1. Each word is pronounced one after another.
2. Each word takes a certain amount of time to pronounce with durations independent of each other.
3. The sum of the time to pronounce all the words in a sequence should equal to the final audio length.

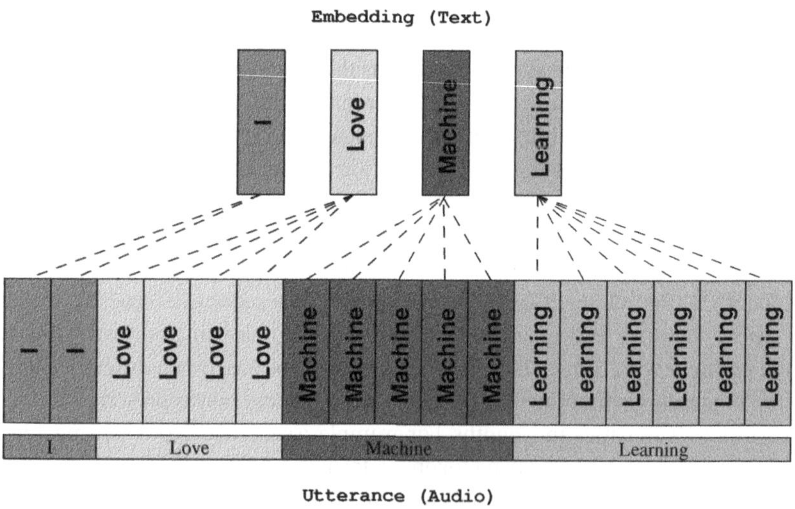

Figure 5.16 Example effect of Length Regulator module on the four-word sentence "I love machine learning," assuming we know that the first word "I" occupies two frames, the second word "love" occupies four frames, the third word "machine" occupies five frames, and the last word "learning" occupies six frames. Each token in the sentence is first passed through an embedding layer to create an embedding vector. Then, the embedding vectors are repeated based on the number of frames the token is supposed to occupy in the utterance (audio).

We also assume that each token is represented by one embedding vector, but may correspond to multiple (or at least 1) time-steps in the mel-spectrogram which we know from the dataset labeling process. With these observations and assumptions, the Length Regulator module serves as an adaptor to repeat/tile the token embedding vectors over the period of pronunciation in the mel-spectrogram. See Figure 5.16 for an example.

For a batch of variable-length text inputs, the Length Regulator module will then return a tensor of shape `[batch_size, time_length, num_channels]` and a corresponding binary mask for masking out the paddings later on. We will leverage the knowledge of indexing and scattering to implement the vectorized Length Regulator module. The following code snippet implements the Length Regulator module as a `keras.Layer` subclass.

```
import tensorflow as tf
from keras import Input, Model
from keras.layers import Layer

# Import functions that we implemented previously
from .utils import tf_ragged_range


class LengthRegulator(Layer):
    """FastSpeech Length Regulator"""
    def __init__(
        self,
        num_channels=256,
        max_len=-1,
        name="length_regulator",
    ):
        super(LengthRegulator, self).__init__(name=name)
        # embedding size
        self.num_channels = num_channels
        # maximum length of the audio signal
        # if -1, determine the audio signal from inputs dynamically
        self.max_len = max_len

    def call(
        self,
        token_embeddings,
        spect_durations,
        training=False,
    ):
        """
        Layer call method
        * token_embeddings : (batch_size, seq_len, num_channels)
        * spect_durations : (batch_size, seq_len), where the value of
            the durations are expressed as number of audio sample frames
        * training : bool, optional, by default False

        The length of the returned sequence after length regulator is
        the maximum of the sum of the durations across the batch.
        For example, [[2, 8, 0], [3, 1, 4], [2, 0, 0]].  Then the
↪   durations
```

```
        are [10, 8, 2], and the output sequence length is 10.
        The mask will indicate the valid positions of the output.
        """
        # Preprocess durations
        spect_durations = tf.cast(tf.round(spect_durations), tf.int64)
        durations = tf.math.maximum(spect_durations, 1)
        durations = tf.cast(durations, tf.int64)

        # Save some shapes
        batch_size = tf.shape(token_embeddings)[0]
        seq_len = tf.shape(durations)[1]
        batch_duration = tf.reduce_sum(durations, axis=1)
        max_durations = tf.maximum(tf.reduce_max(batch_duration),
    ↪   self.max_len)
        # reshaping to (batch_size x seq_len, num_channels)
        hidden_states = tf.reshape(
            token_embeddings, (-1, self.num_channels)
        )
        durations = tf.reshape(durations, (-1,))

        # repeat: (sum(spec_durations), hidden_size)
        hidden_states = tf.repeat(hidden_states, durations, axis=0)

        # scatter the repeated hidden states to the right positions
        row_indices = tf.repeat(
            tf.repeat(tf.range(batch_size, dtype=tf.int64), seq_len),
            durations,
        )
        col_indices = tf_ragged_range(batch_duration)
        indices = tf.stack([row_indices, col_indices], axis=1)
        output = tf.scatter_nd(
            indices,
            hidden_states,
            shape=(batch_size, max_durations, self.num_channels),
        )
        # create output mask
        mask_batch_duration = tf.reduce_sum(spect_durations, axis=1)
        mask_values = tf.ones(
            [tf.reduce_sum(mask_batch_duration), 1],
            dtype=tf.float32,
        )
        mask_row_indices = tf.repeat(
            tf.range(batch_size, dtype=tf.int64), mask_batch_duration
        )
        mask_col_indices = tf_ragged_range(mask_batch_duration)
        mask_indices = tf.stack([mask_row_indices, mask_col_indices],1)
        mask = tf.scatter_nd(
            mask_indices,
            mask_values,
            shape=(batch_size, max_durations, 1),
        )

        # Further truncate any potential padding zeros
        tight_size = tf.cast(
```

```
93                 tf.reduce_max(tf.reduce_sum(mask, axis=1)), tf.int64
94             )
95         output = output[:, :tight_size, :]
96         mask = mask[:, :tight_size, :]
97 
98         return output, mask
99 
100     @property
101     def model(self):
102         text_embeddings = Input((None, self.num_channels), tf.float32)
103         spect_durations = Input((None,), dtype=tf.float32)
104         model = Model(
105             inputs=[text_embeddings, spect_durations],
106             outputs=self.call(
107                 text_embeddings, spect_durations, training=False
108             ),
109         )
110         return model
```

- `__init__` **method**: Initializes the layer subclass
  - `num_channels` stores the embedding dimension size, since later in the `call` method, we will need to reshape the embedding tensors. It is indeed also possible to determine the embedding dimension size from the input dynamically.
  - `max_len` stores the maximum audio sequence length. If the user specified a value greater than zero, then the output will have this sequence length, instead of being dynamically determined from the inputs.
- The `call` method implements the main logic of the Length Regulator module, given a batch of token sequence embeddings, as well as the durations of each token pronunciation as the number of mel-spectrogram frames.
  - `token_embedding` has the shape `[batch_size, seq_len, num_channels]` and has dtype `tf.float32`. The padding token should correspond to 0 valued embedding vectors.
  - `spect_durations` has shape `[batch_size, seq_len]`. Padding tokens have 0 durations.
  - Line 44 first rounds off the durations to integers. This is because, during inference time, another module of the FastSpeech2 model will also make a prediction of the duration of the tokens, which will not necessarily be integers. This line makes sure that in the subsequent operations, we are dealing with the number of mel-spectrogram frames expressed as integers.
  - Lines 45 and 46 compute how many times each token embedding needs to be repeated. This is again an inference-time consideration: we need to repeat any token in the sequence at least once. We will encounter two scenarios where the duration can be 0 during inference time:

1. The duration of pronunciation is predicted to be short so we rounded to 0 durations at inference time.
2. We are dealing with paddings at the end of the sequence. In this case, the values of the embedding vectors for the padding token should be all zeros. Even if we leave the zero paddings around, the model should be able to learn not to generate any audio on top of it. Toward the end of the `call` function, we will do extra cleaning to remove these paddings.

- Lines 49–52 store some shapes which we will leverage later on to reshape our inputs. This also includes pre-computing the expected duration of the audio (`max_durations`).
- Lines 54–56 reshape the batch of token sequence embeddings into shape `[batch_size x seq_len, num_channels]`. This turns the 3D tensor into a stack of embedding vectors, which we will then use the `tf.repeat` operation to repeat/upsample the vectors.
- Line 57 reshapes the batch of durations into a 1D array of `[batch_size x seq_len]`. This new array will indicate how many times each piece of embedding vectors from `hidden_states` (Lines 54–56) needs to be repeated.
- Line 60 performs the tiling/repetition operation. We now should have a stack of embedding vectors of shape `[batch_size x time_len, num_channels]`.
- Lines 63–73 implement a scattering logic to reshape the stack of repeated embedding vectors back into the shape of `[batch_size, time_len, num_channels]`. Because our repeated embedding vectors are organized in the shape of `[batch_size x time_len, num_channels]`, we need to reindex the first dimension and split them into two separate dimensions, i.e. `[batch_size, time_len]`. Since this is a 2D matrix, we use the terminology `row` to indicate the batch dimension and `col` to indicate `time_len` here:
  * `row_indices`: This is the index of the batch dimension. It is constructed with two steps of `tf.repeat` operations. We first construct the batch index `[0, 1, …, batch_size]`. We then repeat each batch index by the sequence length. That way, we are assigning a batch index to each token embedding, which becomes `[0, 0, …, 1, 1, …, batch_size, batch_size, …]`. Finally, since each token embedding vector is repeated `duration` times (Line 60), we also repeat the batch index by that many times corresponding to each token. This should give us an array whose length is identical to the second dimension (`axis=1`) of `hidden_states` that we obtained from Line 60.
  * `col_indices`: This is the index of the time dimension, which is constructed by our helper function `tf_ragged_range`. For each

batch, we enumerate the vectors one by one, assigning an index `[0, 1, 2, 3, ..., batch_0_duration, 0, 1, 2, 3, ...., batch_1_duration, ...]`. The `batch_duration` variable we computed earlier tells us how many frames the current sequence should have.

  * Lines 68–73 utilize the indices we computed to re-create the tensor of shape `[batch_size, time_len, num_channels]`. The scatter operation will automatically leave zero paddings to accommodate the representation of variable length sequences.

- Lines 75–89 implement the same scattering logic, except now it is operating on a tensor of shape `[batch_size, time_len, 1]`.
- Lines 92–94 remove any extraneous padding zeros beyond the expected duration as indicated by `spect_durations`.

- **The `model` property**: I often implement this method to quickly unit test my layer implementation. After initializing the layer, I do `print(layer.model.summary())` to print out a summary of the layer for me to examine. If there are errors thrown, then it is very likely that this Tensorflow/Keras model will not train under the graph mode, even if we have verified the model can generate outputs in eager mode. Here is an output of the `LengthRegulator` layer summary:

```
Model: "model_8"
________________________________________________________________________
Layer (type)              Output Shape   Param # Connected to
========================================================================
input_24                  [(None,None)]  0       []
(InputLayer)

tf.math.round_11          (None, None)   0       ['input_24[0][0]']
(TFOpLambda)

tf.cast_47                (None, None)   0       ['tf.math.round_11[0][0]']
(TFOpLambda)

tf.math.reduce_sum_36     (None,)        0       ['tf.cast_47[0][0]']
(TFOpLambda)

tf.math.maximum_22        (None, None)   0       ['tf.cast_47[0][0]']
(TFOpLambda)

tf.cast_50                (None,)        0       ['tf.math.reduce_sum_36[0][0]']
(TFOpLambda)

tf.cast_48                (None, None)   0       ['tf.math.maximum_22[0][0]']
(TFOpLambda)
```

```
tf.math.reduce_max_39 ()              0        ['tf.cast_50[0][0]']
(TFOpLambda)

tf.math.reduce_sum_35 (None,)         0        ['tf.cast_48[0][0]']
(TFOpLambda)

tf.range_40           (None,)         0        ['tf.math.reduce_max_39[0][0]']
(TFOpLambda)

tf.cast_49            (None,)         0        ['tf.math.reduce_sum_35[0][0]']
(TFOpLambda)

tf.expand_dims_38     (1, None)       0        ['tf.range_40[0][0]']
(TFOpLambda)

tf.expand_dims_39     (None, 1)       0        ['tf.cast_50[0][0]']
(TFOpLambda)

input_23              [(None, None, 0          []
(InputLayer)           256)]

tf.math.reduce_max_38 ()              0        ['tf.cast_49[0][0]']
(TFOpLambda)

tf.math.greater_     ((None, None)    0        ['tf.expand_dims_38[0][0]',
equal_19 TFOpLambda)                            'tf.expand_dims_39[0][0]']

tf.compat.v1.shape_22 (3,)            0        ['input_23[0][0]']
(TFOpLambda)

tf.range_38          (None,)          0        ['tf.math.reduce_max_38[0][0]']
(TFOpLambda)

tf.where_19          (None, None)     0        ['tf.math.greater_equal_19[0]
(TFOpLambda)                                   [0]','tf.expand_dims_38[0][0]']

tf.__operators__.    ()               0        ['tf.compat.v1.shape_22[0][0]']
getitem_38
(SlicingOpLambda)

tf.expand_dims_36    (1, None)        0        ['tf.range_38[0][0]']
(TFOpLambda)

tf.expand_dims_37    (None, 1)        0        ['tf.cast_49[0][0]']
(TFOpLambda)

tf.reshape_74        (None,)          0        ['tf.where_19[0][0]']
(TFOpLambda)

tf.math.             (None, None)     0        ['tf.expand_dims_36[0][0]',
greater_equal_18                                'tf.expand_dims_37[0][0]']
(TFOpLambda)
```

```
tf.range_39            (None,)          0          ['tf.__operators__.getitem_38
(TFOpLambda)                                        [0][0]']

tf.__operators         (None,)          0          ['tf.reshape_74[0][0]']
__.ne_17
(TFOpLambda)

tf.compat.v1           (2,)             0          ['tf.cast_48[0][0]']
.shape_23(TFOpLambda)

tf.where_18            (None, None)     0          ['tf.math.greater_equal_18[0][
(TFOpLambda)                                       0]', 'tf.expand_dims_36[0][0]']

tf.math.reduce         ()               0          ['tf.math.reduce_sum_35[0][0]']
_max_37 (TFOpLambda)

tf.repeat_44           (None,)          0          ['tf.range_39[0][0]',
(TFOpLambda)                                        'tf.math.reduce_sum_36[0][0]']

tf.compat.v1           (None,)          0          ['tf.reshape_74[0][0]','
.boolean_mask_17                                    'tf.__operators__.ne_17[0][0]
(SlicingOpLambda)

tf.math.reduce         ()               0          ['tf.math.reduce_sum_36[0][0]']
_sum_37 (TFOpLambda)

tf.range_37            (None,)          0          ['tf.__operators__.getitem_38
(TFOpLambda)                                        [0][0]']

tf.__operators         ()               0          ['tf.compat.v1.shape_23[0][0]']
__.getitem_39
(SlicingOpLambda)

tf.reshape_72          (None,)          0          ['tf.where_18[0][0]']
(TFOpLambda)

tf.math.maximum_23     ()               0          ['tf.math.reduce_max_37[0][0]']
(TFOpLambda)

tf.stack_17            (None, 2)        0         ['tf.repeat_44[0][0]','tf.compat
(TFOpLambda)                                      .v1.boolean_mask_17 [0][0]']

tf.ones_8              (None, 1)        0          ['tf.math.reduce_sum_37[0][0]']
(TFOpLambda)

tf.repeat_42           (None,)          0          ['tf.range_37[0][0]',
(TFOpLambda)                                        'tf.__operators__.getitem
                                                     _39[0][0]']

tf.reshape_69          (None,)          0          ['tf.cast_48[0][0]']
(TFOpLambda)
```

```
tf.__operators__          (None,)          0       ['tf.reshape_72[0][0]']
.ne_16 (TFOpLambda)

tf.scatter_nd_17          (None, None,     0       ['tf.stack_17[0][0]','tf.__
(TFOpLambda)               1)                       operators__.getitem_38[0][0]',
                                                    'tf.math.maximum_23[0][0]',
                                                    'tf.ones_8[0][0]']

tf.repeat_43              (None,)          0       ['tf.repeat_42[0][0]',
(TFOpLambda)                                        'tf.reshape_69[0][0]']

tf.compat.v1              (None,)          0       ['tf.reshape_72[0][0]',
.boolean_mask_16                                   'tf.__operators__.ne_16[0][0]']
(SlicingOpLambda)

tf.reshape_68             (None, 256)      0       ['input_23[0][0]']
(TFOpLambda)

tf.math.reduce            (None, 1)        0       ['tf.scatter_nd_17[0][0]']
_sum_38 (TFOpLambda)

tf.stack_16               (None, 2)        0      ['tf.repeat_43[0][0]','tf.compat
(TFOpLambda)                                       .v1.boolean_mask_16[0][0]']

tf.repeat_41              (None, 256)      0       ['tf.reshape_68[0][0]',
(TFOpLambda)                                        'tf.reshape_69[0][0]']

tf.math.reduce            ()               0       ['tf.math.reduce_sum_38[0][0]']
_max_40(TFOpLambda)

tf.scatter_nd_16          (None, None,     0       ['tf.stack_16[0][0]','tf.__
(TFOpLambda)               256)                     operators__.getitem_38[0][0]',
                                                    'tf.math.maximum_23[0][0]',
                                                    'tf.repeat_41[0][0]']

tf.cast_51                ()               0       ['tf.math.reduce_max_40[0][0]']
(TFOpLambda)

tf.__operators__          (None, None,     0       ['tf.scatter_nd_16[0][0]',
__.getitem_40              256)                     'tf.cast_51[0][0]']
(SlicingOpLambda)

tf.__operators__          (None, None, 1) 0        ['tf.scatter_nd_17[0][0]',
__.getitem_41                                       'tf.cast_51[0][0]']
(SlicingOpLambda)

==================================================================================
Total params: 0 (0.00 Byte)
Trainable params: 0 (0.00 Byte)
Non-trainable params: 0 (0.00 Byte)
__________________________________________________________________________________
```

We can verify the effect of the Length Regulator module we just implemented:

```
# Initialize the model layer
layer = LengthRegulator(num_channels=256)

# Make up two inputs
# First input has 4 words, and second input has 7 words
spect_durations = tf.constant([
    [2, 4, 5, 6, 0, 0, 0],
    [1, 3, 6, 2, 8, 4, 1],
])
# Faking the token embedding vectors with random values
# Masking out the part that should not correspond to any utterance
input_mask = tf.cast(spect_durations > 0, tf.float32)[:, :, None]
token_embeddings = tf.random.uniform((2, 7, 256)) * input_mask

# Pass the inputs through the layer
output, mask = layer(token_embeddings, spect_durations)

# Check the batch size are kept
assert output.shape[0] == token_embeddings.shape[0]
# Check that the sequence length is the sum of durations from
 ↪  spec_durations
assert output.shape[1] == tf.reduce_max(tf.reduce_sum(spect_durations,1))
# Check number of channels is the same
assert output.shape[2] == token_embeddings.shape[2]

# Checking masking indicates valid indices
assert tf.experimental.numpy.allclose(
    tf.reduce_sum(mask, axis=(1, 2)).numpy(), # float
    tf.reduce_sum(spect_durations, axis=1).numpy() # int
)
```

Bibliography

Martın Abadi, Ashish Agarwal, Paul Barham, Eugene Brevdo, Zhifeng Chen, Craig Citro, Greg S. Corrado, Andy Davis, Jeffrey Dean, Matthieu Devin, Sanjay Ghemawat, Ian Goodfellow, Andrew Harp, Geoffrey Irving, Michael Isard, Yangqing Jia, Rafal Jozefowicz, Lukasz Kaiser, Manjunath Kudlur, Josh Levenberg, Dan Mane, Rajat Monga, Sherry Moore, Derek Murray, Chris Olah, Mike Schuster, Jonathon Shlens, Benoit Steiner, Ilya Sutskever, Kunal Talwar, Paul Tucker, Vincent Vanhoucke, Vijay Vasudevan, Fernanda Viegas, Oriol Vinyals, Pete Warden, Martin Wattenberg, Martin Wicke, Yuan Yu, and Xiaoqiang Zheng. Tf.pad. https://www.tensorflow.org/api_docs/python/tf/pad, 2023.

Jacob Devlin, Ming-Wei Chang, Kenton Lee, and Kristina Toutanova. BERT: Pre-training of deep bidirectional transformers for language understanding, May 2019.

Alexey Dosovitskiy, Lucas Beyer, Alexander Kolesnikov, Dirk Weissenborn, Xiaohua Zhai, Thomas Unterthiner, Mostafa Dehghani, Matthias Minderer, Georg Heigold, Sylvain Gelly, Jakob Uszkoreit, and Neil Houlsby. An image is worth 16x16 words: Transformers for image recognition at scale, June 2021.

Anmol Gulati, James Qin, Chung-Cheng Chiu, Niki Parmar, Yu Zhang, Jiahui Yu, Wei Han, Shibo Wang, Zhengdong Zhang, Yonghui Wu, and Ruoming Pang. Conformer: Convolution-augmented transformer for speech recognition, May 2020.

Charles R. Harris, K. Jarrod Millman, Stéfan J. van der Walt, Ralf Gommers, Pauli Virtanen, David Cournapeau, Eric Wieser, Julian Taylor, Sebastian Berg, Nathaniel J. Smith, Robert Kern, Matti Picus, Stephan Hoyer, Marten H. van Kerkwijk, Matthew Brett, Allan Haldane, Jaime Fernández del Río, Mark Wiebe, Pearu Peterson, Pierre Gérard-Marchant, Kevin Sheppard, Tyler Reddy, Warren Weckesser, Hameer Abbasi, Christoph Gohlke, and Travis E. Oliphant. Np.pad. https://numpy.org/doc/stable/reference/generated/numpy.pad.html, 2023.

Jinglin Liu, Chengxi Li, Yi Ren, Feiyang Chen, and Zhou Zhao. DiffSinger: Singing voice synthesis via shallow diffusion mechanism, 2021.

Aaron van den Oord, Sander Dieleman, Heiga Zen, Karen Simonyan, Oriol Vinyals, Alex Graves, Nal Kalchbrenner, Andrew Senior, and Koray Kavukcuoglu. WaveNet: A generative model for raw audio, September 2016.

Adam Paszke, Sam Gross, Francisco Massa, Adam Lerer, James Bradbury, Gregory Chanan, Trevor Killeen, Zeming Lin, Natalia Gimelshein, Luca Antiga, Alban Desmaison, Andreas Köpf, Edward Yang, Zach DeVito, Martin Raison, Alykhan Tejani, Sasank Chilamkurthy, Benoit Steiner, Lu Fang, Junjie Bai, and Soumith Chintala. Torch.nn.functional.pad. https://pytorch.org/docs/stable/generated/torch.nn.functional.pad.html, 2023.

Yi Ren, Yangjun Ruan, Xu Tan, Tao Qin, Sheng Zhao, Zhou Zhao, and Tie Yan Liu. FastSpeech: Fast, robust and controllable text to speech. *Advances in Neural Information Processing Systems 32(NeurIPS)*, 2019. ISSN 10495258.

Yi Ren, Chenxu Hu, Xu Tan, Tao Qin, Sheng Zhao, Zhou Zhao, and Tie-Yan Liu. FastSpeech 2: Fast and high-quality end-to-end text to speech, pages 1–*15*, 2020.

Ashish Vaswani, Noam Shazeer, Niki Parmar, Jakob Uszkoreit, Llion Jones, Aidan N. Gomez, Łukasz Kaiser, and Illia Polosukhin. Attention is all you need. *Advances in Neural Information Processing Systems*, 2017-Decem (Nips), pages 5999–6009, 2017. ISSN 10495258.

Heiga Zen, Takashi Nose, Junichi Yamagishi, Shinji Sako, Takashi Masuko, Alan W. Black, and Keiichi Tokuda. The HMM-based Speech Synthesis System (HTS) Version 2.0. 2007.

6

String Processing

In our journey on vectorization so far, we have only been dealing with numerical data types (integers, float, complex, etc.). However, string data are also becoming more common especially when dealing with natural language processing (NLP) problems. In other applications of machine learning, we sometimes need to deal with categorical features, which often come as string values in the raw data form. Therefore, in this chapter, we will briefly examine vectorization operations on string values.

6.1 String Data Types

6.1.1 NumPy String, Bytes, and Object

NumPy has three data types that can be used to encode string values in its `np.ndarray`: `'S'` for byte string (corresponding to `np.bytes_` dtype), `'U'` for unicode characters (`np.str_`), and `'O'` or `object` for encoding objects in general (corresponding to `np.object_` dtype), including strings and bytes. We can construct an array of strings in NumPy as the following:

```
import numpy as np
value = ["a", "b", "c"]
A_string = np.array(value, dtype=np.str_)
# array(['a', 'b', 'c'], dtype='<U1')

A_bytes = np.array(value, dtype=np.bytes_)
# array([b'a', b'b', b'c'], dtype='|S1')

A_object = np.array(value, dtype=np.object_)
# array(['a', 'b', 'c'], dtype=object)
```

Vectorization: A Practical Guide to Efficient Implementations of Machine Learning Algorithms,
First Edition. Edward DongBo Cui.

Notice for string and byte type arrays (`A_string` and `A_bytes` respectively), the `dtype` of the displayed array has additional information. Using the string array `A_string` as an example, other than `np.str_` corresponds to the notation of `U` (and a < prefix), the `1` following the dtype notation indicates the maximum length of the strings in the array. For example, let us construct an array with strings of variable lengths:

```
A_string = np.array(["1", "12", "123", "1234", "12345"])
# array(['1', '12', '123', '1234', '12345'], dtype='<U5')
```

Notice now that `dtype` becomes <`U5`, which indicates the maximum length of the strings in the array is 5.

The data types of the array can be inter-converted via `.astype` method from `np.ndarray`. For example, to convert an array of `object` into an array of `bytes`, we can do

```
A_object = np.array(["a", "b", "c", "d"], dtype=np.object_)
# array(['a', 'b', 'c', 'd'], dtype=object)
A_object.astype(np.bytes_)
# array([b'a', b'b', b'c', b'd'], dtype='|S1')
```

6.1.2 Pandas String

Pandas [McKinney, 2010] can handle strings either as an `object` type or as its native implementation of string data type `pd.StringDtype` [McKinney, 2023b]. There are several ways to initialize a Pandas column with string data.

```
import pandas as pd
feature = ["a", "b", "c", "d"]
df = pd.DataFrame({
    "feature1": pd.Series(feature, dtype=pd.StringDtype),
    "feature2": pd.Series(feature, dtype="string"),
    "feature3": pd.Series(feature, dtype=str),
    "feature4": pd.Series(feature, dtype="str"),
    "feature4": pd.Series(feature, dtype=object),
    "feature5": pd.Series(feature, dtype="object")
})
```

Inspecting the DataFrame information with `df.info()`, we have

```
<class 'pandas.core.frame.DataFrame'>
RangeIndex: 4 entries, 0 to 3
```

```
Data columns (total 6 columns):
 #   Column    Non-Null Count  Dtype
---  ------    --------------  -----
 0   feature1  4 non-null      string
 1   feature2  4 non-null      string
 2   feature3  4 non-null      object
 3   feature4  4 non-null      object
 4   feature5  4 non-null      object
 5   feature6  4 non-null      object
dtypes: object(4), string(2)
memory usage: 320.0+ bytes
```

Notice that there are two ways to specify a column of string type, which is to use either `pd.StringDtype` (`feature1`) or `"string"` (`feature2`) as the `dtype` argument. There are several ways to initialize `object` data type columns, using `str` (`feature3`), the string `"str"` (`feature4`), `object` (`feature5`), or the string `"object"` (`feature6`). It is interesting that `pd.Series` interprets the Python class `str` as well as `"str"` as `object`. This may lead to one to believe that any value in the `dtype` argument may be interpreted as an `object`, but try something else, like the name of the author of this book, and you will see this is not the case.

When casting NumPy string arrays to Pandas series, the data type becomes `object` by default, that is

```
import numpy as np
import pandas as pd
A_string = np.array(["a", "b", "c", "d"])
# array(['a', 'b', 'c', 'd'], dtype='<U1')

A_series = pd.Series(A_string)
# 0    a
# 1    b
# 2    c
# 3    d
# dtype: object
```

Pandas handles null values in `string` and `object` columns differently. `string` columns have a special placeholder `<NA>` for missing values, while `object` uses `None`.

```
import pandas as pd
feature = ["a", "b", "c", None, "d"]
df = pd.DataFrame({
    "feature1": pd.Series(feature, dtype="string"),
    "feature2": pd.Series(feature, dtype="object")
```

```
})
# display the table
#    feature1 feature2
# 0        a        a
# 1        b        b
# 2        c        c
# 3     <NA>     None
# 4        d        d

# Check the two missing values
print(df["feature1"][3] is None) # False
print(df["feature2"][3] is None) # True
```

6.1.3 Tensorflow Bytes

Tensorflow has the `tf.string` data type, but all the data stored in `tf.Tensor` becomes bytes. For example,

```
import tensorflow as tf
string_array = ["lorem", "ipsum", "dolar"]
tf_string_array = tf.constant(string_array)
# tf.Tensor([b'lorem' b'ipsum' b'dolar'], shape=(3,), dtype=string)
```

The `tf.Tensor.numpy` method can convert Tensorflow's `tf.Tensor` to a NumPy array. For float and integer types, the conversions typically have no issues retaining the data as float and integer of the corresponding precisions (i.e. `tf.int64 -> np.int64, tf.int32 -> np.int32, tf.float32 -> np.float32, tf.float64 -> np.float64`). But when converting a byte string to a NumPy array, we will get an array of bytes in NumPy instead of typical string representations (`U` or `S`). This can sometimes be inconvenient if we were to implement further string processing on the text, where some methods are specifically implemented against string types instead of byte types. For example, suppose we would like to replace values using a dictionary, where the keys of the dictionary are specified using strings. However, the values we would like to replace are specified in bytes. If caution is not exercised, the program may not be able to achieve the goal of replacing values at all.

```
import tensorflow as tf
mapping = {"A": "1", "B": "2", "C": "5"}
values = tf.constant(["A", "C", "B"]) # should result ["1", "5", "2"]
```

```
# alghouth the following does not throw the error
replaced_values = [mapping.get(v, "0") for v in values.numpy()]
# the replaced results are incorrect
# ["0", "0", "0"]
```

Therefore, additional casting is needed. Typically, we can use the `.astype(str)` method to cast from bytes to string. To illustrate,

```
tf_string_array = tf.constant(["A", "B", "C"], dtype=tf.string)
# <tf.Tensor: shape=(3,), dtype=string, numpy=array([b'A', b'B', b'C'],
↪ dtype=object)>

# Numpy array of bytes (object dtype)
np_string_array_bytes = tf_string_array.numpy()
# array([b'A', b'B', b'C'], dtype=object)

# Casting to string
np_string_array_str = tf_string_array.numpy().astype(str)
# array(['A', 'B', 'C'], dtype='<U1')
```

This may seem inconvenient, but Tensorflow's `bytes` data type can be quite versatile. In the last section of the chapter (Section 6.7), we will examine how Tensorflow can represent structured data using byte strings.

6.1.4 PyTorch

PyTorch's `torch.Tensor` does not handle string or bytes data types, so it relies on external libraries like NumPy and Pandas to prepare and transform the string data into numerical representations first before passing it to the main model. PyTorch has an extension, `torchtext`, which hosts a collection of routines to prepare text for NLP tasks.

6.2 String Operations

Table 6.1 provides a list of string tensor operations that help vectorize string preprocessing during machine learning tasks. It can be seen that typical string processing tasks in native Python have corresponding operations implemented in Pandas and Tensorflow, thus saving the users from writing for loops when processing a batch of string values.

Table 6.1 List of string transformation operations in Python, NumPy, Pandas, and Tensorflow.

Operation	Python	NumPy	Pandas	Tensorflow
Strip space/newline	strip	np.char.strip	pd.Series.str.strip	tf.strings.strip
Substring	slicing	Use np.char.find, then slicing	pd.Series.str.slice	tf.strings.substr
Upper casing	upper	np.char.upper	pd.Series.str.upper	tf.strings.upper
Lower casing	lower	np.char.lower	pd.Series.str.lower	tf.stirngs.lower
Join	join	np.char.join	pd.Series.str.join	tf.strings.join, tf.strings.reduce:join
Split	split	np.char.split	pd.Series.str.split	tf.strings.split
Length	len	np.char.str_len	pd.Series.str.len	tf.strings.length
Formatting	format, f-string	np.char.mod	pd.Series.map, pd.Series.apply	tf.strings.format
Convert to string	str, format, f-string	np.ndarray.astype(str)	pd.Series.astype(str)	tf.strings.as_string
String to number	int, float	np.ndarray.astype(<int, float>)	pd.to_numeric, pd.Series.astype(dtype)	tf.string.to_number
Replace	replace, regex	np.char.replace	pd.Series.str.replace	tf.strings.regex_replace
Contains, is substring	in, regex	np.char.count(...) > 0	pd.Series.str.contains, pd.Series.str.match, pd.Series.str.fullmatch	tf.strings.regex_full_match

6.3 Case Study: Parsing DateTime from String Representations

We may bc familiar with Python's `datetime` library which allows us to handle timestamps easily via intuitive application programming interfaces (APIs). For example, we would like to convert a local timestamp to a Coordinated Universal Time (UTC) timestamp like the following:

```
import datetime
datetime_local = datetime.datetime.strptime(
    "2023-05-07 17:32:25-08:00", '%Y-%m-%d %H:%M:%S%z'
)
datetime_utc = datetime_local.astimezone(datetime.timezone.utc)

# datetime.datetime(2023, 5, 8, 1, 32, 25, tzinfo=datetime.timezone.utc)
```

Pandas also provides vectorized ways to manipulate timestamps easily as well with its `pd.Timestamp` class [McKinney, 2023a]. However, deep-learning libraries like Tensorflow and PyTorch do not have a built-in timestamp class (at the time of writing), so it is more typical to handle timestamp features using strings when attempting to model time series data. The following function implements a way to convert an array of timestamp strings in the format of `year-month-date hour:minute:second-timezone`, or using `datetime.strftime` expression `%Y-%m-%d %H:%M:%S%Z`, into a series of integers representing `year, month, date, hour, minute, seconds, zone`, respectively.

```
def _parse_timestamps(
        timestamps_array: tf.Tensor, utc: bool = False
):
        """
        Parse timestamp strings, assuming %Y-%m-%d %H:%M:%S%Z.
        For example: 2023-05-07 17:32:25-08:00
        """
        # use substring to extract time components
        year = tf.strings.to_number(
            tf.strings.substr(timestamps_array, 0, 4),
            out_type=tf.int64
        )
        month = tf.strings.to_number(
            tf.strings.substr(timestamps_array, 5, 2),
            out_type=tf.int64
```

```
    )
    day = tf.strings.to_number(
        tf.strings.substr(timestamps_array, 8, 2),
        out_type=tf.int64
    )
    hour = tf.strings.to_number(
        tf.strings.substr(timestamps_array, 11, 2),
        out_type=tf.int64
    )
    minute = tf.strings.to_number(
        tf.strings.substr(timestamps_array, 14, 2),
        out_type=tf.int64
    )
    seconds = tf.strings.to_number(
        tf.strings.substr(timestamps_array, 17, 2),
        out_type=tf.int64
    )
    zone = tf.strings.to_number(
        tf.strings.substr(timestamps_array, 19, 3)
    )
    zone = tf.cast(zone, tf.int64)
    # retroactively offset year/ month / day based on zone
    if not utc:
        return year, month, day, hour, minute, seconds, zone

    # offsets [-1, 0, +1] day
    day += (hour - zone) // 24
    num_days = tf.constant(
        [31, 28, 31, 30, 31, 30, 31, 31, 30, 31, 30, 31],
        dtype=day.dtype
    )
    is_leap_year = tf.math.logical_and(
        year % 4 == 0, year % 400 != 0
    )
    # deals with month offsets
    month_days = tf.gather(num_days, month - 1)
    month_days = tf.where(
        tf.math.logical_and(month == 2, is_leap_year),
        month_days + 1,
        month_days
    )
    month -= (month_days - day) // month_days
    # deals with year offset
    year += (month - 1)  // 12

    # Go back to correct Month
    month = (month - 1) % 12 + 1
    # Correct day: day > max number of days for the month
    day = tf.where(
```

```
65              day > month_days, day % month_days, day
66          )
67          # month_days of the correct month
68          month_days = tf.gather(num_days, month - 1)
69          month_days = tf.where(
70              tf.math.logical_and(month == 2, is_leap_year),
71              month_days + 1,
72              month_days
73          )
74
75          # Correct day: case when day = 0
76          day = tf.where(
77              day < 1, month_days, day
78          )
79
80          # Correct hour
81          hour = (hour - zone) % 24
82
83          # make utc zone, which are 0s
84          zone = tf.zeros_like(seconds)
85
86          return year, month, day, hour, minute, seconds, zone
```

- Lines 9–36 extract the values of the year, month, date, hour, minute, seconds, and timezone information from the string using the substring method and convert each into an integer. A comment on `zone`: at the time of writing, `tf.strings.to_number` will throw an error if the string to be converted is prefixed by + and `out_type=tf.int64` (confirmed in Tensorflow version 2.11, 2.13, and 2.14 on my MacBook Pro 2021). However, using the default `out_type=tf.float32` has no problem. To ensure reproducibility, I have used `tf.cast` to convert the parsed number from float to integer.
- Lines 38 and 39 return the parsed time information if we want to retain the timestamp as is without converting back to UTC timestamp, by setting `utc=True`. The rest of the code implements the logic to convert from local timezone to the UTC timestamp.
- Line 42: We first offset the `day` based on the hour and timezone. `(hour - zone) // 24` computes the offset value:
  - If `hour - zone >= 24`, then we need to add a day. In this case, it is possible we are exceeding the maximum number of days of the month, for example `day = 32`. We will handle it after computing all the offsets for month and year and come back to correct this later on.
  - If `hour - zone < 0`, we need to subtract a day. In this case, we may be going to get `day = 0`. Again, we will handle it after computing all the offsets for month and year and come back to correct this later on.
  - Otherwise, there is no need to offset the `day`.

- Lines 43–46 set up the `num_days` vector, which encodes the number of days of each month (for a non-leap year, we have additional logic to handle leap years below).
- Lines 47–49 determine if the timestamp occurs on a leap year.
- Lines 51–56 compute how many days the month is supposed to have, that is `month_days`. We use `tf.gather` to look up the expected number of days for a non-leap year first. We then use `tf.where` to replace any leap month with 29 days instead of 28 days.
- Line 57 offsets the month.
  - If `month_days > day`, then we do not need to offset the `day`. However, if `day = 0`, then we need to offset the month backward by 1. For now, we may encounter cases where the `month = 0` after the offset. This is okay since we will correct it later on after first correcting the `year`.

    If `month_days <= day`, then we need to add a month. For now, we may encounter cases where `month = 13` after the offset. Again, this is okay, since we will correct it later on after first correcting the `year`.
- Line 59 offsets the year.
  - If `month > 12` (i.e. `month = 13`), then we need to add a year.
  - If `month < 1` (i.e. `month = 0`), then we need to subtract a year.
  - Otherwise, the year does not change.
- Line 62 corrects the `month`
  - If `month = 13`, then we need to change it to `month = 1`.
  - If `month = 0`, then we need to change it to `month = 12`.
- Lines 64–78 handle the date correction.
  - Lines 64–66: If `day > month_days` exceeds the maximum number of days for the `month`, then we need to reset starting from the previous month and count the number of days exceeded.
  - Lines 68–73 compute `month_days` (which is the last day of the month) before handling the case where `day = 0`. Because we offset the `month` backward, sometimes to February which can potentially be a leap month, we need to compute which day we need to set back to (as the last day of the previous month) corresponding to the corrected `month` (in this case, a month is subtracted).
  - Lines 76–78 correct the `day` when `day = 0`, by setting the date to the previous month's last day.
- Line 81 corrects the hour based on zone.
  - If `hour - zone > 24`, e.g. `hour - zone = 25`, then we need to reset the value to `(hour - zone) % 24 = 1`.
  - If `hour - zone < 0`, e.g. `hour - zone = -2`, then we need to reset the value to `(hour - zone) % 24 = 22`.

- Line 84 makes the UTC zone which is filled with zeros. This is just trying to be consistent with our return signature `year, month, day, hour, minute, seconds, zone`.
- Line 86 returns the outputs.

Let's write several test cases to ensure that the implementation above is indeed working as expected

```
import datetime
import numpy as np
import tensorflow as tf
timestamps_array = tf.constant(
    [
        "2023-05-07 05:32:25-08:00", # utc: same date
        "2023-05-07 17:32:25-08:00", # utc: next day
        "2023-05-07 05:32:25+11:00", # utc: previous day
        "2023-08-31 22:15:37-08:00", # utc: next month
        "2023-06-01 02:23:15+08:00", # utc: previous month
        "2023-12-31 23:32:15-04:00", # utc: next year
        "2023-01-01 02:25:10+04:00", # utc: last year
        "2024-02-28 23:59:11-02:00", # utc: leap year next day (29th)
        "2024-03-01 01:38:17+04:00", # utc: leap year last day (29th)
    ]
)
# parse with our function
year, month, day, hour, minute, seconds, zone = _parse_timestamps(
    timestamps_array, utc=True
)
output_tf = tf.stack(
    [year, month, day, hour, minute, seconds], axis=1
)

# parse each with datetime
output_dt = np.zeros((timestamps_array.shape[0], 6), dtype=int)
for ii, tstamp in enumerate(timestamps_array.numpy().astype(str)):
    datetime_local = datetime.datetime.strptime(
        tstamp, '%Y-%m-%d %H:%M:%S%z'
   )
    datetime_utc = datetime_local.astimezone(datetime.timezone.utc)
    output_dt[ii][0] = datetime_utc.year
    output_dt[ii][1] = datetime_utc.month
    output_dt[ii][2] = datetime_utc.day
    output_dt[ii][3] = datetime_utc.hour
    output_dt[ii][4] = datetime_utc.minute
    output_dt[ii][5] = datetime_utc.second

assert np.allclose(output_dt, output_tf)
```

Note

Fingers crossed that this would no longer be needed in the future.

6.4 Mapping Strings to Indices

When handling categorical features, there is a need to translate the string representations of the features into numerical representations. One-hot encoding and embedding techniques are commonly used to represent string features. To create these types of numerical representations, we first need to determine a list of ***vocabulary*** of the categorical feature (i.e. all valid values encountered during training time) and assign an ordinal number to each value of the feature. We call this the ***index*** of the categorical feature value. Much like the concept of indexing that we have seen in Chapter 3, we can use the vocabulary and the index to look up the corresponding category. From the index, we can then further create the one-hot encoding or look up a corresponding learned vector representation from an embedding table. Suppose we would like to look up integer values to some letters of the alphabet, we can do the following in Python:

```
# create a list to store all the alphabet
alphabet_lookup = [chr(ii+97) for ii in range(26)]
values = ["b", "a", "c", "d", "z"]
# using list.index to look
index = [alphabet_lookup.index(v) for v in values]
# [1, 0, 2, 3, 25]

# Look up letter based on index
retrieved_values = [alphabet_lookup[ii] for ii in index]
# the two values list should be the same
assert values == retrieved_values
```

Various libraries for machine learning implement a so-called "lookup" method (e.g. `tensorflow.lookup`, `pandas.map`), which allows the users to map from string values to integer indices via vectorization when processing a batch of categorical features.

6.4.1 NumPy `np.unique`

NumPy's `np.unique` can determine the unique set of values from a given array of strings (as well as other dtypes). When specifying the argument `return_inverse=True`, the method can also return an integer value corresponding to the lookup index of the unique set. For instance,

```
import numpy as np
feature = ["a", "b", "c", "b", "a"]
```

```
unique, index = np.unique(feature, return_inverse=True)
# unique = array(['a', 'b', 'c'], dtype='<U1')
# index = array([0, 1, 2, 1, 0])
```

Here, the `unique` array would be the vocabulary of the categorical feature, while the `index` would be the corresponding index. It can be easy to verify that using the `unique` vocabulary and `index` of the category, we can reconstruct the feature

```
# code continues from above
recovered_feature = np.take(unique, index)
assert np.all(feature == recovered_feature)
```

The disadvantage of using `np.unique`, however, is that the indices determined may not be reproducible: if the order of the original list is changed, then it is no longer guaranteed the same string value will be mapped to the same integer index value. For instance,

```
import numpy as np
feature_a = ["a", "b", "c", "b", "a"]
feature_b = ["c", "b", "a", "b", "b"]

_, index_a = np.unique(feature_a, return_inverse=True)
# array([0, 1, 2, 1, 0])
_, index_b = np.unique(feature_b, return_inverse=True)
# array([2, 1, 0, 1, 1])
```

Pandas also has the method `pd.factorize` which can achieve similar effects.

6.4.2 Pandas `pd.Categorical`

Instead of using `pd.factorize`, an alternative way to encode string values in Pandas is to use `pd.Categorical` dtype to represent the feature column. We can create a Pandas DataFrame with a categorical column as the following:

```
import pandas as pd
df = pd.DataFrame({
    "user_id": pd.Series([1, 2, 3, 4, 5, 6], dtype=int),
    "age": pd.Series([
        "18-25", "65+", "26-30", "30-40", "40-50", "50-65"
```

```
    ], dtype="category")
})
```

Display the table,

```
   user_id    age
0        1  18-25
1        2    65+
2        3  26-30
3        4  30-40
4        5  40-50
5        6  50-65
```

We can use `df.info(memory_usage="deep")` to examine the data types as well as the size of the table in the memory:

```
# df.info(memory_usage="deep")
<class 'pandas.core.frame.DataFrame'>
RangeIndex: 6 entries, 0 to 5
Data columns (total 2 columns):
 #   Column    Non-Null Count  Dtype
---  ------    --------------  -----
 0   user_id   6 non-null      int64
 1   age       6 non-null      category
dtypes: category(1), int64(1)
memory usage: 724.0 bytes
```

One advantage of using `pd.Categorical` is to save the memory usage of the DataFrame. Suppose we have a much larger table with two categorical features:

```
import pandas as pd
df = pd.DataFrame({
"user_id": pd.Series(np.arange(6000), dtype=int),
"age": pd.Series([
    "18-25", "65+", "26-30", "30-40", "40-50", "50-65"
] * 1000, dtype="category"),
"item_id": pd.Series([
    "qbirvk9xs6lhtsgbtoc782jda9sopijt9m4132s5xftromiq8cuabc9z2y",
    "cu0lw3aotlipgelkuu2q25sbxkqlaiivskkp631iqbtsqn9bmbkszynt5mg8",
    "glg7hb0a0tii3huq7ixx3rdmt36sz0869hw439o7iwdtxslcglx6whqtjl6",
```

```
        "4w58m7a91s4q3968oc8xn1om4p89zh14vt014atlif86bclh4hf7xnaugu3",
        "cbjqeyvrvaomkpkrebftrffnpivbnzpiikyjmgsoelsbulobgfbmdewsozrv",
        "qbirvk9xs61htsgbtoc782jda9sopijt9m4132s5xftromiq8cuabc9z2y0"
    ] * 1000, dtype="category")
})

# make a copy of the table, and set the "age" and "item_id" columns to string
df_non_category = df.copy()
df_non_category["age"] = df_non_category["age"].astype(str)
df_non_category["item_id"] = df_non_category["item_id"].astype(str)
```

We can then compare the memory usage of the two tables created above, one using `pd.Categorical` and one not using it to store the two string features:

```
# df.info(memory_usage="deep")
<class 'pandas.core.frame.DataFrame'>
RangeIndex: 6000 entries, 0 to 5999
Data columns (total 3 columns):
 #   Column   Non-Null Count  Dtype
---  ------   --------------  -----
 0   user_id  6000 non-null   int64
 1   age      6000 non-null   category
 2   item_id  6000 non-null   category
dtypes: category(2), int64(1)
memory usage: 60.0 KB

# df_non_category.info(memory_usage="deep")
<class 'pandas.core.frame.DataFrame'>
RangeIndex: 6000 entries, 0 to 5999
Data columns (total 3 columns):
 #   Column   Non-Null Count  Dtype
---  ------   --------------  -----
 0   user_id  6000 non-null   int64
 1   age      6000 non-null   object
 2   item_id  6000 non-null   object
dtypes: int64(1), object(2)
memory usage: 1.1 MB
```

The table with non-categorical columns occupies 1.1 MB of memory. In contrast, only 60.0 KB of memory was used when storing the two columns as categorical variables, saving memory usage by almost 20 times!

More recent versions of tree-based model libraries like LightGBM and XGBoost can handle `pd.Categorical` features directly without the need to create further index lookups. This is another advantage of using `pd.Categorical` when working with machine learning models.

6.4.3 Scikit-learn `sklearn.preprocessing.LabelEncoder`

None of the encoding methods we have discussed so far allow us to easily reuse the mapping between the string value and the indices, a requirement when making inferences on new data during production. The `sklearn.preprocessing .LabelEncoder` class allows us to create and save the mapping for later use.

```
from sklearn.preprocessing import LabelEncoder
vocabulary = [chr(ii+97) for ii in range(26)]
feature_train = ["b", "a", "c", "d", "z"]
# Create the encoder
feature_lookup = LabelEncoder()
feature_lookup.fit(vocabulary)

# Transform during training
index_train = feature_lookup.transform(feature_train)
# array([ 1,  0,  2,  3, 25])

# New data during inference
feature_test = ["f", "g", "a", "b", "h"]
index_test = feature_lookup.transform(feature_test)
# array([5, 6, 0, 1, 7])
```

Notice that both `feature_train` and `feature_test` features have the string value `a` and `b` in them. The corresponding `index_train` and `index_test` indices are both `0` and `1`, respectively. This demonstrates that `sklearn.preprocessing.LabelEncoder` can (re)produce consistent string value to index mappings.

6.4.4 Tensorflow `tf.lookup`

Tensorflow's `tf.lookup` offers yet another suite of tools to convert string values of features into indices. `tf.lookup` has methods that build hash tables, so it can be used for a more general purpose not limited to string to integer mapping. One key feature of `tf.lookup`, in my opinion, is the ability to handle new feature values that are out of vocabulary (OOV). This is something missing from the three libraries we have discussed so far. In my everyday practice, I often reserve index `0` for OOV vocabularies and start the index from `1`. Another option is to still use `0` as the first index and use the index `len(vocabulary)` as the OOV indices (which we can also use `-1` as an equivalent alternative). To illustrate the usage, suppose we have a list of the alphabet that we would like to map to integer indices:

```
import tensorflow as tf
# Create a list to store all the alphabet
alphabet_lookup = tf.constant([chr(ii+97) for ii in range(26)])
# Last value in the list is oov
values = tf.constant(["b", "a", "c", "d", "z", "1"])

# Create the lookup hash table
table = tf.lookup.StaticHashTable(
    tf.lookup.KeyValueTensorInitializer(
        keys=alphabet_lookup,
        # index starts from 1
        values=list(range(1, len(alphabet_lookup)+1)),
        key_dtype=tf.string,
        value_dtype=tf.int64,
    ),
    default_value=0, # oov
)

# Map the values
index = table.lookup(values)
# <tf.Tensor: shape=(6,), dtype=int64, numpy=array([ 2,  1,  3,  4, 26,
↪  0])>
```

`tf.lookup` can also create lookup tables by loading from a text file.

```
import os
import tempfile
import tensorflow as tf

# Create a list to store all the alphabet and save to a file
alphabet_lookup = [chr(ii+97) for ii in range(26)]
_, file_path = tempfile.mkstemp()
with open(file_path, "w") as fid:
    fid.write("\n".join(alphabet_lookup))

# Last value in the list is oov
values = tf.constant(["z", "a", "c", "d", "h", "3"])

# Create the vocabulary lookup
table = tf.lookup.StaticHashTable(
    tf.lookup.TextFileInitializer(
        filename=file_path,
        # each line of the text file is a feature value
        key_index=tf.lookup.TextFileIndex.WHOLE_LINE,
        # use the line number as the index of the value
```

```
        value_index=tf.lookup.TextFileIndex.LINE_NUMBER,
        key_dtype=tf.string,
        value_dtype=tf.int64,
        value_index_offset=1, # index starts at 1
    ),
    default_value=0, # oov
)

# Apply the lookup
index = table.lookup(values)
# <tf.Tensor: shape=(6,), dtype=int64, numpy=array([26,  1,
    3,  4,  8,  0])>

# Clean up the temp file
os.remove(file_path)
```

When using Keras with Tensorflow, `keras.layers.StringLookup`, building on top of `tf.lookup`, offers additional ways to incorporate string to integer indices lookup tables inside a Keras model. It also provides further post-processing to transform the integer indices into common labeling schemes such as one-hot encoding or Term Frequency-Inverse Document Frequency (TF-IDF).

6.4.5 TorchText `torchtext.vocab`

Since PyTorch `torch.Tensor` does not handle strings and bytes, the library itself does not provide a way to handle the transformation from string values to integer indices and relies on custom preprocessing. `TorchText` is a library currently in pre-release (at the time of writing, the most recent version is 0.16) that implements text pre-processing utilities for NLP tasks. `torchtext.vocab.vocab` is the method to convert string values to integer indices based on a lookup table from a predefined dictionary which maps between the string value and the frequency of occurrence of the value. It sorts the order of the index based on the frequency, with the lowest frequency as the starting index.

```
# !pip install torchtext==0.16.0
import torchtext
# A dictionary that maps between vocabulary and frequency of
    occurrence
# This creates a->0, b->1, c->2
vocab = {"a":1, "b": 2, "c":10}
# Special tokens to insert at the beginning of the
    vocabulary
```

```
# The mapping now becomes <pad>->0, a->1, b->2, c->3
vocab_lookup = torchtext.vocab.vocab(
    vocab, specials=["<pad>"],  special_first=True
)
# handle oov, anything else, use -1
vocab_lookup.set_default_index(-1)
# look up values: "d" is oov, "<pad>" is special which maps
↪ to index 0
values = ["a", "c", "b", "d", "<pad>"]
# Look up a single value
index = vocab_lookup["a"] # 1
# Loook up an array of values
indices = vocab_lookup.lookup_indices(values)
# [1, 3, 2, -1, 0]
```

A nice feature that `torch.vocab` has is the ability to build the vocabulary iteratively, instead of predefining the vocabulary by summarizing the data (cf. `tensorflow_transform.compute_and_apply_vocabulary`, a Tensorflow extension library that allows iterative vocabulary building but requires the mastery of Tensorflow Extended [TFX]). For instance,

```
from torchtext.vocab import build_vocab_from_iterator

def get_feature(feature_batch):
    for feature in feature_batch:
        yield feature

feature_batch = [["a", "b", "c"], ["a", "d", "c"], ["f", "g"]]
vocab_lookup = build_vocab_from_iterator(
    get_feature(feature_batch), specials=["<pad>"]
)
vocab_lookup.set_default_index(-1) # oov

# look up values
values = ["a", "g", "2", "<pad>"]
index = vocab_lookup.lookup_indices(values)
# [1, 6, -1, 0]
```

6.5 Case Study: Factorization Machine

Factorization Machine (FM), proposed by Steffen Rendle in 2010 [Rendle, 2010], is a supervised machine learning algorithm that can be used for classification, regression, and ranking tasks. It is a combination of linear regression and matrix factorization. It also uses a polynomial kernel. Thus, FM is an extension of basic matrix factorization, since it can model not only user and item ids as inputs but also contextual features such as item attributes, user characteristics, and user behavior contexts. FM is used as a strong baseline model in building recommendation engines in e-commerce, media streaming, and advertising industries. The strengths of FM are:

1. It can model d-way feature interaction, where d is the order of polynomial, i.e. typically set to $d = 2$.
2. A fast optimization algorithm for FM can reduce polynomial computation time to linear complexity.

Let's first examine the mathematical formulation of FM.

6.5.1 Factorization Machine Model

Let $x \in \mathbb{R}^n$ denote a d-dimensional feature vector of a sample, and y denote the label data, which can be real values (in regression models) or class labels. The model for a Factorization Machine of degree 2 is

$$\hat{y}(x) = w_0 + \sum_{i=1}^{n} w_i x_i + \sum_{i=1}^{n} \sum_{j=i+1}^{n} \langle \mathbf{v}_i, \mathbf{v}_j \rangle x_i x_j$$

where

- $w_0 \in \mathbb{R}$ is the global mean
- $w_i \in \mathbb{R}^n$ is the weights of the ith feature
- $\mathbf{V} \in \mathbb{R}^{nxk}$ represents the feature embeddings
- $\mathbf{v}_i$ represents the ith row of $\mathbf{V}$
- k is the number of latent factors
- $\langle \cdot, \cdot \rangle$ is the dot product between two vectors. Thus, $\langle \mathbf{v}_i, \mathbf{v}_j \rangle$ models feature interactions

The third term is the feature interaction term. Notice that we are modeling interactions for $j > i$. Although FM can be generalized to higher orders than order 2, the numerical stability is not great. The generalization to d-way FM is

$$\hat{y}(x) = w_0 + \sum_{i=1}^{n} w_i x_i + \sum_{l=2}^{d} \sum_{i_1=1}^{n} \cdots \sum_{i_l=i_{l-1}+1}^{n} \left(\prod_{j=1}^{l} x_{i_j} \right) \left(\sum_{f=1}^{k_l} \prod_{j=1}^{l} v_{i_j,f}^{(l)} \right)$$

with time complexity $O(kn^d)$, where

- l is the index of the interactions, from 2 to d
- f is the index of latent factors, from 1 to k_l

6.5.2 More Efficient Optimization Criterion

The third term adds a nonlinearity to the model, but it is the difficult part for optimization ($O(kn^2)$, n^2 since all pair-wise interactions of features have to be computed, and the length of the latent factors is k). Rearrangement of the third term can reduce the time complexity to $O(kn)$, which is linear.

$$
\begin{aligned}
&\sum_{i=1}^{n}\sum_{j=i+1}^{n}\langle \mathbf{v}_i, \mathbf{v}_j\rangle x_i x_j \\
&= \frac{1}{2}\sum_{i=1}^{n}\sum_{j=i+1}^{n}\langle \mathbf{v}_i, \mathbf{v}_j\rangle x_i x_j - \frac{1}{2}\sum_{i=1}^{n}\langle \mathbf{v}_i, \mathbf{v}_i\rangle x_i x_i \quad \text{(upper triangle – diagonal entry)} \\
&= \frac{1}{2}\left(\sum_{i=1}^{n}\sum_{j=1}^{n}\sum_{f=1}^{k} v_{i,f}v_{j,f}x_i x_j - \sum_{i=1}^{n}\sum_{f=1}^{k} v_{i,f}v_{i,f}x_i x_i\right) \quad \text{(expanding the dot product)} \\
&= \frac{1}{2}\sum_{f=1}^{k}\left(\left(\sum_{i=1}^{n} v_{i,f}x_i\right)\left(\sum_{j=1}^{n} v_{j,f}x_j\right) - \sum_{i=1}^{n} v_{i,f}^2 x_i^2\right) \quad \text{(factor out the feature summation)} \\
&= \frac{1}{2}\sum_{f=1}^{k}\left(\left(\sum_{i=1}^{n} v_{i,f}x_i\right)^2 - \sum_{i=1}^{n} v_{i,f}^2 x_i^2\right)
\end{aligned}
$$

For sparse features, only nonzero elements need to be computed so that the overall complexity is linear to the number of nonzero features.

The final model can be written as

$$
\hat{y}(x) = w_0 + \sum_{i=1}^{n} w_i x_i + \frac{1}{2}\sum_{f=1}^{k}\left(\left(\sum_{i=1}^{n} v_{i,f}x_i\right)^2 - \sum_{i=1}^{n} v_{i,f}^2 x_i^2\right) \tag{6.1}
$$

6.5.3 Implementation of Deep Factorization Machine in Tensorflow

Warning

I used the following versions of the Python packages on M1 MacBookPro 2021 to build the exercise on DeepFM. Specifically, older versions of `pandas` may throw errors when loading the data or may handle the unicode characters incorrectly.

```
python==3.10.6
numpy==1.24.3
pandas==2.1.1
tensorflow==2.13.0
keras==2.13.1
```

We use Keras API in Tensorflow to implement the Deep Factorization Machine (DeepFM) [Guo et al., 2017] algorithm. This is a simple modification of the FM algorithm by adding a deep neural network component to help predict the final output. Recall that a dense feedforward neural network is implementing the following equation:

$$z = \sigma(xw + b)$$

where σ is some activation function such as the sigmoid function. x typically represents a vector input in batches, while w and b are weight matrices and bias vectors, respectively.

The class `KerasDeepFactorizationMachine` contains a set of methods to implement the mathematical formulations of DeepFM and returns a Keras Model object in the `build_model` method via Keras' functional API. Of course, one can also choose to subclass the `keras.Model` class for its implementation and create another build function separately in order to take advantage of the `model.fit` and `model.evaluate` APIs later on.

```
from typing import List

from keras import Model, Input
import keras.backend as K
from keras.layers import (
    Concatenate,
    Flatten,
    Lambda,
    Embedding,
    Reshape,
    Dense,
    Dropout,
    Add,
    Activation,
)
from keras.regularizers import l2

class KerasDeepFactorizationMachine:
    """
    Parameters:
    ----------
    * feature_name: name of the features. For the purpose of
        distinguishing between feature names in each keras steps
    * feature_size: a list of size of each features, in the correct
        order. This is the number of unique categories of the feature.
        If the feature is continuous, use 0.
    * num_label: a list of integers, in the correct order of the
        features, that indicates the number of labels applied for a
        multilabel categorical feature.
        - For non-multilabel categorical features, use 1;
        - For continuous features, use 1;
        - Otherwise, specify the maximum number of multi-labels
    * emb_size: embedding size
    * hidden_layer_dim: list of hidden layer dimensions for
        Multiplayer perceptron part of the Deep Factorization Machine.
        If None or empty [], DFM becomes FM.
```

```
    * drop_rate: dropout rate of MLP hidden layers
    * output_activation: activation function of the output layer.
        Default to 'linear' (no activation) for regression task.
    For binary classification task, use "sigmoid".
    * loss: type of loss to use. Default to 'mse'
    * metrics: list of validation metrics. Default to ['mse']
    * l2_fm: embedding regularization
    * l2_dnn: regularization of the MLP layer
    * l2_out: l2 regularization of the output layer
    """

    def __init__(
        self,
        feature_name: List[str],
        feature_size: List[int],
        num_label: int = None,
        emb_size: int = 64,
        hidden_layer_dim: List[int] = [16, 16],
        drop_rate: float = 0.1,
        output_activation: str = "linear",
        loss: str = "mse",
        metrics: List[str] = ["mse"],
        l2_fm: float = 0.0,
        l2_dnn: float = 0.0,
        l2_out: float = 0.0,
    ):
        self.feature_name = feature_name
        self.feature_size = feature_size
        self.num_label = num_label or ([1] * len(feature_size))
        self.emb_size = emb_size
        self.hidden_layer_dim = hidden_layer_dim
        self.drop_rate = drop_rate
        self.output_activation = output_activation
        self.loss = loss
        self.metrics = metrics
        self.l2_fm = l2_fm
        self.l2_dnn = l2_dnn
        self.l2_out = l2_out
        # count number of inputs
        self.dim_input = len(self.feature_size)

        # Build the model
        self.model = self.build_model()

    def get_embed(
        self,
        x_input,
        x_size,
        k_latent,
        n_label,
        name,
        keepdims=False,
    ):
        """Get embedding layer depending on type of variable"""
```

```
        if x_size > 0:  # cateogry
            embed = Embedding(
                x_size,
                k_latent,
                input_length=1,
                mask_zero=n_label > 1,
                name=f"embedding_{name}",
                embeddings_regularizer=l2(self.l2_fm),
            )(x_input)
            if n_label > 1:
                # Average the multilabel embedding
                embed = Lambda(
                    lambda x: K.mean(x, axis=1, keepdims=keepdims),
                    name=f"multilabel_pool_{name}",
                )(embed)
            else:  # single label
                if keepdims:
                    embed = Reshape(
                        (1, k_latent), name=f"reshape_{name}"
                    )(embed)
                else:
                    embed = Flatten(name=f"flat_{name}")(embed)
        else:  # continuous variable, x_size=0
            embed = Dense(
                k_latent,
                kernel_regularizer=l2(self.l2_fm),
                name=f"continuous_{name}",
            )(x_input)
            if keepdims:
                embed = Reshape((1, k_latent), name=f"rehape_{name}")(
                    embed
                )
        return embed

    def fm_1d(self):
        """Linear (1st order) term of Factorization Machine"""
        # 1D terms
        biases = [
            self.get_embed(x, size, 1, n, name + "_1d")
            for (x, size, n, name) in zip(
                self.input_x,
                self.feature_size,
                self.num_label,
                self.feature_name,
            )
        ]
        # Add all terms
        y_fm_1d = Add(name="fm_1d_output")(biases)

        return y_fm_1d

    def fm_2d(self):
        """Interaction (2nd order) term of Factorization Machine"""
        # (None, 1, k) embeddings for each feature
```

```
        interactions = [
            self.get_embed(
                x, size, self.emb_size, n, name + "_2d", keepdims=True
            )
            for (x, size, n, name) in zip(
                self.input_x,
                self.feature_size,
                self.num_label,
                self.feature_name,
            )
        ]
        # Concatenate all 2D embed layers => (None, n_features, k)
        embed_2d = Concatenate(axis=1, name="concat_embed_2d")(
            interactions
        )

        # Calculate the interaction term sum(x)^2 - sum(x^2)
        square_of_sum = Lambda(
            lambda x: K.square(K.sum(x, axis=1, keepdims=False)),
            name="square_of_sum",
        )(embed_2d)
        sum_of_square = Lambda(
            lambda x: K.sum(K.square(x), axis=1, keepdims=False),
            name="sum_of_square",
        )(embed_2d)
        second_order_term = Lambda(
            lambda s: 0.5 * K.sum(s[0] - s[1], axis=1),
            name="sq_sum-sum_sq",
        )((square_of_sum, sum_of_square))
        y_fm_2d = Reshape((1,), name="fm_2d_output")(second_order_term)

        return y_fm_2d, embed_2d

    def deep_mlp(self):
        """
        Deep Multilayer Perceptron part of the
        Deep Factorization Machine
        """
        # flat embed layers from 3D to 2D tensors
        y_mlp = Flatten(name="flat_embed_2d")(self.embed_2d)
        for i, dim in enumerate(self.hidden_layer_dim):
            if self.drop_rate > 0:
                y_mlp = Dropout(self.drop_rate)(y_mlp)
            y_mlp = Dense(
                dim,
                activation="relu",
                name=f"mlp_hidden_{i+1}",
                kernel_regularizer=l2(self.l2_dnn),
            )(y_mlp)
        # Output of the MLP layer
        y_mlp = Dense(1, activation="relu", name="mlp_output")(y_mlp)

        return y_mlp
```

```
199         def build_model(self):
200             # Make 1 input for each feature
201             self.input_x = [
202                 Input(
203                     shape=(self.num_label[i],),
204                     name=f"input_{self.feature_name[i]}",
205                 )
206                 for i in range(self.dim_input)
207             ]
208             # Build each part of the model
209             self.y_fm_1d = self.fm_1d()
210             self.y_fm_2d, self.embed_2d = self.fm_2d()
211             # Decide whether or not add the deep MLP part
212             if self.hidden_layer_dim:
213                 self.y_mlp = self.deep_mlp()
214                 y_hat = Concatenate()(
215                     [self.y_fm_1d, self.y_fm_2d, self.y_mlp]
216                 )
217             else:
218                 y_hat = Concatenate()([self.y_fm_1d, self.y_fm_2d])
219 
220             y_hat = Dense(
221                 1, name="deepfm_output", kernel_regularizer=l2(self.l2_out)
222             )(y_hat)
223             y_hat = Activation(self.output_activation)(y_hat)
224 
225             # Make the model
226             model = Model(inputs=self.input_x, outputs=[y_hat])
227             model.compile(
228                 optimizer="adam", loss=self.loss, metrics=self.metrics
229             )
230 
231             return model
```

- `__init__` **method (Lines 48–62)**:
  - Lines 63–76 keep a record of the model configurations as class attributes
  - Line 79 calls the model building routine `self.build_model`, which leverages the methods to build FM and deep feedforward layers.
- `get_embed` **method (Lines 81–123)**: This is a helper method to create embedding layers for categorical features or dense feedforward layers for numerical value features. It can handle categorical features that are represented as a list (multivalue attributes, such as genres of a movie) by taking the average of the embeddings of each individual element. The arguments of this method will help the method decide which type of feature the embedding needs to be created for:
  - `x_input`: This is the feature itself to be used for embedding lookup
  - `x_size`: This tells the method how many unique values the feature has. Typically for a categorical feature, there is at least 1 unique value the

feature can take. We use `x_size=0` to indicate the feature is a continuous feature.
- `k_latent`: This is the latent dimension size of the embeddings. For the FM model, the latent dimensions should be always identical among the features.
- `n_label`: This tells the method if the feature is a list or a single-value feature. We assume that a categorical feature can be either a single-value feature or a list feature, but all continuous features are single-value features.
- `name`: This provides a name for the embedding. Used for Keras model summary to keep track of the flow of the data.
- `keepdims`: If `True`, then the method returns `(batch_size, 1, emb_size)` embeddings. This can be useful for later concatenation operations when building the model.

- `fm_1d` **(Lines 125–140)**: This is the method that implements the linear part of Equation 6.1. It iterates through the features and creates the linear or the bias terms of the features, using the embedding of size 1. Then, the bias terms are all summed together.
- `fm_2d` **(Lines 142–176)**: This is the method that implements the second-order part of Equation 6.1. It first iterates through the features to create embeddings with the size of K and uses the reformulated optimization criterion to compute the sum of pair-wise second-order interactions between the features.
- `deep_mlp` **(Lines 178–197)**: This method implements the dense feedforward layer given the feature embeddings obtained from `fm_2d` as inputs. The model first concatenates the embeddings of the features and uses it to predict a single score that will later be combined together with the scores from `fm_1d` and `fm_2d`.
- `build_model` **(Lines 199–231)**: This method creates the model. It first creates the `Input` layers for each of the features (Lines 201–207). It then proceeds to call the three terms (`fm_1d`, `fm_2d`, `deep_mlp`) to compute predicted scores for the current batch of examples (Lines 209–210). The scores are combined into an array via `Concatenate`, and the array is sent to another feedforward layer to obtain the final score (Lines 212–223). Finally, the method builds the `Keras.Model` object and returns it (Lines 226–231).

6.5.4 Training DeepFM on MovieLens 1M Dataset

In this section, we train the DeepFM model using the MovieLens [Harper and Konstan, 2016] 1M dataset, which can be downloaded from https://grouplens.org/datasets/movielens/1m.

Let's first prepare the data for our model

```python
import numpy as np
import pandas as pd
from IPython.display import display
from keras.preprocessing.text import Tokenizer
from keras.preprocessing.sequence import pad_sequences


def load_ratings():
    COL_NAME = ["uid", "mid", "rating", "timestamp"]
    df = pd.read_csv(
        "./ml-1m/ratings.dat",
        sep="::",
        header=None,
        engine="python",
        names=COL_NAME,
    )
    return df


def load_movies():
    COL_NAME = ["mid", "movie_name", "movie_genre"]
    df = pd.read_csv(
        "./ml-1m/movies.dat",
        sep="::",
        header=None,
        engine="python",
        names=COL_NAME,
        encoding="iso-8859-1",  # deal with unicode characters
    )
    return df


def load_users():
    COL_NAME = ["uid", "gender", "age", "occupation", "zipcode"]
    df = pd.read_csv(
        "./ml-1m/users.dat",
        sep="::",
        header=None,
        engine="python",
        names=COL_NAME,
    )
    return df


def text2seq(text, n_genre):
    """using tokenizer to encoded the multi-level categorical feature"""
    tokenizer = Tokenizer(
        lower=True, split="|", filters="", num_words=n_genre
    )
    tokenizer.fit_on_texts(text)
    seq = tokenizer.texts_to_sequences(text)
    seq = pad_sequences(seq, maxlen=3, padding="post")
    return seq
```

```

n_genre = 15
ratings = load_ratings()
movies = load_movies()
users = load_users()
# Preview the data
print("====== rating.dat ======")
display(ratings.head())
print("===== movies.dat ======")
display(movies.head())
print("====== users.dat ======")
display(users.head())
movies["movie_genre"] = text2seq(
    movies.movie_genre.values, n_genre=n_genre
).tolist()

ratings = ratings.join(movies.set_index("mid"), on="mid", how="left")

ratings = ratings.join(users.set_index("uid"), on="uid", how="left")

ratings.to_csv("./ml-1m/combined_data.csv", index=False)

print("====== preprocessed data =======")
```

We then need to transform the string categorical variables into indices so that they can be used for embedding lookups in the model.

```
# Transformation and feature engineering
from sklearn.model_selection import train_test_split
from ast import literal_eval

df = pd.read_csv("./ml-1m/combined_data.csv")
df["gender"] = df["gender"].replace({"F": 0, "M": 1})
df["movie_genre"] = df["movie_genre"].apply(literal_eval)

# Reindex the categorical variables
# Recode uid, mid, age, such that they are starting from 0, and
# the max(category) = len(category)-1
df = df.sort_values("uid")
df["uid"] = pd.factorize(df["uid"])[0]
df = df.sort_values("mid")
df["mid"] = pd.factorize(df["mid"])[0]
df = df.sort_values("age")
df["age"] = pd.factorize(df["age"])[0]
# occupation, movie_genre are fine

# Train test split
df_train, df_test = train_test_split(
    df, test_size=0.2, random_state=42, shuffle=True
)

```

```
for field in ["uid", "mid", "gender", "age", "occupation"]:
    # For each categorical feature, if the value in df_test is not
    # in df_train, then we move the entries contain that new value
    # to trainig.
    # This will ensure that we won't have out of vocabulary cases
    # to be handled during evaluation. However, in real production,
    # we need to address the so-called "out-of-vocabulary" issue.
    test_set = set(df_test[field])
    train_set = set(df_train[field])
    diff_id = list(test_set.difference(train_set))
    index = df_test[field].isin(diff_id)
    df_train = pd.concat([df_train, df_test.loc[index, :]]).reset_index(
        drop=True
    )
    df_test = df_test.loc[~index, :]
    # Checking again
    assert set(df_test[field]).issubset(set(df_train[field])), field

df_train.to_pickle("./ml-1m/train.pkl", compression="gzip")
df_test.to_pickle("./ml-1m/test.pkl", compression="gzip")
```

To load the data in batches from the above created.pickle files, we make our own data loader using `keras.utils.Sequence`.

```
import numpy as np
from keras.utils import Sequence


class DFMData(Sequence):
    def __init__(self, x_set, y_set, batch_size, shuffle=True):
        self.x, self.y = x_set, y_set
        self.batch_size = batch_size
        self.shuffle = True
        self.index = np.arange(len(self.y))
        self.on_epoch_end()

    def __len__(self):
        return int(np.ceil(len(self.y) / float(self.batch_size)))

    def __getitem__(self, index):
        start_index = index * self.batch_size
        end_index = (index + 1) * self.batch_size
        if end_index > len(self.y):
            end_index = None

        batch_x = [x[start_index:end_index] for x in self.x]
        batch_y = self.y[start_index:end_index]

        return batch_x, batch_y

    def on_epoch_end(self):
        if self.shuffle:
            np.random.shuffle(self.index)
```

```
            self.x = [
                x[self.index] if x.ndim < 2 else x[self.index, :]
                for x in self.x
            ]
            self.y = self.y[self.index]


# Prepare data loader
df_train = pd.read_pickle("./ml-1m/train.pkl", compression="gzip")
df_test = pd.read_pickle("./ml-1m/test.pkl", compression="gzip")

# Make the dataset provider
batch_size = 128
train_dataset = DFMData(
    [
        df_train["uid"].values,
        df_train["mid"].values,
        df_train["gender"].values,
        df_train["age"].values,
        df_train["occupation"].values,
        np.vstack(df_train["movie_genre"].values),
    ],
    df_train["rating"].values,
    batch_size=batch_size,
    shuffle=True,
)

test_dataset = DFMData(
    [
        df_test["uid"].values,
        df_test["mid"].values,
        df_test["gender"].values,
        df_test["age"].values,
        df_test["occupation"].values,
        np.vstack(df_test["movie_genre"].values),
    ],
    df_test["rating"].values,
    batch_size=batch_size,
    shuffle=True,
)
```

Finally, let's train the model.

```
import json
from keras.callbacks import EarlyStopping


def compute_steps(num_examples, batch_size):
    return num_examples // batch_size + int(
        num_examples % batch_size > 0
    )

```

```

# Build the model
dfm = KerasDeepFactorizationMachine(
    feature_name=[
        "user",
        "movie",
        "gender",
        "age",
        "occupation",
        "genre",
    ],
    feature_size=[6040, 3706, 2, 7, 21, 15],
    num_label=[1, 1, 1, 1, 1, 3],
    emb_size=25,
    hidden_layer_dim=[64, 64],
)

history = dfm.model.fit(
    x=train_dataset,
    validation_data=test_dataset,
    epochs=30,
    steps_per_epoch=compute_steps(df_train.shape[0], batch_size),
    validation_steps=compute_steps(df_test.shape[0], batch_size),
    callbacks=[EarlyStopping(monitor="val_mse", patience=0, verbose=1)],
)

# Save model, including the weights
dfm.model.save("./ml-1m/keras_dfm_model.h5")

# save history
with open("./ml-1m/keras_dfm_history.json", "w") as f:
    json.dump(history.history, f)
```

We can further plot the training history (as illustrated in Figure 6.1).

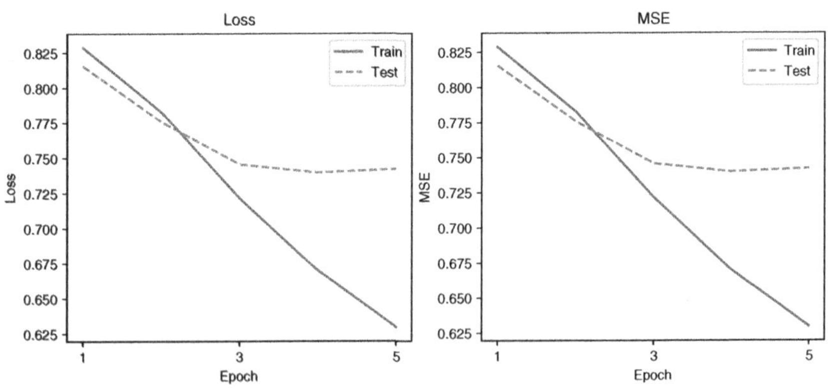

Figure 6.1 Training results of Deep Factorization Machine.

```
import json
from collections import namedtuple
import matplotlib.pyplot as plt

plt.rcParams["font.family"] = "TeX Gyre Heros"
history = namedtuple("history", ["history"])
with open("./ml-1m/keras_dfm_history.json", "r") as fid:
    history.history = json.load(fid)

fig, axs = plt.subplots(1, 2, figsize=(10, 4))
n_epochs = len(history.history["loss"])
epochs_arr = np.arange(1, n_epochs)
ax = axs[0]
ax.plot(epochs_arr, history.history["loss"][1:], label="train")
ax.plot(
    epochs_arr, history.history["val_loss"][1:], ls="--", label="test"
)
ax.legend()
# ax.set_yscale('log')
ax.set_xlabel("epoch")
ax.set_ylabel("Loss")
ax.set_title("Loss")
ax.set_xticks(epochs_arr[::2])
ax = axs[1]
ax.plot(epochs_arr, history.history["mse"][1:], label="train")
ax.plot(
    epochs_arr,
    history.history["val_mse"][1:],
    ls="--",
    label="test",
)
ax.legend()
# ax.set_yscale('log')
ax.set_xlabel("epoch")
ax.set_ylabel("MSE")
ax.set_title("MSE")
_ = ax.set_xticks(epochs_arr[::2])
plt.savefig("dfm-train.svg")
```

6.6 Regular Expressions (Regex)

Aside from simple native Python string methods and corresponding vectorized methods implemented by NumPy, Pandas, and Tensorflow, regular expression offers a versatile and elegant way to process large amounts of text data. Speaking of my personal experiences using regex: when I started my PhD program in 2014, I rotated in one of the genetics labs whose main focus of the study was human cancer genomics. The human genome has about 3 billion base pairs, and since DNA is a double helix, this is about 6 billion letters of "A," "T," "C," and "G" stored in the text file, making each data file gigabytes in size! Trying to search for specific occurrences of cancer genes in a text file like this cannot rely on for loops in Python (which I naively attempted once at the time), as it will take years to finish the search. At least at the time, the common way to parse the genome text file is to use Linux's `awk` command. Still, this took days to finish parsing the files I would need to parse. Then eventually, I came across Perl which is well known for its capability of text processing using regex. The adoption of Perl with regex cut down the processing time significantly, down to several hours from the original several days.

Let's examine several basic patterns of using regex in Python in the following examples. But I would like to encourage interested readers to refer to official Python documentation on regex (such as this one "Regular Expression HOWTO" [Kuchling, 2023]) to learn more about this topic.

```
1  import re
2  # Example 1: remove punctuation
3  text = "How is it going? I'm good!"
4  result = re.sub(r'''[\.\?\!\,\:\;\"\']''', '', text)
5  # 'How is it going Im good'
6
7  # Example 2: tokenize: spaces and newline
8  text = """The quick    brown fox  jumps
9  over the lazy dog"""
10 result = re.split(r'\s+', text)
11 # ['The', 'quick', 'brown', 'fox', 'jumps', 'over',
12 #  'the', 'lazy', 'dog']
13
14 # Example 3: find occurrences of the pattern
15 text = "The human genome has 3 billion base pairs, "
16     "making the text file storing it as large as "
17     "12 GB on disk."
18 result = re.findall(r'\d+', text)
19 # ['3', '12']
```

- **Lines 2–5, Example 1**: In this example, we attempt to remove punctuations characters, including period `.`, question mark `?`, exclamation mark `!`, comma `,`, colon `:`, semicolon `;`, double quote `"`, and apostrophe `'`. We achieve this by replacing (`re.sub` or substituting) the character in our list with an empty string. The backslash character \ is called the escape character, which will make sure the regex treats the character following the backslash as a string character, rather than a special pattern-matching metacharacter, such as `.` (matching any character other than newline \n), + (matching at least 1 occurrence), `*` (matching any number of occurrences, including 0 occurrences), and ? (matching 0 or 1 occurrences of the preceding pattern character, but no more than 1). Alternatively, one can also use `string.punctuations` from Python's built-in `string` package for the list of punctuation characters, i.e. `re.sub(rf'''[{string.punctuation}]''', '', text)`.

- **Lines 7–12, Example 2**: In this example, we attempt to locate all white space characters (which may occur consecutively) and split by the white space characters. The pattern `\s` indicates that we are trying to find white space characters (newlines as well as spaces and tabs). Then the + metacharacter tells `res.split` that it is possible to have 1 or more than 1 white space characters occurring consecutively to delimit the tokens in the sentence, so we would like to split the sentence such that all white space characters are removed from the final list of tokens. An alternative to this regex pattern is to use `re.findall(r'([\w]*\w)', text)` instead. Here, `\w` will match an alphanumerical character. Therefore, `[\w]*\w` will match a string that starts and ends with alphanumerical characters, which effectively breaks the original long string into a list of shorter strings by removing any non-alphanumerical characters such as spaces and newlines.
- **Lines 14–19, Example 3**: In this example, we try to extract some patterns from a sentence. Specifically, we want to find all occurrences of numbers in the sentence, i.e. `"3"` and `"12"`. The `\d+` pattern indicates that what we would like to find are one or more (+) digits (`\d`) that are grouped together. `re.findall` will return all the occurrences that match the pattern.

6.7 Data Serialization and Deserialization

The current section is not regarding how to manipulate string values in the data, but rather how to turn an entire dataset into a string or byte representation, which we call data "**serialization**." Within the context of machine learning data management, serialization is the process of converting structured data into a stream of byte strings so it can be further compressed and consumed by any downstream processes. In production machine learning systems that use microservices, data hosting services are separate from the model inference services. Hence, there is a need to transfer data from one service to another service via the network. In cloud environments, model training needs to load data from the cloud disk to the virtual machine instance which has access to costly resources such as GPUs. Therefore, the service needs to be able to load the data from the remote disk quickly, so that the training instance can fully utilize the GPUs for training rather than waiting for data to be downloaded. On the other hand, the process of reversing the serialization process, or turning byte strings back into structured data, is called **deserialization**.

Tensorflow uses Protocol Buffers [Google, 2023] (protobuf) to serialize the structured data into a sequence of bytes. The serialized byte strings can then be compressed and stored in the TFRecord format, a row-based data storage format that supports distributed data processing, transformation, shuffling, and batching via the `tf.data` API. Suppose we have the following batch of data with three features:

```
import os
import tempfile
import tensorflow as tf
inputs = [
    {"student_id": "a-153248", "num_classes_taken": 5, "gpa": 3.97},
    {"student_id": "b-132429", "num_classes_taken": 3, "gpa": 3.51},
    {"student_id": "a-293841", "num_classes_taken": 4, "gpa": 4.00},
    {"student_id": "c-912431", "num_classes_taken": 4, "gpa": 3.91},
    {"student_id": "c-924145", "num_classes_taken": 3, "gpa": 3.20},
]
```

We recognize that `student_id` is a string feature, `num_classes_taken` is an integer feature, and `gpa` is a float feature. We can then put this data, one row at a time, into the `tf.train.Example` protobuf,

```
# Create the Features protobuf
examples = []
for x in inputs:
    # Create the dictionary
    feature = {
        "student_id": tf.train.Feature(
            bytes_list=tf.train.BytesList(
                value=[bytes(x["student_id"], "utf-8")]
            )
        ),
        "num_classes_taken": tf.train.Feature(
            int64_list=tf.train.Int64List(value=[x["num_classes_taken"]])
        ),
        "gpa": tf.train.Feature(
            float_list=tf.train.FloatList(value=[x["gpa"]])
        )
    }

    # Create the Examples protobuf
    example = tf.train.Example(features=tf.train.Features(feature=feature))
    examples.append(example)

# print the last example
print(example)

# features {
#   feature {
```

```
#       key: "student_id"
#       value {
#         bytes_list {
#           value: "c-924145"
#         }
#       }
#     }
#     feature {
#       key: "num_classes_taken"
#       value {
#         int64_list {
#           value: 3
#         }
#       }
#     }
#     feature {
#       key: "gpa"
#       value {
#         float_list {
#           value: 3.2
#         }
#       }
#     }
# }
```

In the example, the protobuf for the byte string needs to encode the data types using separate protobuf definitions, i.e. string features need to be cast to bytes first and then encoded by `tf.train.BytesList`, integer features need to be encoded by `tf.train.Int64List`, and float features need to be encoded by `tf.train.FloatList`. Notice that the value assigned for each feature is a list with a single element. This is because these protobuf types actually encode a list of values of a certain type rather than a single value. Suppose `{num_classes_taken: [5, 3, 4]}`, then we can encode this feature as `int64_list=tf.train.Int64List(value=[5, 3, 4])`.

To write our string representations of the data into a TFRecord, we can first serialize the protobuf string into bytes, and then write to a compressed TFRecord. To understand how exactly TFRecord is formatted to store the data in a serialized format, see the implementation from the Ray Data library [Moritz et al., 2023].

```
# Create a temporary TFRecord file
base_dir = tempfile.TemporaryDirectory()
file_path = os.path.join(base_dir.name, "temp.tfrecord")
options = tf.io.TFRecordOptions(compression_type="GZIP") # gzip compressed
with tf.io.TFRecordWriter(file_path, options=options) as writer:
    for example in examples:
        serialized_example = example.SerializeToString()
        writer.write(serialized_example)
```

```

# print last serialized example
print(serialized_example)
# b'\nI\n\x1a\n\x11num_classes_taken\x12\x05\x1a\x03\n\x01\x03\n\x0f
    ↪ \n\x03gpa
    ↪ \x12\x08\x12\x06\n\x04\xcd\xccL@\n\x1a\n\nstudent_id\x12\x0c\n\n\n
    ↪ \x08c-924145'
```

Lastly, we can then attempt to read it back

```
dataset = tf.data.TFRecordDataset(
    tf.data.TFRecordDataset.list_files(file_path),
    compression_type="GZIP", # make sure to decompress before decoding
)

# print a record
print(next(iter(dataset)))
# tf.Tensor(b'\nI\n\x1a\n\x11num_classes_taken\x12\x05\x1a\x03\n\x01
    ↪ \x05\n
    ↪ \x0f\n\x03gpa\x12\x08\x12\x06\n\x04{\x14~@\n\x1a\n\nstudent_id\x12
    ↪ \x0c\n\n\n\x08a-153248', shape=(), dtype=string)
```

But this reads only the serialized byte string, instead of the original structured data. To parse the structured data back from the serialized byte strings, we need to do the following:

```
# Define the feature schema
feature_spec = {
    "student id": tf.io.FixedLenFeature((1, ), dtype=tf.string),
    "num_classes_taken": tf.io.FixedLenFeature((1, ), dtype=tf.int64),
    "gpa": tf.io.FixedLenFeature((1, ), dtype=tf.float32)
}

# Parsing a batch of data back
serialized_data = next(iter(dataset.batch(4)))
parsed_data = tf.io.parse_example(serialized_data, feature_spec)

print(parsed_data)
# {'gpa': <tf.Tensor: shape=(4, 1), dtype=float32, numpy=
#  array([[3.97],
#         [3.51],
#         [4.  ],
#         [3.91]], dtype=float32)>,
# 'num_classes_taken': <tf.Tensor: shape=(4, 1), dtype=int64, numpy=
#  array([[5],
#         [3],
#         [4],
#         [4]])>,
#  'student_id': <tf.Tensor: shape=(4, 1), dtype=string, numpy=
```

```
105  #   array([[b'a-153248'],
106  #          [b'b-132429'],
107  #          [b'a-293841'],
108  #          [b'c-912431']], dtype=object)>}
```

Between Lines 83 and 87, the `feature_spec` variable defines the schema of the data to be parsed. We used `tf.io.FixedLenFeature` to indicate that we are parsing a tensor with known shapes. We specify the shape argument (first argument) as `(1,)` to indicate the tensor to be parsed has shape `(batch_size, 1)`. Indeed, we can see that the recovered data matches the shape we have specified. On the other hand, the `dtype` argument specifies what data type to treat the parsed data as. This is matching to the data types we have seen in the parsed data.

Bibliography

Google. Protocol Buffers. https://protobuf.dev/, 2023.

Huifeng Guo, Ruiming Tang, Yunming Ye, Zhenguo Li, and Xiuqiang He. DeepFM: A factorization-machine based neural network for CTR prediction. *IJCAI International Joint Conference on Artificial Intelligence*, pages 1725–1731, 2017. ISSN 10450823. doi: 10.24963/ijcai.2017/239.

F. Maxwell Harper and Joseph A. Konstan. The MovieLens datasets: History and context. *ACM Transactions on Interactive Intelligent Systems*, 5(4):1–19, January 2016. ISSN 2160-6455, 2160-6463. doi: 10.1145/2827872.

A. M. Kuchling. Regular Expression HOWTO. https://docs.python.org/3/howto/regex.html, 2023.

Wes McKinney. Data structures for statistical computing in python. In *Python in Science Conference*, pages 56–61, Austin, Texas, 2010. doi: 10.25080/Majora-92bf1922-00a.

Wes McKinney. Pandas Timestamp. https://pandas.pydata.org/docs/reference/api/pandas.Timestamp.html, 2023a.

Wes McKinney. Pandas: Working with text data. https://pandas.pydata.org/docs/user_guide/text.html, 2023b.

Philipp Moritz, Robert Nishihara, Stephanie Wang, Alexey Tumanov, Richard Liaw, Eric Liang, Melih Elibol, Zongheng Yang, William Paul, Michael I. Jordan, and Ion Stoica. Ray Data Write TFRecord. https://github.com/ray-project/ray/blob/master/python/ray/data/datasource/tfrecords_datasource.py#L414C1-L424, 2023.

Steffen Rendle. Factorization machines. *Proceedings - IEEE International Conference on Data Mining, ICDM*, pages 995–1000, 2010. ISSN 15504786. doi: 10.1109/ICDM.2010.127.

7

Sparse Matrix

As a data scientist who developed the recommender system for a media streaming platform, I often worked with sparse data. In the most classic examples of recommender systems such as Netflix or MovieLens dataset, we have data on the ratings of specific items by the users. For example, a customer may give a 4.5-star rating on a movie. However, given there are millions or even billions of users, and tens of thousands of titles on Netflix (or millions to billions in the context of e-commerce like Amazon and Alibaba), it is not possible for every user to give a rating on every single piece of content. Oftentimes, we will have most of the users rating on several items, and most of the items will only have some tens to hundreds of users to rate them. If we imagine this as a matrix where the rows represent the user IDs, whereas the columns represent the item ID, then we have a very large matrix, but only with a small percentage of it filled in. Most of the entries in this matrix will be empty (see Figure 7.1, for example). This matrix, with most of the values missing or empty, is called a **sparse matrix**. The task of a recommender system is often to predict the ratings on these missing entries in order to make recommendations to users on new items.

We can, in theory, fill these missing values with zeros and use `numpy.ndarray` or `pandas.DataFrame` to represent it. But every 0 we store in our memory or disk is probably a waste of memory or storage space. Also, if we fill these missing values with 0s, then we would not be able to distinguish cases where a user literally rates a book with 0 stars (although most rating systems probably do not allow it, perhaps also consider other types of data where a value of 0 is distinguished from missing values). Therefore, it makes more sense to store only the entries of the matrix where it contains actual data collected from the users. Sparse matrix data structures are designed exactly for this purpose. Sparse matrices can be implemented in various ways, using different indexing and data value storage logics.

Vectorization: A Practical Guide to Efficient Implementations of Machine Learning Algorithms,
First Edition. Edward DongBo Cui.

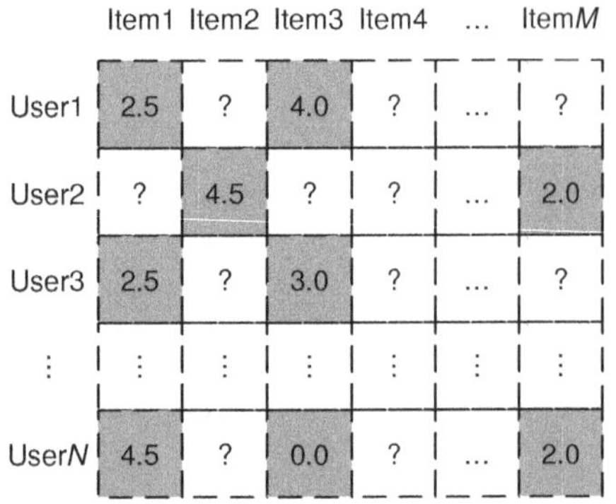

Figure 7.1 Example user-item rating matrix in a typical recommendation system.

In this chapter, we will first examine the seven variants of sparse matrices implemented in the Scipy library. To help further advance our knowledge in practical machine learning implementations, we will also look briefly into the sparse data structure classes implemented by deep learning libraries such as Tensorflow and PyTorch. Lastly, we will close the chapter with a case study on implementing matrix factorization with the alternating least square (ALS) method using the sparse data structures we learned from this chapter.

7.1 `scipy`'s Sparse Matrix Classes

`scipy` [Virtanen et al., 2020] implements seven different sparse matrix data structures under its `scipy.sparse` API [Virtanen et al., 2023]. Each sparse matrix data structure has its pros and cons. All seven types can be constructed from dense arrays (with 0s assumed to be empty for numerical data and an empty string “ ” assumed to be empty-valued for string data). All seven types are inter-convertible from each other. Some are good for slicing the matrix and doing arithmetic operations, while others are better used for constructing the data matrix from other data structures. In the following sections, we will analyze each one of them.

7.1.1 Coordinate Sparse Matrix (`coo_matrix`)

`coo_matrix` is the best and fastest format for constructing a new sparse matrix using large arrays and row/column indices. This is probably the most intuitive way to store a sparse matrix: by storing its row/column coordinates, as well as the data at these corresponding coordinates. Once the sparse matrix is constructed, one can easily cast it from `coo_matrix` to other sparse matrix formats.

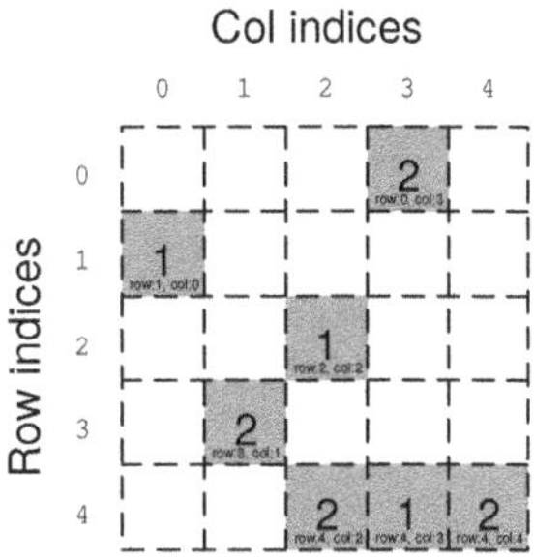

Figure 7.2 `coo_matrix` example.

A feature of the `coo_matrix` format when constructing a sparse matrix is that it allows repeated `(row, col)` coordinates. When converting to other sparse matrix formats such as `csc_matrix` or `csr_matrix` (which we will introduce next), the duplicated `(row, col)` entries will be summed together. This is especially helpful when constructing finite element matrices. To construct a sparse matrix without summing the duplicated entries, one can use `dok_matrix` (to be introduced in a later section).

For the example matrix shown in Figure 7.2, the `coo_matrix` can be constructed by

```
from scipy.sparse import coo_matrix
A = coo_matrix([
    [0, 0, 0, 2, 0],
    [1, 0, 0, 0, 0],
    [0, 0, 1, 0, 0],
    [0, 2, 0, 0, 0],
    [0, 0, 2, 1, 2],
])
```

The underlying data structure is

```
A.data = [2, 1, 1, 2, 2, 1, 2]
A.row  = [0, 1, 2, 3, 4, 4, 4]
A.col  = [3, 0, 2, 1, 2, 3, 4]
```

7.1.2 Compressed Sparse Column Matrix (`csc_matrix`)

`csc_matrix` is good for column-wise slicing and fast arithmetics including addition, subtraction, multiplication, division, power, and matrix multiplication (+, −, `*`, `/`, `**` and@, respectively). Neither `csc_matrix` is efficient when adding a

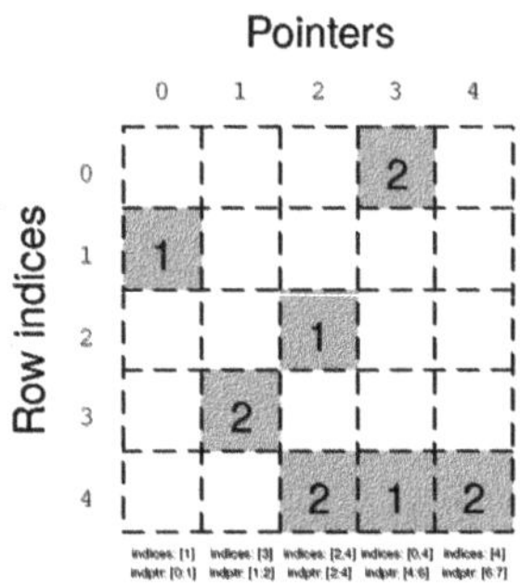

Figure 7.3 `csc_matrix` example.

new element to the data nor it is the best choice for constructing a new sparse matrix.

`csc_matrix` stores three pieces of information regarding the sparse matrix, the data itself as an array (`data`), the indices of the *rows* (`indices`), and the indirect register pointer (`indptr`) with respect to the `indices`. As an example, suppose we have the sparse `csc_matrix` shown in Figure 7.3.

Then, its `csc_matrix` representation is constructed as

```
from scipy.sparse import csc_matrix
A = csc_matrix([
    [0, 0, 0, 2, 0],
    [1, 0, 0, 0, 0],
    [0, 0, 1, 0, 0],
    [0, 2, 0, 0, 0],
    [0, 0, 2, 1, 2],
])
```

The underlying properties are

```
A.data = [1, 2, 1, 2, 2, 1, 2]
A.indices = [1, 3, 2, 4, 0, 4, 4]
A.indptr = [0, 1, 2, 4, 6, 7]
```

where

- `data` is simply a column-wise flattened version of the matrix.
- `indices` is the row indices for the corresponding elements of the `data`, for example the first element in the data is a 1, and it is located in the row index 1 (second row); the second element in the data is a 2, the row index is 3 (fourth row), etc.

- `indptr` is a pointer for `data` and `indices`, an array of length `col + 1`, with a maximum of its element as the length of the `data` or `indices`. It represents the range of column indices that each element in the data belongs to. A convenient heuristic to remember is that `indptr[i] - indptr[i-1]` can tell us how many elements are present within column `i-1`. In the above example, we have

  1. Column 0 has `data` and `indices` between indices `A.indptr[0]=0` and `A.indptr[0+1]=1`, i.e. `A.data[0:1]=1` and `A.indices[0:1]=1`
  2. Column 1 has `data` and `indices` between indices `A.indptr[1]=1` and `A.indptr[1+1]=2`, i.e. `A.data[1:2]=2` and `A.indices[1:2]=3`
  3. Column 2 has `data` and `indices` between indices `A.indptr[2]=2` and `A.indptr[2+1]=4`, i.e. `A.data[2:4]=[1,2]` and `A.indices[2:4]=[2,4]`
  4. Column 3 has `data` and `indices` between indices `A.indptr[3]=4` and `A.indptr[3+1]=6`, i.e. `A.data[4:6]=[2,1]` and `A.indices[4:6]=[0,4]`
  4. Column 4 has `data` and `indices` between indices `A.indptr[4]=6` and `A.indptr[4+1]=7`, i.e. `A.data[6:7]=2` and `A.indices[6:7]=2`

7.1.3 Compressed Sparse Row Matrix (`csr_matrix`)

`csr_matrix` is good for row-wise slicing and like `csc_matrix`, fast for arithmetic operations. `csr_matrix` is also not efficient when adding a new element to the data.

The data structure of the csr matrix is similar to that of the csc matrix, except that the `indices` indicate the indices of the *columns*. For the same example matrix (Figure 7.4), the `csr` representation is constructed as

```
from scipy.sparse import csr_matrix
A = csr_matrix([
    [0, 0, 0, 2, 0],
    [1, 0, 0, 0, 0],
    [0, 0, 1, 0, 0],
    [0, 2, 0, 0, 0],
    [0, 0, 2, 1, 2],
])
```

The underlying properties are

```
A.data    = [2, 1, 1, 2, 2, 1, 2]
A.indices = [3, 0, 2, 1, 2, 3, 4]
A.indptr  = [0, 1, 2, 3, 4, 7]
```

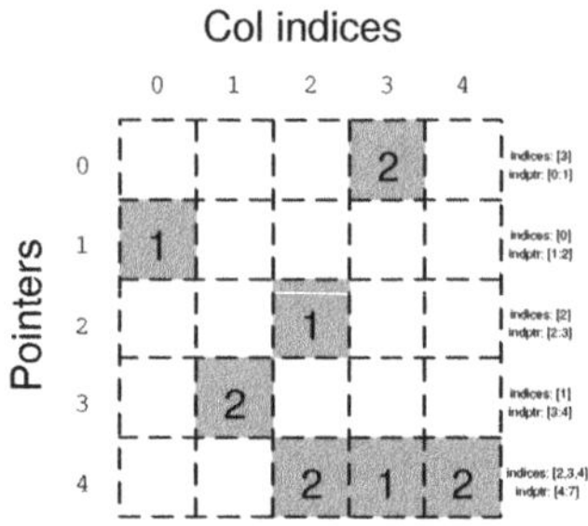

Figure 7.4 `csr_matrix` example.

where

- `data` is a row-wise flattened array
- `indices` are indices of the columns
- `indptr` is a pointer for `data` and `indices`, an array of cumulative length of rows, with the max of its element as the length of the `data` or `indices`

7.1.4 Block Sparse Row Matrix (`bsr_matrix`)

`bsr_matrix` has all the advantages of `csr_matrix`. In addition, when data is dense at certain regions of the matrix, arithmetic operations can be even more efficient than `csr_matrix`. This dense region of the matrix is called a "block," which is a dense matrix of arbitrary size (R, C). To efficiently store the sparse matrix in the `bsr` format, it is required that the block size are all the same and are divisible by the number of rows and columns, i.e. if the block size is (R, C) and matrix size is (M, N), then it is required that `M % R = 0` and `N % C = 0`.

The `bsr` sparse matrix representation has the following structure:

- `data`: A $K \times R \times C$ matrix, where K is the number of blocks, and R and C are the dimensions of each block
- `indices`: Length K array, column indices of the blocks
- `indptr`: Length $K + 1$ array, pointers of block `data` and `indices`

For the example matrix shown in Figure 7.5, the `bsr` representation can be constructed as

```
from scipy.sparse import bsr_matrix
A = bsr_matrix([
    [0, 0, 0, 0, 3, 3],
    [0, 0, 0, 0, 3, 3],
    [1, 1, 0, 0, 0, 0],
    [1, 1, 0, 0, 0, 0],
    [0, 0, 2, 2, 0, 0],
    [0, 0, 2, 2, 0, 0]]
)
```

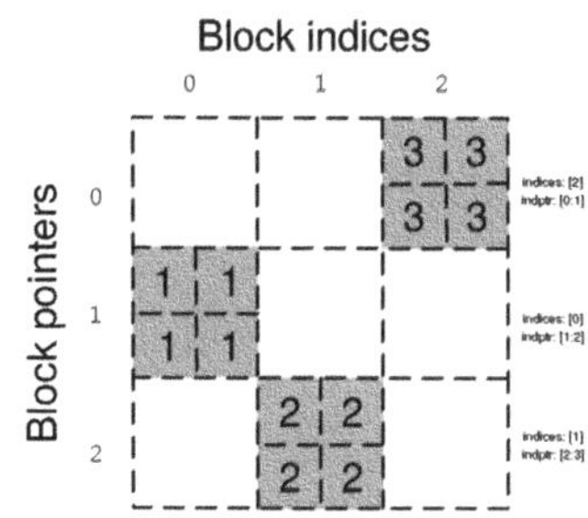

Figure 7.5 `bsr_matrix` example.

Its data structure is

```
A.data = [
    [
        [3, 3],
        [3, 3],
    ],
    [
        [1, 1],
        [1, 1],
    ],

    [
        [2, 2],
        [2, 2],
    ],
]
A.indices = [2, 0, 1]
A.indptr = [0, 1, 2, 3]
```

7.1.5 Dictionary of Keys Sparse Matrix (`dok_matrix`)

`dok_matrix` is the best format to use when we try to update/replace the elements in a sparse matrix. It is basically a dictionary of the format { `(row, col): data, ...`}.

- `keys()`: This method will return the (row, col) tuples which index the data
- `values()`: This method will return the data stored

For the matrix shown in Figure 7.6, a `dok_matrix` can be constructed as

```
from scipy.sparse import dok_matrix
A = dok_matrix([
    [0, 0, 0, 2, 0],
    [1, 0, 0, 0, 0],
```

```
        [0, 0, 1, 0, 0],
        [0, 2, 0, 0, 0],
        [0, 0, 2, 1, 2],
    ])
```

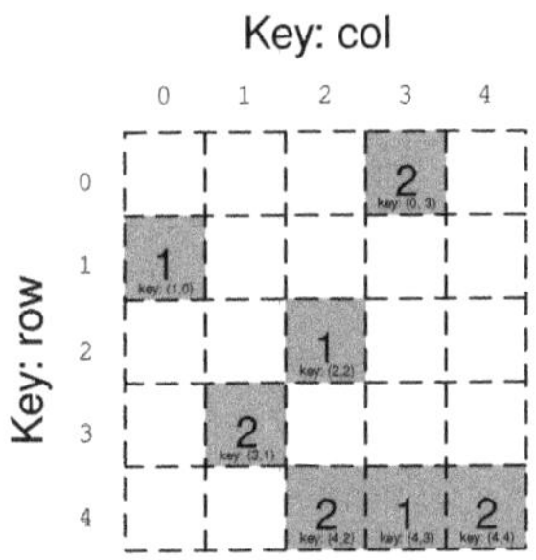

Figure 7.6 `dok_matrix` example.

The `dok_matrix`'s equivalent `dict` structure is

```
{
    (1, 0): 1,
    (3, 1): 2,
    (2, 2): 1,
    (4, 2): 2,
    (0, 3): 2,
    (4, 3): 1,
    (4, 4), 2,
}
```

7.1.6 Row-Based List of List Sparse Matrix (`lil_matrix`)

`lil_matrix` is also a good format to construct a sparse matrix, but only if the matrix does not involve a lot of data. For larger data construction, consider using `coo_matrix`. `lil_matrix` stores the following two pieces of information about the sparse matrix

- `rows`: Lists of column indices for each row.
- `data`: Lists of list of row elements.

For the example matrix shown in Figure 7.7, the `lil` sparse matrix can be constructed as

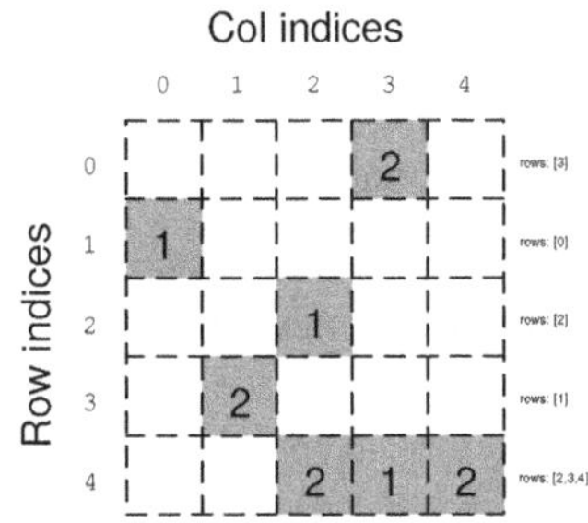

Figure 7.7 `lil_matrix` example.

```
from scipy.sparse import lil_matrix
A = lil_matrix([
    [0, 0, 0, 2, 0],
    [1, 0, 0, 0, 0],
    [0, 0, 1, 0, 0],
    [0, 2, 0, 0, 0],
    [0, 0, 2, 1, 2],
])
```

The data structures are

```
A.rows = [ [3], [0], [2], [1], [2,3,4] ]
A.data = [ [2], [1], [1], [2], [2,1,2] ]
```

`lil_matrix` offers a very simple and flexible interface to slice data. A key advantage of using `lil` format to slice the matrix is that it allows row and column indices to be broadcasted.

Suppose we have an array of row indices `rows` and column indices `cols`, to slice out a subset of the sparse matrix for every `rows` and `cols`, we can do

```
rows = rows[:, np.newaxis] # column vector
cols = cosl[np.newaxis, :] # row vector
mat_lil[rows, cols] += np.ones(
    (rows.size, cols.size)
) # add one to these slices
mat_lil[rows, cols] = ... # assigning these subset of slices
```

For very large matrices, however, it is recommended to construct them using `coo_matrix`.

7.1.7 Diagonal Storage Sparse Matrix (dia_matrix)

dia_matrix is the best format to use when constructing diagonal matrices, or off-diagonal matrices. It stores two pieces of information about the sparse matrix.

- offsets: An array of diagonal offsets. 0 means at the main diagonal of the matrix; a negative value means below the main diagonal; a positive value means above the main diagonal. The array is usually sorted ascendingly (for example below the main diagonal, main diagonal, above the main diagonal)
- data: $K \times D$ matrix, where K is the number of diagonal terms, and D is the size of the main diagonal. Note that if the diagonal is at the off-diagonal (i.e. not at the main diagonal), then
  1. If this diagonal is below the main diagonal (offset < 0), then trailing zeros will be added to the array to match the size of the main diagonal array.
  2. If this diagonal is above the main diagonal (offset > 0), then leading zeros will be added to the array to match the size of the main diagonal array.
  3. These zero values can actually be any other values, but they will not be part of the matrix. In the scipy.sparse.dia_matrix document example, the dia_matrix was constructed using three identical length vectors, placed at the diagonals [-1, 0, 1]. Although, when accessing the data via A.data, it was shown that there were no zeros filled right after the initial construction, casting first to coo_matrix then back to dia_matrix and access data with A.data again will show these leading and lagging zeros.

Figure 7.8 shows an example dia_matrix.

We can construct the dia_matrix using

```
from scipy.sparse import dia_matrix
A = dia_matrix([
    [1, 0, 0, 0, 0],
    [0, 1, 0, 0, 0],
    [0, 0, 1, 0, 0],
    [2, 0, 0, 1, 0],
    [0, 2, 0, 0, 1],
])
```

In the example matrix shown in Figure 7.8, the two attributes of the matrix are

```
A.data = [
    [2, 2, 0, 0, 0],
    [1, 1, 1, 1, 1],
]
A.offsets = [-3, 0]
```

Figure 7.8 `dia_matrix` example.

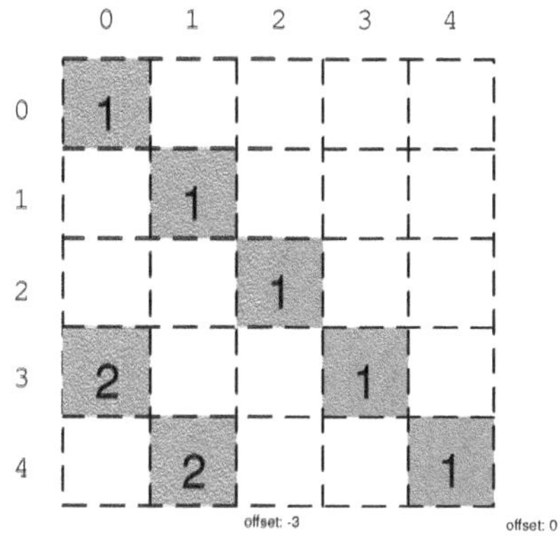

7.1.8 Comparisons Between Different Sparse Matrix Formats

Table 7.1 summarizes the ways to construct and update the seven sparse matrices. In addition to constructing from dense matrices with 0s being treated as the empty entries, there are more efficient ways to construct these matrices by directly specifying the indices and the corresponding data values (as well as other needed properties) with respect to each data structure. For example, to construct a `coo_matrix`, one can directly specify the `row`, `col`, `data`, and `shape` of the sparse matrix, i.e. `coo_matrix((data, (rows, cols)), [shape=(M,N)])`. Moreover, for `dok_matrix`, `lil_matrix`, and `dia_matrix`, I have included some in-place methods to update the matrix, which are not found in `scipy`'s documentation.

In addition to using `.to<format>` methods to convert between different sparse matrix formats, we can also leverage the `.asformat()` method. This can be convenient if we would like to implement a function that does some operation in another format but still would like to retain the original format when returning the result. For example,

```
def function_retaining_sparse_format(A):
    # Convert to csr_matrix
    B = A.tocsr()  # this makes a copy
    # Do something on B ...

    # Convert back to the original format
    return B.asformat(A.format)
```

Table 7.2 makes a comparison of the performance of several types of operations for each sparse matrix. In general, `csc_matrix`, `csr_matrix`, and

Table 7.1 Construction, update, and inter-conversion of seven sparse matrix formats.

Type	From Dense Matrix	Inter-conversion	Empty Matrix	From Array and Row/Col Indices	Format-Specific Construction/Updates
`csc`	`csc_matrix(D)`	`.tocsc()`	`csc_matrix((M,N), [dtype])`	`csc_matrix((data, (rows, cols)), [dtype])`	indices & indptr construction: `csc_matrix((data, indices, indptr), [shape=(M,N)])`
`csr`	`csr_matrix(D)`	`.tocsr()`	`csr_matrix((M,N), [dtype])`	`csr_matrix((data, (rows, cols)), [dtype])`	indices & indptr construction: `csr_matrix((data, indices, indptr), [shape=(M,N)])`
`bsr`	`bsr_matrix(D,[ blocksize = (R,C)])`	`.tobsr([ blocksize = (R,C)])`	`bsr_matrix((M,N), [blocksize=(R,C), dtype])`	`bsr_matrix((data, (rows, cols)), [blocksize=(R,C), shape=(M,N)])`	indices & indptr construction: `bsr_matrix((data, indices, indptr), [shape=(M,N)])`
`coo`	`coo_matrix(D)`	`.tocoo()`	`coo_matrix((M,N), [dtype])`	`coo_matrix((data, (rows, cols)), [shape=(M,N)])`	Add new data: `mat_coo.data = np.r_[mat_coo.data, data], mat_coo.row = np.r_[mat_coo.row, rows], mat_coo.col = np.r_[mat_coo.col, cols]`
`dok`	`dok_matrix(D)`	`.todok()`	`dok_matrix((M,N), [dtype])`	N/A	row & col update: `dict.update(mat_dok, zip(zip(rows, cols), data))`
`lil`	`lil_matrix(D)`	`.tolil()`	`lil_matrix((M,N), [dtype])`	N/A	row & col update: `mat_lil[rows, cols] = data`
`dia`	`dia_matrix(D)`	`.todia()`	`dia_matrix((M,N), [dtype])`	`dia_matrix((data, offsets), shape=(M,N))`	`mat_dia.setdiag(data, offset)`

Table 7.2 Performance comparisons of all seven sparse matrix data structures on different operations.

Type	Construction	Add New Element	Slicing	Arithmetic Operations
csc	slow	very slow	efficient col slicing	efficient
csr	slow	very slow	efficient row slicing	efficient
bsr	slow	very slow	efficient row slicing	Can sometimes be more efficient than csr_matrix
coo	fast	efficient	no native support	no native support
dok	efficient	efficient	no native support, $O(1)$ access of individual elements	slow
lil	efficient, convenient	efficient	flexible slicing, but slow col slicing	slow
dia	efficient	efficient	no native support	efficient

bsr_matrix are great for arithmetic operations, while others are better for matrix construction.

7.2 Sparse Matrix Broadcasting

There are many cases where we want to add, subtract, or normalize over columns, rows, or the entire matrix, without affecting the missing data. In this section, we are going to consider several cases of sparse matrix broadcasting.

7.2.1 Scalar Broadcasting

Unlike numpy, simply adding a scalar on the sparse matrix with mat_csc + 1 will give us the error.

```
NotImplementedError: adding a nonzero scalar to a sparse
matrix is not supported
```

Fortunately, the actual data of the matrix is stored in .data attribute as a numpy.ndarray. Therefore, we can simply modify the .data attributes,

```
mat_csc.data += 1
```

7.2.2 Row-wise Broadcasting

Suppose we have the `csr` sparse matrix `mat_csr` of size $M \times N$ and a row vector `array_r` of length M. We want to add (subtract, multiply, or divide) each row of the sparse matrix by the row vector elementwise. Recall in Chapter 2, in NumPy with dense matrices (`mat_dense`), we can simply do this via broadcasting

```
mat_dense + array_r[np.newaxis, :]
```

With the Sparse matrix, we again can access the `.data` attribute. We can also access the column indices of each data point via the `.indices` attribute (see Figure 7.9), then the equivalent sparse matrix row-wise broadcasting is then

```
mat_csr.data += np.take(array_r, mat_csr.indices)
```

For other types of sparse matrices, first, convert to `csr_matrix` with `tocsr()`, make the broadcasting calculation, and then convert it back to the original format if needed. This can be implemented as

```
def R_plus_rv(R, rv):
    """Sparse matrix plus a row vector"""
    R = R.tocsr()
    R.data += np.take(rv, R.indices)
    return R
```

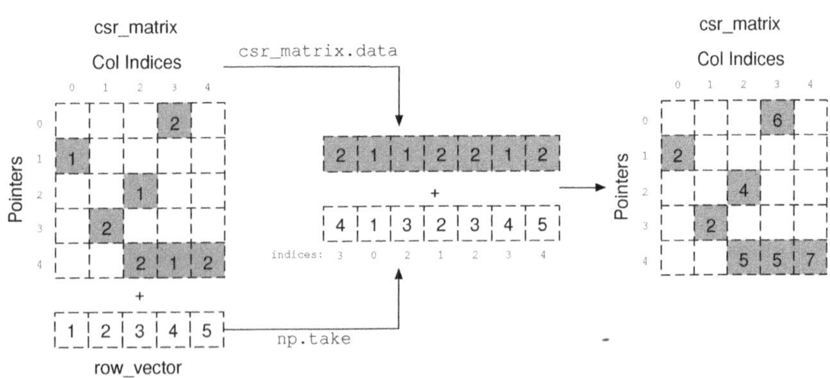

Figure 7.9 Row-wise broadcasting example.

7.2.3 Column-wise Broadcasting

Similarly, for column-wise broadcasting, in NumPy using dense matrices, this can be done with broadcasting

```
mat_dense + array_c[:, np.newaxis]
```

On sparse matrices, suppose we have a `csc` matrix `mat_csc` of size $M \times N$ and a vector `array_c` of size N. If we want to broadcast the vector to all the columns of the sparse matrix (see also Figure 7.10), we can do

```
mat_csc.data += np.take(array_c, mat_csc.indices)
```

Again, for other types of sparse matrices, first convert to `csc_matrix` with `tocsc()`, make the broadcasting calculation, and then convert it back to the original format if needed. This can be implemented as

```
def R_plus_cv(R, cv):
    """Sparse matrix plus a column vector"""
    R = R.tocsc()
    R.data += np.take(cv, R.indices)
    return R
```

7.2.4 Multiplication on Sparse Indices

In matrix factorization problems, we often need to compute a loss function after the prediction by multiplying the two factorized dense matrices. But since the loss

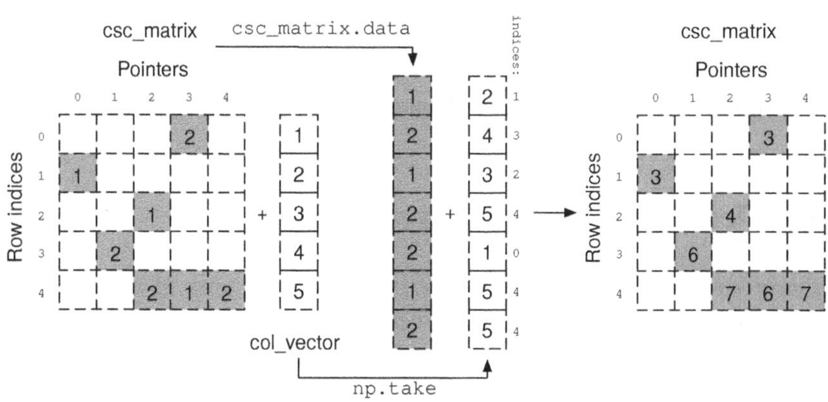

Figure 7.10 Column-wise broadcasting example.

function should only be calculated where the data is available on the large sparse matrix, it is only necessary to execute at a small subset of combinations of rows, instead of the full matrix multiplication.

Suppose we have two dense matrices $U \in \mathbb{R}^{N\times K}$ and $W \in \mathbb{R}^{M\times K}$ and suppose N and M can be very large, but K is very small. We have a user-item rating sparse matrix R, and we are trying to use the product

$$\hat{R} = UW^T$$

to predict R. Since R is sparse, it is better if predicted $\hat{R}$ is also sparse to avoid unnecessary computations.

A trick is to use `coo_matrix` to determine which combinations of rows and columns of R need to be predicted. Then, we can slice out only the necessary rows from the two dense matrices (for necessary rows of U, they would be the rows of R; for necessary rows of W, they would be the columns of R). After selecting only the necessary rows, we can do dot products on these sliced sets of row vectors. The following snippets implement this idea:

```
def multiply_U_W(U, W, R):
    """Compute product of U and W.T,
    but only at the position where R is available"""
    iU, iW = R.tocoo().row, R.tocoo().col # np.nonzero(R)
    # values = np.sum(U[iU, :] * W[iW, :], axis=1)
    values = np.einsum('ij,ij->i', U[iU, :], W[iW, :])
    # Recreate the coo matrix, making sure its shape
    # is the same as R
    result = coo_matrix(
        (values, (iU, iW)), shape=R.shape
   ).tocsc()
    return result
```

This piece of code will soon make more sense once we introduce the matrix factorization algorithm later in the chapter.

7.3 Tensorflow's Sparse Tensors

Tensorflow (as of the time of writing) supports two types of sparse matrix structures, `coo` format whose constructions and operations are handled by `tf.sparse.SparseTensor`, and `csr` format whose operations are handled by `tensorflow.python.ops.linalg.sparse.sparse_csr_matrix_ops`.

7.3.1 SparseTensor Class

The major sparse tensor API, `tf.sparse.SparseTensor` (or simply `tf.SparseTensor`) uses the `coo` format, and a sparse tensor can be constructed as

```
import tensorflow as tf

tf.SparseTensor(
    indices=tf.constant([
        [0, 3], [1, 0], [2, 2], [3, 1], [4, 2],[4, 3], [4, 4]],
        dtype=tf.int64
    ),
    values=tf.constant([2, 1, 1, 2, 2, 1, 2], dtype=tf.int32),
    dense_shape=(5, 5),
)
```

Tensorflow's `tf.SparseTensor` can support more than two-dimensional data, whereas Scipy's `coo_matrix` can only support two-dimensional matrices. To represent an N-dimensional sparse tensor, we need to expand the indices to represent N-dimensional coordinates, for example

```
tf.SparseTensor(
    indices=tf.constant([
        [0, 3, 0], [1, 0, 0], [2, 2, 1], [3, 1, 2],
        [4, 2, 3],[4, 3, 4], [4, 4, 5]],
        dtype=tf.int64
    ),
    values=tf.constant([2, 1, 1, 2, 2, 1, 2], dtype=tf.int32),
    dense_shape=(5, 5, 6),
)
```

Like Scipy's implementation, a `tf.SparseTensor` can also be constructed by converting the data from a dense tensor.

```
tf.sparse.from_dense(
    tf.constant([
        [0, 0, 0, 2, 0],
        [1, 0, 0, 0, 0],
        [0, 0, 1, 0, 0],
        [0, 0, 1, 0, 0],
        [0, 2, 0, 0, 0],
        [0, 0, 2, 1, 2],
    ])
)
```

There are several caveats to pay attention to when working with Tensorflow `tf.SparseTensor`:

1. The `indices` of the `SparseTensor` must be of `dtype=tf.int64`
2. Some operations assume the indices are ordered ascendingly. For example, `[[1, 0], [0, 1], [0, 0], [2, 1]]` needs to be ordered as `[[0, 0], [0, 1], [1, 0], [2, 1]]` such that each index dimension is always monotonically increasing. If the indices are not ordered already, use `tf.sparse.reorder` to reorder the indices.
3. The `dense_shape` of the `SparseTensor` must be static and known beforehand. If it needs to be dynamically computed, use `tf.sparse.reset_shape` to specify the shape. This is to make sure that the `SparseTensor` can be correctly constructed in graph mode of Tensorflow runtime (for example during training or serving using TFServing).

7.3.2 Sparse CSR Matrix

The other sparse tensor type, `CSRSparseMatrix`, is not very well known and is part of the raw operations in Tensorflow. It can support only two-dimensional matrices but can operate under batch mode (i.e. representing a series of matrices using a shape of `(batch_size, n_rows, n_cols)`). It also only supports float and complex data types (i.e. `tf.float32`, `tf.float64`, `tf.complex64`, `tf.complex128`). The `csr` matrix operations can be imported using `from tensorflow.python.ops.linalg.sparse import sparse_csr_matrix_ops`. For the rest of the section, we will briefly demonstrate some common sparse matrix operations using `CSRSparseMatrix`.

Suppose we have the dense tensor like before

```
import tensorflow as tf
X = tf.constant([
    [0, 0, 0, 2, 0],
    [1, 0, 0, 0, 0],
    [0, 0, 1, 0, 0],
    [0, 0, 1, 0, 0],
    [0, 2, 0, 0, 0],
    [0, 0, 2, 1, 2],
], dtype=tf.float32)
```

To construct a matrix from this dense tensor [Abadi et al., 2023a]

```
from tensorflow.python.ops.linalg.sparse import sparse_csr_matrix_ops
csr_sparse = sparse_csr_matrix_ops.CSRSparseMatrix(
    X,
    indices=tf.where(X > 0),  # optional
)
```

or via the operation `dense_to_csr_sparse_matrix`

```
# needs the indices of the non-missing values
csr_sparse = sparse_csr_matrix_ops.dense_to_csr_sparse_matrix(
    X, indices=tf.where(X > 0)
)
```

To convert from `csr` to `coo`, use

```
# Convert from csr to coo
coo_sparse = sparse_csr_matrix_ops.csr_sparse_matrix_to_sparse_tensor(
    csr_sparse, type=tf.float32,
)
# Casting to Sparse Tensor class instead of intermediate
 ↪ representation
coo_sparse = tf.SparseTensor(
    coo_sparse.indices, coo_sparse.values, coo_sparse.dense_shape
)
```

To convert from `coo` back into `csr` matrix, use

```
# Convert from coo to csr
csr_from_coo = sparse_csr_matrix_ops.sparse_tensor_to_csr_sparse_matrix(
    coo_sparse.indices, coo_sparse.values, coo_sparse.dense_shape
)
```

This can also be done by directly constructing from the `CSRSparseMatrix` class [Abadi et al., 2023b],

```
csr_from_coo = sparse_csr_matrix_ops.CSRSparseMatrix(coo_sparse)
```

`tf.SparseTensor` can perform a multiplication between a sparse and a dense tensor using `tf.sparse.sparse_dense_matmul`. Like `tf.SparseTensor`, `CSRSparseTensor` also has an operation `sparse_matrix_mat_mul` [Abadi et al., 2023a] to multiply a dense vector with the sparse matrix:

```
b_mat = tf.constant([[0, 1, 2, 4, 1]], tf.float32)
w = sparse_csr_matrix_ops.sparse_matrix_mat_mul(
    csr_sparse,
    b_mat,
    transpose_b=True,
)
```

We can easily verify that the results are identical to the dense operation `X.numpy() @ b_mat.numpy().T`

```
array([[ 8.],
       [ 0.],
       [ 2.],
       [ 2.],
       [ 2.],
       [10.]])
```

An operation that `tf.SparseTensor` cannot perform is the sparse-to-sparse matrix multiplication. This can be done with the `csr` matrix, using `sparse_csr_matrix_ops.sparse_matrix_sparse_mat_mul` [Abadi et al., 2023c]. For example, let's compute XX^T.

```
res = sparse_csr_matrix_ops.sparse_matrix_sparse_mat_mul(
    a=csr_sparse, b=csr_sparse, type=tf.float32, transpose_b=True,
)
# converting back to dense
res = sparse_csr_matrix_ops.csr_sparse_matrix_to_dense(
    res, type=tf.float32
)
```

We can easily verify that the result is the same as using the dense matrix multiplication procedure `X.numpy() @ X.numpy().T`:

```
array([[4., 0., 0., 0., 0., 2.],
       [0., 1., 0., 0., 0., 0.],
       [0., 0., 1., 1., 0., 2.],
       [0., 0., 1., 1., 0., 2.],
       [0., 0., 0., 0., 4., 0.],
       [2., 0., 2., 2., 0., 9.]], dtype=float32)
```

7.4 PyTorch's Sparse Matrix

PyTorch currently supports `coo`, `csr`, `csc`, `bsr`, and `bsc` (`bsc` to `bsr` is as `csc` to `csr`, where `bsc` uses columns to represent block indices instead of rows) [Paszke et al., 2023]. The construction of these matrices is similar to Scipy's implementations, where they take indices/pointers, and data as arguments. Just like Tensorflow, only the `coo` format supports multi-dimensional Sparse Tensors, while the other four types (collectively named compressed sparse tensor, see Figure 7.11) allow sparse format that can be arranged into row-by-column

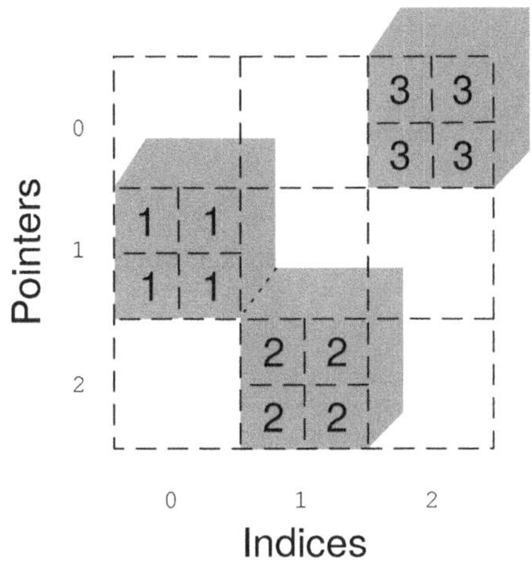

Figure 7.11 Generalized compressed sparse tensor.

sparse structures, with the addition of batch dimension to collectively represent a series of such sparse tensors. In general, these sparse tensors have shape `(batch_size, 2, m, n, ...)`, where `m, n, ....` are dense shapes of the "blocks" of the tensors scattered on the plane of the two sparse dimensions.

The pointers and indices are used to generically represent the two sparse index dimensions and, depending on which compressed tensor specification is used, they can carry different meanings.

In Torch's `coo` matrix construction, the indices are arranged as row vectors instead of column vectors as in Tensorflow's `tf.SparseTensor`. For example, in Tensorflow, we would construct a sparse matrix

```
import tensorflow as tf

X = tf.constant([
    [0, 0, 0, 2, 0],
    [1, 0, 0, 0, 0],
    [0, 0, 1, 0, 0],
    [0, 0, 1, 0, 0],
    [0, 2, 0, 0, 0],
    [0, 0, 2, 1, 2],
], dtype=tf.float32)
```

as

```
tf.SparseTensor(
    indices=tf.constant(
        [[0, 3], [0, 1], [1, 2], [2, 2], [3, 1], [4, 2], [4, 3], [4, 4]],
```

```
            tf.int64
        ),
        values=tf.constant([2, 1, 1, 1, 2, 2, 1, 2], tf.float32),
        dense_shape=(6, 5),
    )
```

In Torch's coo matrix, it would be

```
import torch
coo_sparse = torch.torch.sparse_coo_tensor(
    # instead of torch.sparse_coo_tensor!
    indices=[
        [0, 0, 1, 2, 3, 4, 4, 4], [3, 1, 2, 2, 1, 2, 3, 4],
    ],
    values=[2, 1, 1, 1, 2, 2, 1, 2],
    size=(6, 5),
    dtype=torch.float32,
)
# tensor(indices=tensor([[0, 0, 1, 2, 3, 4, 4, 4],
#                        [3, 1, 2, 2, 1, 2, 3, 4]]),
#        values=tensor([2., 1., 1., 1., 2., 2., 1., 2.]),
#        size=(6, 5), nnz=8, layout=torch.sparse_coo)
```

Interconversions between dense and sparse tensors are similar to those in Scipy, using methods such as `coo_sparse_tensor.to_dense()` to convert from sparse to dense tensors, and `dense_tensor.to_sparse()` to convert from dense to sparse (default to coo format). Conversion between each sparse tensor format is similar, using `to_sparse_{format}`, even directly from dense tensors. Table 7.3 details the comparisons of these five supported sparse matrices in PyTorch:

Table 7.3 PyTorch Sparse tensor comparisons.

Type	Construction	Interconversion	Sparse Dimensions
coo	`torch.sparse_coo_tensor(indices, values, size, nnz)`	`.to_sparse_coo()`	>= 2
csr	`torch.sparse_csr_tensor(crow_indices, col_indices, values)`	`.to_sparse_csr()`	2
csc	`torch.sparse_csc_tensor(ccol_indices, row_indices, values)`	`.to_sparse_csc()`	2
bsr	`torch.sparse_bsr_tensor(crow_indices, col_indices, values)`	`.to_sparse_bsr()`	2
bsc	`torch.sparse_bsc_tensor(ccol_indices, row_indices, values)`	`.to_sparse_bsc()`	2

7.5 Sparse Matrix in Other Python Libraries

In addition to the sparse matrix implementations in Scipy, Tensorflow, and PyTorch, other Python libraries also have sparse data structure implemented, including but not limited to:

- `pandas`: Implemented as types `SparseArray` and `SparseDataFrame`, similar to `coo`, and interfaces with `scipy.sparse.coo_matrix` very well.
  - conversion from pandas to `scipy.sparse.coo_matrix` with `df.sparse.to_coo()`
  - conversion to pandas from `scipy.sparse.coo_matrix` with `pd.Series.sparse.from_coo()`
- `pyspark`: `pyspark.mllib.linalg.SparseMatrix`, default `csc` format. But when the matrix is transposed by setting the argument `isTransposed=True`, this sparse matrix can be viewed as `csr` format.
- `pyarrow`: `pyarrow` supports `coo` (`pyarrow.SparseCOOTensor`), `csr` (`pyarrow.SparseCSRTensor`), `csc` (`pyarrow.SparseCSCTensor`), and `csf` (`pyarrow.SparseCSFTensor`, or Compressed Sparse Fiber [Nisa et al., 2019]) format. `pyarrow` is an efficient library for big data operations and integrates tightly with `numpy` and `pandas`.

7.6 When (Not) to Use Sparse Matrix

Data sparsity is the key to deciding if we should be using a sparse matrix or a dense matrix to represent our data. In general, the space complexity of a multidimensional sparse tensor is $O((d + 1)N)$, where N is the size of the data and d is the number of dimensions of the tensor. In the case of a two-dimensional matrix, we need to store three times the information about the data. We need $2N$ of spaces to store the row and column indices, and another space of size N to store the data itself. As a rule of thumb, when the data sparsity is <33% (i.e., only less than one-third of the matrix is filled with non-missing values), then it is probably more efficient to represent the data using a sparse matrix rather than a dense tensor.

7.7 Case Study: Sparse Matrix Factorization with ALS

In this case study, we will implement the classic recommender system algorithm matrix factorization. The matrix factorization problem is solved numerically by using the method called **alternating least squares (ALS)** [Bell and Koren, 2009].

7.7.1 Matrix Factorization

Let the user-item rating matrix R be of shape $N \times M$, where there are N users and M items. R can be approximated by $\hat{R}$, which is a product of two smaller matrices U ($N \times K$) and W ($M \times K$), i.e.

$$R \approx \hat{R} = UW^T$$

where U is usually called the user matrix (encoding N users), whereas W is the item matrix (encoding M items). K is the number of latent features, which is usually between 10 and 50. *Note that our data is arranged as row vectors rather than column vectors that are more commonly used in mathematical derivations.*

At prediction time, suppose we want to find the rating for ith user and jth item, then we only need to calculate

$$\hat{r}_{ij} = \vec{u}_i \cdot \vec{w}_j^T$$

where $\vec{u}_i, \vec{w}_j$ are length K vectors of user i and item j, respectively.

Interpretation

Suppose we have $K = 5$ explicit features, where

- $K = 1$ Action/adventure
- $K = 2$ Comedy
- $K = 3$ Romance
- $K = 4$ Horror
- $K = 5$ Animation

then,

- $u_{i,1}$ means how much user i likes Action/adventure
- $u_{i,2}$ means how much user i likes Comedy
- $w_{j,1}$ means how much movie j contains Action/adventure
- $w_{j,2}$ means how much movie j contains Comedy

Then, the dot product

$$\hat{r}_{ij} = \vec{u}_i \cdot \vec{w}_j^T = \|\vec{u}_i\| \ \|\vec{w}_j\| \cos\theta \propto \text{Similarity}(i, j)$$

means how much user i is correlated with movie j.

However, in practice, the features are usually **latent**.

7.7.2 Parameter Updates with ALS

The loss function for this regression problem is the sum-squared error

$$J = \sum_{i,j \in \Omega} (r_{ij} - \hat{r}_{ij})^2 = \sum_{i,j \in \Omega} (r_{ij} - \vec{u}_i \vec{w}_j^T)^2$$

We want to minimize J. By setting the gradients as zero, we have

$$\frac{\partial J}{\partial u_i} = 2\sum_{j\in\Psi_i}(r_{ij} - \vec{u}_i\vec{w}_j^T)(-\vec{w}_j) = 0$$

$$\sum_{j\in\Psi_i} r_{ij}(-\vec{w}_j) - \sum_{j\in\Psi_i}(\vec{u}_i\vec{w}_j^T)(-\vec{w}_j) = 0$$

$$\sum_{j\in\Psi_i}(\vec{u}_i\vec{w}_j^T)\vec{w}_j = \sum_{j\in\Psi_i} r_{ij}\vec{w}_j$$

$$\sum_{j\in\Psi_i}\vec{u}_i(\vec{w}_j^T\vec{w}_j) = \sum_{j\in\Psi_i} r_{ij}\vec{w}_j$$

$$\vec{u}_i\left(\sum_{j\in\Psi_i}\vec{w}_j^T\vec{w}_j\right) = \sum_{j\in\Psi_i} r_{ij}\vec{w}_j$$

Notice that $\vec{w}_j^T\vec{w}_j$ is an outer product. Also, notice that the last step is in the form of $AX = b$ (if X is column vector, or $XA = b$ is X is a row vector).

We can invert the matrix $A = \sum_{j\in\Psi_i}\vec{w}_j^T\vec{w}_j$ part to solve for $\vec{u}_i$, yielding

$$\vec{u}_i = \sum_{j\in\Psi_i} r_{ij}\vec{w}_j\left(\sum_{j\in\Psi_i}\vec{w}_j^T\vec{w}_j\right)^{-1}$$

Numerically, we can take advantage of `numpy`'s `np.linalg.solve(A, b)` function to solve for $\vec{u}_i$.

To solve for w_j, we yield similar results

$$\vec{w}_j = \sum_{i\in\Omega_j} r_{ij}\vec{u}_i\left(\sum_{i\in\Omega_j}\vec{u}_i^T\vec{u}_i\right)^{-1}$$

We apply the isolated u_i and w_j updates iteratively to minimize the loss function J. This is called **alternating least squares (ALS)**. We first initialize u_i and w_j randomly, then update according to the equations we obtained.

7.7.3 Adding Bias Terms to Matrix Factorization

Recall that in our collaborative filtering, we have the similarity measure as

$$s(i,j) = \bar{r}_i + \frac{\sum_{i'\in\Omega_j} w_{ii'}(r_{i'j} - \bar{r}_{i'})}{\sum_{i'\in\Omega_j}\|w_{ii'}\|}$$

where we have the bias term $\bar{r}_i$. Thus, we revised the rating as

$$\hat{r}_{ij} = \vec{u}_i \cdot \vec{w}_j^T + b_i + c_j + \mu$$

where

- b_i is user bias
- c_j is item bias
- μ is the global average rating (average of the data, therefore, no need to update during training)

Then, the updates for $\vec{u}_i$ and $\vec{w}_j$ from before now become

$$\vec{u}_i := \sum_{j\in\Psi_i}(r_{ij} - b_i - c_j - \mu)w_j\left(\sum_{j\in\Psi_i}\vec{w}_j^T\vec{w}_j\right)^{-1}$$

$$\vec{w}_j := \sum_{i\in\Omega_j}(r_{ij} - b_i - c_j - \mu)\vec{u}_i\left(\sum_{i\in\Omega_j}\vec{u}_i^T\vec{u}_i\right)^{-1}$$

We also need to update b_i and c_j. Taking the partial derivative with respect to b_i, we have

$$\frac{\partial J}{\partial b_i} = 2\sum_{j\in\Psi_i}(r_{ij} - \vec{u}_i\cdot\vec{w}_j^T - b_i - c_j - \mu)(-1) = 0$$

$$\sum_{j\in\Psi_i}(r_{ij} - \vec{u}_i\cdot\vec{w}_j^T - c_j - \mu) = \sum_{j\in\Psi_i}b_i$$

$$\frac{1}{|\Psi_i|}\sum_{j\in\Psi_i}(r_{ij} - \vec{u}_i\cdot\vec{w}_j^T - c_j - \mu) = b_i$$

where $|\Psi_i|$ is the cardinality of the set, which represents the number of items that user i rated. It can be seen that the **bias b_i is the average deviation between target and modeling prediction, if the model did not involve the term b_i.**

Similarly, taking the partial derivative with respect to c_j, we have

$$\frac{\partial J}{\partial c_j} = 2\sum_{i\in\Omega_j}(r_{ij} - \vec{u}_i\cdot\vec{w}_j^T - b_i - c_j - \mu)(-1) = 0$$

$$\sum_{i\in\Omega_j}(r_{ij} - \vec{u}_i\cdot\vec{w}_j^T - b_i - \mu) = \sum_{i\in\Omega_j}c_j$$

$$\frac{1}{|\Omega_j|}\sum_{i\in\Omega_j}(r_{ij} - \vec{u}_i\cdot\vec{w}_j^T - b_i - \mu) = c_j$$

The update rules for b_i and c_j are

$$b_i := \frac{1}{|\Psi_i|}\sum_{j\in\Psi_i}(r_{ij} - \vec{u}_i\cdot\vec{w}_j^T - c_j - \mu)$$

$$c_j := \frac{1}{|\Omega_j|}\sum_{i\in\Omega_j}(r_{ij} - \vec{u}_i\cdot\vec{w}_j^T - b_i - \mu)$$

7.7.4 Adding Regularization Term

Starting with a simple linear regression, we have the model as

$$\hat{y} = Aw$$

where w is a column weight vector. The L2 regularized loss function is

$$J = \sum_{i}^{N} (y_i - \hat{y}_i)^2 + \lambda \|w\|_2^2$$

where $\|\cdot\|_2$ means the second norm. This is known as Ridge Regression [Wikipedia, 2024], and w has the known solution

$$w = (\lambda I + A^T A)^{-1} A^T y$$

For matrix factorization, we modify our loss function to be regularized as

$$J = \sum_{i,j \in \Omega} (r_{ij} - \hat{r}_{ij})^2 + \lambda(\|U\|_F^2 + \|W\|_F^2 + \|b\|_2^2 + \|c\|_2^2)$$

where $\|\cdot\|_F$ is the Frobenius Norm, which is the L2 norm for a matrix.

Now to solve for the updates for the four parameters, starting with u_i by taking the partial derivative, we have

$$\frac{\partial J}{\partial u_i} = 2 \sum_{j \in \Psi_i} (r_{ij} - \vec{u}_i \cdot \vec{w}_j^T - b_i - c_j - \mu)(-w_j) + 2\lambda \vec{u}_i = 0$$

$$\vec{u}_i \left(\sum_{j \in \Psi_i} \vec{w}_j^T \vec{w}_j + \lambda I \right) = \sum_{j \in \Psi_i} (r_{ij} - b_i - c_j - \mu) \vec{w}_j$$

$$\vec{u}_i = \sum_{j \in \Psi_i} (r_{ij} - b_i - c_j - \mu) \vec{w}_j \left(\sum_{j \in \Psi_i} \vec{w}_j^T \vec{w}_j + \lambda I \right)^{-1}$$

Similarly, for w_j, we have

$$\vec{w}_j = \sum_{i \in \Omega_j} (r_{ij} - b_i - c_j - \mu) \vec{u}_i \left(\sum_{i \in \Omega_j} \vec{u}_i^T \vec{u}_i + \lambda I \right)^{-1}$$

Taking the partial derivative with respect to b_i, we have

$$\frac{\partial J}{\partial b_i} = 2 \sum_{j \in \Psi_i} (r_{ij} - \vec{u}_i \cdot \vec{w}_j^T - b_i - c_j - \mu)(-1) + 2\lambda b_i = 0$$

$$\sum_{j \in \Psi_i} (r_{ij} - \vec{u}_i \cdot \vec{w}_j^T - c_j - \mu) = \sum_{j \in \Psi_i} b_i + \lambda b_i$$

$$\frac{1}{|\Psi_i| + \lambda} \sum_{j \in \Psi_i} (r_{ij} - \vec{u}_i \cdot \vec{w}_j^T - c_j - \mu) = b_i$$

Similarly, we can update c_j as

$$c_j = \frac{1}{|\Omega_j| + \lambda} \sum_{i \in \Omega_j} (r_{ij} - \vec{u}_i \cdot \vec{w}_j^T - b_i - \mu)$$

7.7.5 Implementing ALS

Let's first import all the libraries and define some helper functions for sparse matrix operations.

```
import numpy as np
import scipy.sparse
from scipy.sparse import csc_matrix, coo_matrix, dok_matrix
from scipy.sparse import load_npz, save_npz
from tqdm import tqdm

def multiply_U_W(U, W, R):
    """Compute product of U and W.T,
    but only at the position where R is available"""
    iU, iW = R.tocoo().row, R.tocoo().col # np.nonzero(R)
    # values = np.sum(U[iU, :] * W[iW, :], axis=1)
    values = np.einsum('ij,ij->i', U[iU, :], W[iW, :])
    # Recreate the coo matrix, making sure its shape
    # is the same as R
    result = coo_matrix(
        (values, (iU, iW)), shape=R.shape
    ).tocsc()
    return result


def R_plus_cv(R, cv):
    """Sparse matrix plus a column vector"""
    R = R.tocsc()
    R.data += np.take(cv, R.indices)
    return R


def R_plus_rv(R, rv):
    """Sparse matrix plus a row vector"""
    R = R.tocsr()
    R.data += np.take(rv, R.indices)
    return R
```

Let us unwrap the first function `multiply_U_W` furthermore and explain it using the context of matrix factorization that we discussed in this section. Suppose we have a rating matrix R with shape (N, M), where N is the number of users, and M is the number of items in the catalog. Using the matrix factorization algorithm, we want to factorize R into two smaller matrices U and W, which uses latent features to represent user and item characteristics. See Figure 7.12a for this setup. In particular, we have four users, and three items, with latent dimension size $K = 5$. Therefore, we have a sparse `coo` matrix of shape (4, 3), and there are total $L = 6$ ratings in this sparse matrix. If we multiply U and W together, we will need to perform dot products on all 12 pairs of user and item vectors. Therefore, the goal is to compute the dot product only on the six entries that have ratings. This is a 50% reduction in the number of dot product operations needed.

- Line 10: We first extract the row and column indices of the entries. Both `iU` and `iW` have a length L that is identical to the number of entries in R. We can treat `iU` as the user indices, and `iW` as the item indices.
- Lines 11 and 12: These lines are two variants of the same implementation. With `U[iU, :]`, we are extracting the vectors from the user matrix U for each of the ratings being made and forming a matrix of (L, K) (see Figure 7.12b). Similarly, `W[iW, :]` extracts vectors from the matrix W on all the items that are rated by any users. Notice that the order of extraction/slicing is the same as the order of iterating the entries in R row-wise: `R[0, 1] = 3`, `R[0, 2] = 4`, ..., `R[3, 2] = 2`. Then, the corresponding rows of the extracted user and item vectors are multiplied and summed together (i.e. same as the dot product), which results in the prediction of each rating. Another way to look at this is that we use a flat index of R to take entries from either U or W. The resulting matrix is supposedly a 3D tensor of (N, M, K), but we are flattening the first two dimensions, so now their shape becomes (L, K) after skipping the empty entries.
- Lines 15–17: This step can be seen as scattering the predictions back to the position of the original sparse rating matrix. The prediction from the above step is a vector of length L, and its ordering is the same as that of `R.data`. Therefore, we can create a new sparse `coo` matrix $\hat{R}$ with `row`, `col`, and `shape` identical to that of R (see Figure 7.12c).

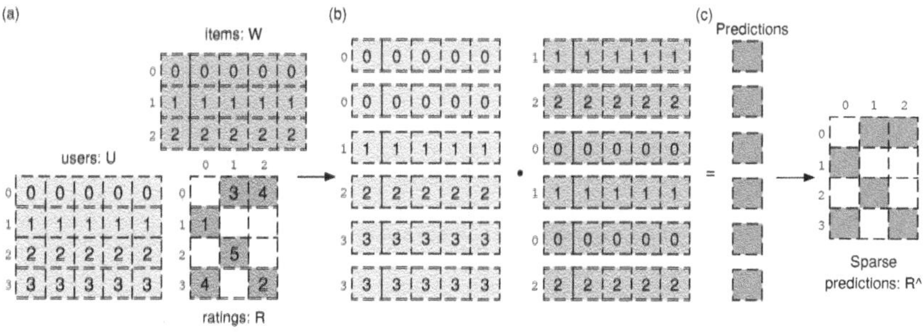

Figure 7.12 Illustration of matrix multiplication on sparse indices.

These three helper functions were introduced at the beginning of the chapter and will allow us to update our parameters more efficiently on the sparse matrix. To recap,

- `multiply_U_W` performs a sparse matrix multiplication between the two dense matrices U and W^T, but it will skip entries where R is missing. We therefore can use this function to skip the predictions of the missing values during training.
- `R_plus_cv` is a broadcasting operation on the sparse matrix R, where we add a constant value to each row of the sparse matrix given a column vector.
- `R_plus_rv` is a broadcasting operation on the sparse matrix R, where we add a constant value to each column of the sparse matrix given a row vector.

Recall that the final update rules for the four parameters, u_i, w_j, b_i, c_j, are the following:

$$\vec{u}_i := \sum_{j\in\Psi_i}(r_{ij} - b_i - c_j - \mu)\vec{w}_j\left(\sum_{j\in\Psi_i}\vec{w}_j^T\vec{w}_j + \lambda I\right)^{-1} \tag{7.1}$$

$$\vec{w}_j := \sum_{i\in\Omega_j}(r_{ij} - b_i - c_j - \mu)\vec{u}_i\left(\sum_{i\in\Omega_j}\vec{u}_i^T\vec{u}_i + \lambda I\right)^{-1} \tag{7.2}$$

$$b_i := \frac{1}{|\Psi_i| + \lambda}\sum_{j\in\Psi_i}(r_{ij} - \vec{u}_i \cdot \vec{w}_j^T - c_j - \mu) \tag{7.3}$$

$$c_j := \frac{1}{|\Omega_j| + \lambda}\sum_{i\in\Omega_j}(r_{ij} - \vec{u}_i \cdot \vec{w}_j^T - b_i - \mu) \tag{7.4}$$

Let's now implement matrix factorization with ALS in the following:

```
def compute_prediction(U_W_prod, mu, b, c):
    """Computing predictions use the learned parameters."""
    R_hat = U_W_prod.copy()   # csc
    R_hat.data += mu   # add mu
    R_hat = R_plus_cv(R_hat, b)   # add b, of length N
    R_hat = R_plus_rv(R_hat, c).tocsc()   # add c, of length M
    return R_hat


def compute_regularized_loss(R, R_hat, U, W, b, c, l2_reg, total_card):
    """Compute mean squared loss with regularization."""
    loss = ((R.data - R_hat.data) ** 2).sum() + l2_reg * (
        np.sum(U**2) + np.sum(W**2) + np.sum(b**2) + np.sum(c**2)
    )
    loss = loss / total_card  # mean squared error

    return loss
```

```


def regularized_matrix_factorization(
    R, R_eval=None, K=25, l2_reg=0.001, maxiter=100
):
    """
    Solving matrix factorization with alternating least square

    Parameters:
    --------
    * R : training rating sparse matrix of size N x M
    * R_eval : evaluation rating sparse matrix of size N x M
    * K : number of features to use
    * l2_reg: L2 regularization size
    * maxiter : max number of iterations
    Returns:
    --------
    U : N x K user matrix
    W : M x K item matrix
    b : length N vector of user bias
    c : length M vector of item bias
    mu: global mean of the user rating matrix
    J : training loss over epochs
    """
    R = R.tocsc()  # make sure it is csc format
    N, M = R.shape
    # Calculate global mean mu
    mu = R.mean()

    # Initialize the bias terms
    b = np.random.randn(N)
    c = np.random.randn(M)

    # Initialize the user and item matrix
    U = np.random.randn(N, K)
    W = np.random.randn(M, K)
    U_W_prod = multiply_U_W(U, W, R)  # sparse (N, M)
    if R_eval is not None:
        R_eval = R_eval.tocsc()
        U_W_prod_eval = multiply_U_W(U, W, R_eval)

    # Counting cardinalities
    card_psi = R.getnnz(axis=1)
    card_omega = R.getnnz(axis=0)
    total_card = R.getnnz()

    # Initialize loss
    J = np.zeros(maxiter)
    J_eval = np.zeros(maxiter)

    # Iterate
    for epoch in tqdm(range(maxiter)):
        # Compute training prediction
        R_hat = compute_prediction(U_W_prod, mu, b, c)
```

```
72          # Compute regularized loss
73          J[epoch] = compute_regularized_loss(
74              R, R_hat, U, W, b, c, l2_reg, total_card
75          )
76
77          if R_eval is not None:
78              # Compute evaluation prediction
79              R_hat_eval = compute_prediction(U_W_prod_eval, mu, b, c)
80              # Compute regularized loss
81              J_eval[epoch] = compute_regularized_loss(
82                  R_eval, R_hat_eval, U, W, b, c, l2_reg, total_card
83              )
84
85          # Compute updates for U and W
86          # r_ij - b_i - c_j - \mu
87          R_b_c_mu = R.copy()
88          R_b_c_mu.data -= mu
89          R_b_c_mu = R_plus_cv(R_b_c_mu, -b)
90          R_b_c_mu = R_plus_rv(R_b_c_mu, -c).tocsc()
91
92          U = np.linalg.solve(
93              W.T @ W + l2_reg * np.eye(K)[np.newaxis, :, :],
94              R_b_c_mu @ W
95          )  # eq1, (N, K) = solve((1, K, K), (N, k))
96          W = np.linalg.solve(
97              U.T @ U + l2_reg * np.eye(K)[np.newaxis, :, :],
98              R_b_c_mu.T @ U
99          )  # eq2, (M, K) = solve((1, K, K), (M, K))
100         U_W_prod = multiply_U_W(U, W, R)
101
102         # Compute updates for b and c
103         # r_ij - u_i w_j
104         R_u_mu = R - U_W_prod
105         # r_ij - u_i w_j - \mu
106         R_u_mu.data -= mu
107         # r_ij - u_i w_j - c - \mu
108         R_u_c_mu = R_u_mu.copy()
109         R_u_c_mu = R_plus_rv(R_u_c_mu, -c).tocsc()
110         R_u_c_mu = np.asarray(R_u_c_mu.sum(axis=1)).flatten()
111         # r_ij - u_i w_j - b - \mu
112         R_u_b_mu = R_u_mu.copy()
113         R_u_b_mu = R_plus_cv(R_u_b_mu, -b)
114         R_u_b_mu = np.asarray(R_u_b_mu.sum(axis=0)).flatten()
115
116         # Update b and c
117         b = 1 / (card_psi   + l2_reg) * R_u_c_mu  # eq3, (N, )
118         c = 1 / (card_omega + l2_reg) * R_u_b_mu  # eq4, (M, )
119
120     return U, W, b, c, mu, J, J_eval
```

- Lines 1–7 define the function to make sparse predictions, using the formula $\hat{r}_{ij} = \vec{u}_i \cdot \vec{w}_j + b_i + c_j + \mu$.

- Lines 10–17 define a function to compute the loss J using mean-squared error (averaged over all non-missing data points).
- Lines 42–53 are initializations of the learning parameters. Here, `R` is assumed to be a sparse matrix. We use the `csc` format for more efficient mathematical operations below.
- Lines 54–57 make a first round of predictions (without the biases) using matrix multiplication $\vec{u}_i \cdot \vec{w}_j$. We compute the product for both training and evaluation data (if it exists).
- Lines 59–61 are computing the normalizing constants $|\Psi|$ and $|\Omega|$ that are used to update b_i and c_j, respectively. The `.getnnz(axis)` function will count the number of nonzero (non-missing) values in the sparse matrix R.
- Line 62 is computing the number of data points, which is used to normalize the total loss.
- Lines 64–66 initialize the loss to be zeros right before training starts. We want to keep a record of all the losses computed for each epoch, for both train and test data splits, which we will plot later.
- Line 69, we start the training process.
- In Lines 70–83, we first call the functions we defined above to compute the predictions and losses for both training and evaluation data.
- Lines 85–90 compute update terms for U and W, which are part of Equation 7.1 and Equation 7.2, respectively.
- Lines 92–99 use `np.linalg.solve` to solve our $AX = b$ linear regression problem that we posed earlier.
- Line 100 updates the matrix multiplication term $\vec{u}_i \cdot \vec{w}_j$.
- Lines 102–114 compute terms for updating b and c. Notice that we are using the broadcasting functions we wrote earlier to perform sparse-to-sparse or sparse-to-dense vector operations.
- Lines 116–118 make the update on b and c according to Equation 7.3 and Equation 7.4, respectively.
- In Line 120, after `maxiter` iterations of training, the results are finally returned.

7.7.6 Training a Model with MovieLens-100k

We can test this implementation using the MovieLens-100k dataset [Harper and Konstan, 2016], which can be downloaded from https://grouplens.org/datasets/movielens/100k/.

```
# Make data
import numpy as np
import pandas as pd
import scipy.sparse
```

```
from scipy.sparse import csc_matrix, load_npz, save_npz
from sklearn.model_selection import train_test_split
# Split train-test set
df = pd.read_csv(
    "./ml-100k/u.data",
    names=["userId", "movieId", "rating", "timestamp"],
    usecols=["userId", "movieId", "rating"],
    header=None, sep="\t",
)
df_train, df_test = train_test_split(
    df, test_size=0.2, random_state=42
)
#df = None # remove the original dataframe

# Convert to sparse matrix
R_train = csc_matrix(
    (df_train["rating"], (df_train["userId"], df_train["movieId"])),
    shape=(944, 1683),
    dtype=np.float16
)
R_test = csc_matrix(
    (df_test["rating"], (df_test["userId"], df_test["movieId"])),
    shape=(944, 1683),
    dtype=np.float16
)

# Save the data
scipy.sparse.save_npz(
    "./ml-100k/R_train_top.npz", R_train
)
scipy.sparse.save_npz(
    "./ml-100k/R_test_top.npz", R_test
)
```

Let's now train the model

```
# train the model
U, W, b, c, mu, J, J_eval = regularized_matrix_factorization(
    R_train, R_test, K=25, l2_reg=0.1, maxiter=30
)

# save results
np.savez(
    './ml-100k/l2_regularized_matrix_factorization.npz',
    U=U, W=W, b=b, c=c, mu=mu, J=J, J_eval=J_eval,
)
```

We can plot out the training loss function over time (as illustrated in Figure 7.13).

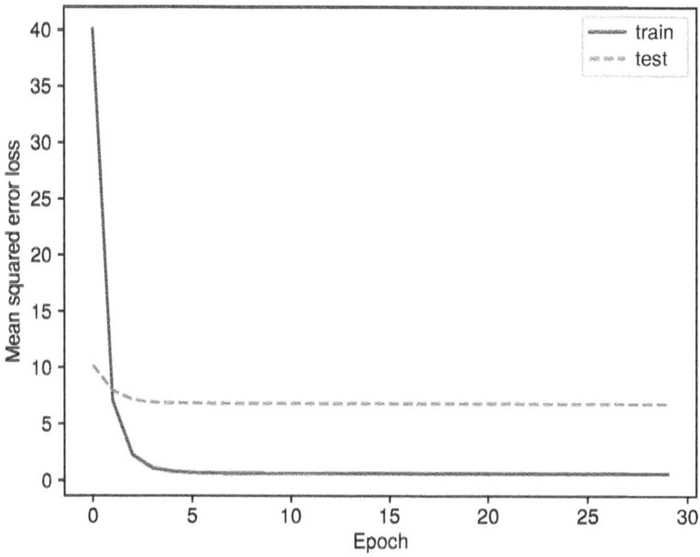

Figure 7.13 ALS training loss.

```
import matplotlib.pyplot as plt
plt.plot(J, label="train")
plt.plot(J_eval, label="test")
plt.legend()
plt.gca().set_xlabel('Epoch')
plt.gca().set_ylabel('Mean Squared Error Loss')
```

Did you know?

ALS can be used for a general task called ***sparse matrix completion*** and is one of the techniques to fill in missing pixels in images. In fact, I was only convinced that my implementation of ALS was correct when I tested the above training function on an image where I randomly blacked out some of the pixels. Feel free to try this out yourself.

Bibliography

Martín Abadi, Ashish Agarwal, Paul Barham, Eugene Brevdo, Zhifeng Chen, Craig Citro, Greg S. Corrado, Andy Davis, Jeffrey Dean, Matthieu Devin, Sanjay Ghemawat, Ian Goodfellow, Andrew Harp, Geoffrey Irving, Michael Isard,

Yangqing Jia, Rafal Jozefowicz, Lukasz Kaiser, Manjunath Kudlur, Josh Levenberg, Dan Mane, Rajat Monga, Sherry Moore, Derek Murray, Chris Olah, Mike Schuster, Jonathon Shlens, Benoit Steiner, Ilya Sutskever, Kunal Talwar, Paul Tucker, Vincent Vanhoucke, Vijay Vasudevan, Fernanda Viegas, Oriol Vinyals, Pete Warden, Martin Wattenberg, Martin Wicke, Yuan Yu, and Xiaoqiang Zheng. Construction of CSRSparseMatrix from a dense tensor (csrsparsematrixtest.py). https://github.com/tensorflow/tensorflow/blob/e65d97b3e338666baea2b2ec2b2d4e38ba8b78f9/tensorflow/python/kernel_tests/linalg/sparse/csr_sparse_matrix_test.py#L65C2-L65C2, 2023a.

Martín Abadi, Ashish Agarwal, Paul Barham, Eugene Brevdo, Zhifeng Chen, Craig Citro, Greg S. Corrado, Andy Davis, Jeffrey Dean, Matthieu Devin, Sanjay Ghemawat, Ian Goodfellow, Andrew Harp, Geoffrey Irving, Michael Isard, Yangqing Jia, Rafal Jozefowicz, Lukasz Kaiser, Manjunath Kudlur, Josh Levenberg, Dan Mane, Rajat Monga, Sherry Moore, Derek Murray, Chris Olah, Mike Schuster, Jonathon Shlens, Benoit Steiner, Ilya Sutskever, Kunal Talwar, Paul Tucker, Vincent Vanhoucke, Vijay Vasudevan, Fernanda Viegas, Oriol Vinyals, Pete Warden, Martin Wattenberg, Martin Wicke, Yuan Yu, and Xiaoqiang Zheng. Construction of CSRSparseMatrix from tf.SparseTensor (csrsparsematrixtest.py). https://github.com/tensorflow/tensorflow/blob/e65d97b3e338666baea2b2ec2b2d4e38ba8b78f9/tensorflow/python/kernel_tests/linalg/sparse/csr_sparse_matrix_test.py#L46, 2023b.

Martín Abadi, Ashish Agarwal, Paul Barham, Eugene Brevdo, Zhifeng Chen, Craig Citro, Greg S. Corrado, Andy Davis, Jeffrey Dean, Matthieu Devin, Sanjay Ghemawat, Ian Goodfellow, Andrew Harp, Geoffrey Irving, Michael Isard, Yangqing Jia, Rafal Jozefowicz, Lukasz Kaiser, Manjunath Kudlur, Josh Levenberg, Dan Mane, Rajat Monga, Sherry Moore, Derek Murray, Chris Olah, Mike Schuster, Jonathon Shlens, Benoit Steiner, Ilya Sutskever, Kunal Talwar, Paul Tucker, Vincent Vanhoucke, Vijay Vasudevan, Fernanda Viegas, Oriol Vinyals, Pete Warden, Martin Wattenberg, Martin Wicke, Yuan Yu, and Xiaoqiang Zheng. CSR Sparse to sparse matrix multiplication. https://github.com/tensorflow/tensorflow/blob/e65d97b3e338666baea2b2ec2b2d4e38ba8b78f9/tensorflow/python/kernel_tests/linalg/sparse/csr_sparse_matrix_ops_test.py#L804-L812, 2023c.

Robert M. Bell and Yehuda Koren. Scalable collaborative filtering with jointly derived neighborhood interpolation weights. *Seventh IEEE International Conference on Data Mining*, February 2009.

F. Maxwell Harper and Joseph A. Konstan. The MovieLens datasets: History and context. *ACM Transactions on Interactive Intelligent Systems*, 5(4):1–19, January 2016. ISSN 2160-6455, 2160-6463. doi: 10.1145/2827872.

Israt Nisa, Jiajia Li, Aravind Sukumaran-Rajam, Prasant Singh Rawat, Sriram Krishnamoorthy, and P. Sadayappan. An efficient mixed-mode representation of sparse tensors. In *Proceedings of the International Conference for High Performance*

Computing, Networking, Storage and Analysis, pages 1–25, Denver, CO, November 2019. ACM. ISBN 978-1-4503-6229-0. doi: 10.1145/3295500.3356216.

Adam Paszke, Sam Gross, Francisco Massa, Adam Lerer, James Bradbury, Gregory Chanan, Trevor Killeen, Zeming Lin, Natalia Gimelshein, Luca Antiga, Alban Desmaison, Andreas Köpf, Edward Yang, Zach DeVito, Martin Raison, Alykhan Tejani, Sasank Chilamkurthy, Benoit Steiner, Lu Fang, Junjie Bai, and Soumith Chintala. PyTorch Sparse Tensors. https://pytorch.org/docs/stable/sparse.html, 2023.

Pauli Virtanen, Ralf Gommers, Travis E. Oliphant, Matt Haberland, Tyler Reddy, David Cournapeau, Evgeni Burovski, Pearu Peterson, Warren Weckesser, Jonathan Bright, Stéfan J. van der Walt, Matthew Brett, Joshua Wilson, K. Jarrod Millman, Nikolay Mayorov, Andrew R. J. Nelson, Eric Jones, Robert Kern, Eric Larson, C. J. Carey, İlhan Polat, Yu Feng, Eric W. Moore, Jake VanderPlas, Denis Laxalde, Josef Perktold, Robert Cimrman, Ian Henriksen, E. A. Quintero, Charles R. Harris, Anne M. Archibald, Antnio H. Ribeiro, Fabian Pedregosa, Paul van Mulbregt, and SciPy 1.0 Contributors. SciPy 1.0: Fundamental algorithms for scientific computing in python. *Nature Methods*, 17:261–272, 2020. doi: 10.1038/s41592-019-0686-2.

Pauli Virtanen, Ralf Gommers, Travis E. Oliphant, Matt Haberland, Tyler Reddy, David Cournapeau, Evgeni Burovski, Pearu Peterson, Warren Weckesser, Jonathan Bright, Stéfan J. van der Walt, Matthew Brett, Joshua Wilson, K. Jarrod Millman, Nikolay Mayorov, Andrew R. J. Nelson, Eric Jones, Robert Kern, Eric Larson, C. J. Carey, İlhan Polat, Yu Feng, Eric W. Moore, Jake VanderPlas, Denis Laxalde, Josef Perktold, Robert Cimrman, Ian Henriksen, E. A. Quintero, Charles R. Harris, Anne M. Archibald, Antnio H. Ribeiro, Fabian Pedregosa, Paul van Mulbregt, and SciPy 1.0 Contributors. Sparse Matrices (scipy.sparse). https://docs.scipy.org/doc/scipy/reference/sparse.html, 2023.

Wikipedia. Ridge regression. https://en.wikipedia.org/wiki/Ridge_regression, May 2024.

8

Jagged Tensors

In various machine learning applications, we may sometimes need to deal with a batch of sequences that vary in length. For example, in natural language processing, sentences, paragraphs, or documents can contain a variable number of tokens. In audio processing, sample clips may come in variable durations. In recommendation systems, users' sequence of catalog interaction events may also vary in size. As we have already seen in Chapter 5, padding with dummy values within a dense tensor is a good way to represent sequences of variable lengths. These tensors still have a typical shape, where each entry along a certain dimension has exactly the same number of elements. For example, a matrix of shape `(4, 5)` has four rows, and each row has uniformly five elements. If we eliminate the padding entries toward the end of the sequence, and then examine the shape of the tensor, we can no longer confidently say that each entry along the dimension has the same number of elements. This tensor without the padding entries has therefore become a so-called **Jagged Tensor** or a **Ragged Tensor**, whose non-uniform dimension is the jagged dimension, and its shape is usually denoted by `None`. For example, a jagged matrix with four rows of data has the shape `(4, None)`.

A jagged tensor differs from a sparse tensor in that the paddings or empty entries of a jagged tensor always occur at the end of the sequence along the jagged dimension, whereas a sparse tensor can have empty entries occurring anywhere within the sequence while having a concrete shape (rather than using `None` to represent the jagged dimension). It can be seen that a sparse tensor is a more general representation of a tensor with padding/filler values. Therefore, we can also use a sparse tensor to represent a jagged tensor. See Figure 8.1 for an illustration of the three ways of representing jagged tensors.

Vectorization: A Practical Guide to Efficient Implementations of Machine Learning Algorithms,
First Edition. Edward DongBo Cui.

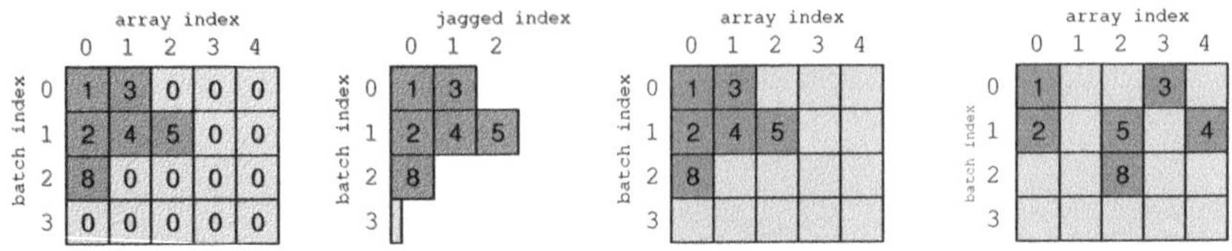

(a) (b) (c) (d)

Figure 8.1 Comparison of three ways of representing a jagged tensor. (a) A dense tensor with padding. Here, we use 0 as padding values. All padding values are to the right of the valid values. The dense tensor has a fixed shape. (b) A jagged tensor where the second dimension is the jagged dimension whose shape is unknown. The batch dimension has a known size. Notice that at row index 3, there are no elements in this row. It does not mean the batch size is unknown. But rather the row with no values in the list is represented as an empty list (this may be considered a type of padding). (c) Use a left-aligned sparse tensor to represent the jagged tensor. Notice that all valid entries are next to each other along the row. A sparse representation of a jagged tensor still needs to specify the size of the presumed jagged dimension, which is usually the expected maximum sequence length. (d) This is not a jagged tensor representation but rather a generic sparse tensor, since the values are not left-aligned.

Native Python simply represents a jagged tensor as a list of variable-length lists, i.e.

```
jagged_array = [[1, 3], [2, 4, 5], [8], []]
```

In contrast, the main NumPy library does not have a nice representation of a jagged tensor; it treats each list within the list as an object, i.e.

```
import numpy as np
jagged_array_np = np.array(
    [[1, 3], [2, 4, 5], [8], []], dtype=object
)
# array([list([1, 3]), list([2, 4, 5]), list([8]), list([])],
↪  dtype=object)
```

But, many vectorization techniques can be applied to an array of objects like such. NumPy/Scipy has an extension library `awkward` [Pivarski et al., 2018] that can handle jagged arrays and tensors, but to the best of my knowledge, this library is not very well adopted.

The main Tensorflow library implements the `tf.ragged` module and the `tf.RaggedTensor` class [Abadi et al., 2023] to represent jagged tensors and allows various vectorizing operations. As we have seen previously, and will see again in this chapter, `tf.ragged.range` will be a very useful function to implement various sequence and set manipulation operations in Tensorflow.

```
import tensorflow as tf
jagged_array_tf = tf.ragged.constant(
    [[1, 3], [2, 4, 5], [8], []]
)
# <tf.RaggedTensor [[1, 3], [2, 4, 5], [8], []]>
```

PyTorch implements the `torch.nested` module and the `torch.nested.nested_tensor` class to represent jagged tensors. Like Tensorflow's implementation, it eliminates the need to represent a variable-length sequence as a dense tensor with paddings, while allowing efficient operations using vectorization.

```
import torch
jagged_array_torch = torch.nested.nested_tensor(
    [[1, 3], [2, 4, 5], [8], []]
)
# nested_tensor([
#   tensor([1, 3]),
#   tensor([2, 4, 5]),
#   tensor([8]),
#   tensor([], dtype=torch.int64)
# ])
```

At the time of writing this chapter, the `torch.nested` module is still in beta (`torch==2.1.0`). So when calling this module, we will see the following warning message:

```
UserWarning: The PyTorch API of nested tensors is in the
↪  prototype stage and will change in the near future.
↪  (Triggered internally at
↪  /Users/runner/work/pytorch/pytorch/pytorch/aten/src/ATen/
↪  NestedTensorImpl.cpp:180.)
```

8.1 Left Align a Sparse Tensor to Represent Ragged Tensor

As mentioned before, other than padded dense tensors and special data structures such as `tf.RaggedTensor`, a jagged tensor can be represented using a sparse tensor whose entries are left aligned. For instance, in Tensorflow, we can convert a `RaggedTensor` to a `SparseTensor`

```
import tensorflow
# RaggedTensor
X = tf.ragged.constant([
    [1, 3, 2], [4, 6], [1], [], [9, 3, 5, 8], []
])
# <tf.RaggedTensor [[1, 3, 2], [4, 6], [1], [], [9, 3, 5, 8], []]>

# Convert to SparseTensor
y = X.to_sparse()

# SparseTensor(indices=tf.Tensor(
# [[0 0]
#  [0 1]
#  [0 2]
#  [1 0]
#  [1 1]
#  [2 0]
#  [4 0]
#  [4 1]
#  [4 2]
#  [4 3]], shape=(10, 2), dtype=int64),
# values=tf.Tensor([1 3 2 4 6 1 9 3 5 8],
# shape=(10,), dtype=int32),
# dense_shape=tf.Tensor([6 4], shape=(2,), dtype=int64))

# Display sparse tensor as dense
tf.sparse.to_dense(y)

# <tf.Tensor: shape=(6, 4), dtype=int32, numpy=
# array([[1, 3, 2, 0],
#        [4, 6, 0, 0],
#        [1, 0, 0, 0],
#        [0, 0, 0, 0],
#        [9, 3, 5, 8],
#        [0, 0, 0, 0]], dtype=int32)>
```

We can make several observations of the sparse tensor that we created:

1. `indices`: The first column of the sparse tensor is the batch dimension index or the row index of each variable-length list. Notice that row index `3` is missing while there is no row index for the last row because the 4th and 6th rows of the original input are both empty. Within each row, the array/column index (i.e. the second column of `indices`) forms `0, 1, 2,`

. . . consecutive integers. Recall that `tf.SparseTensor` uses the coordinate format, corresponding to `coo_matrix` in Scipy's implementation.

2. `values`: This property stores the flattened values of the original input. Its order corresponds to the order of the `indices` along the row dimension (`axis=0`). That is, the number of rows in `indices` tells us how many valid entries are in the sparse matrix. By looking at the `values` only, there is no way we can tell the fourth and sixth rows of the original input are empty. In other words, there is no "padding" value to represent the fourth or sixth row. Empty rows can be a challenge in data preprocessing sometimes. We will look at this issue more carefully in Chapter 9 when discussing segment-wise group-by-aggregate operations.
3. `dense_shape`: The shape of the sparse matrix indicates there should be six rows in the matrix, but based on `indices`, there are only four rows, since the fourth and sixth rows of the matrix are empty. Therefore, it is necessary to look at the `dense_shape` when determining how many rows there are in the matrix, instead of either counting the distinct number of values from `indices[:, 0]` or using the maximum of `indices[:, 1]`.

Since sparse tensor classes are more developed than jagged tensor classes in NumPy/Scipy, Tensorflow, and PyTorch libraries, I commonly use sparse matrices to represent a batch of variable-length lists, which allow me to manipulate the indices of the values directly, while taking advantage of existing implementations of operations on this data structure. Additionally, as one of the early adopters of Tensorflow Extended (TFX) library for data science machine learning operations (MLOps) that supported `tf.Tensor` and `tf.SparseTensor` early on, I had to fully master Tensorflow's `tf.SparseTensor` class to solve various challenges when building recommendation systems. In one of the models I built in the past, I had to preprocess a sequence feature to filter out certain elements based on its values. For instance, given the following data batch,

```
X = [[1, 5], [2, 4, 11], [3, 15, 8], [12, 2, 6, 18, 9], [7, 8, 1]]
```

I would like to create a new set of sequences that filters out any values greater than 10, i.e.

```
X_filtered = [[1, 5], [2, 4], [3, 8], [2, 6, 9], [7, 8, 1]]
```

Tensorflow's `tf.sparse.retain` is a ready-to-use operation that can help filter the data based on values. Therefore, we can do

```
import tensorflow
X = [[1,5], [2,4,11], [3,15,8], [14], [12,2,6,18,9], [7,8,1], [19]]
# Use ragged to sparse to convert to sparse tensor in this example
X = tf.ragged.constant(X).to_sparse()
# Keep values less than or equal to 10
X_filtered = tf.sparse.retain(
    X, X.values <= 10
)

# SparseTensor(indices=tf.Tensor(
# [[0 0]
#  [0 1]
#  [1 0]
#  [1 1]
#  [2 0]
#  [2 2]
#  [4 1]
#  [4 2]
#  [4 4]
#  [5 0]
#  [5 1]
#  [5 2]], shape=(12, 2), dtype=int64),
# values=tf.Tensor([1 5 2 4 3 8 2 6 9 7 8 1],
# shape=(12,), dtype=int32),
# dense_shape=tf.Tensor([7 5], shape=(2,), dtype=int64))
```

If we display the filtered tensor as a dense tensor, we can see that there are now gaps (empty entries) within the matrix, which no longer represents a jagged tensor.

```
tf.sparse.to_dense(X_filtered)
# <tf.Tensor: shape=(7, 5), dtype=int32, numpy=
# array([[1, 5, 0, 0, 0],
#        [2, 4, 0, 0, 0],
#        [3, 0, 8, 0, 0],
#        [0, 0, 0, 0, 0],
#        [0, 2, 6, 0, 9],
#        [7, 8, 1, 0, 0],
#        [0, 0, 0, 0, 0]], dtype=int32)>
```

Looking at `indices` of `X_filtered`, in row `2` (`indices[:, 0] == 2`), the array indices are no longer consecutive integers but skipped index `1`, making the two values on this row having indices `[2, 0]` and `[2, 2]`, skipping `[2, 1]`. This corresponds to the `0` filled gap that we see on the third row of the dense representation `[3, 0, 8, ...]`.

Fortunately, Tensorflow Transform, a supporting library for TFX, implements the sparse tensor transformation function `tft.sparse_tensor_left_align`

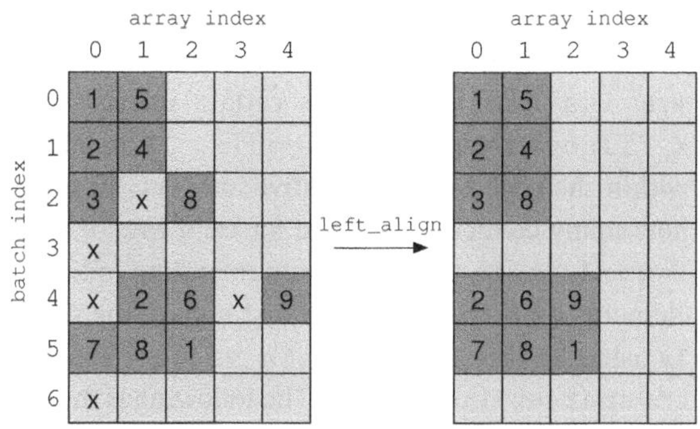

Figure 8.2 Before (left) and after (right) of left aligning the sparse matrix to represent a jagged tensor. The entries marked with x indicate that there used to be a value before filtering.

[Baylor et al., 2023] which removes the gaps between the values and ensures the sparse tensor represents a jagged tensor (see also Figure 8.2 for an illustration).

```
import tensorflow_transform as tft
X_left_aligned = tft.sparse_tensor_left_align(X_filtered)

# SparseTensor(indices=tf.Tensor(
# [[0 0]
#   [0 1]
#   [1 0]
#   [1 1]
#   [2 0]
#   [2 1]
#   [4 0]
#   [4 1]
#   [4 2]
#   [5 0]
#   [5 1]
#   [5 2]], shape=(12, 2), dtype=int64),
# values=tf.Tensor([1 5 2 4 3 8 2 6 9 7 8 1],
# shape=(12,), dtype=int32),
# dense_shape=tf.Tensor([7 5], shape=(2,), dtype=int64))
```

Warning

At the time of writing this chapter, the latest version of `tfx` with `tensorflow_transform` cannot be installed on my MacBook Pro 2021 M1 machine. I am expecting that most readers of the book may still have the same problem running the above code on an Apple Silicon machine. This may be the only piece of code in the entire book that cannot be run easily by the reader. Understanding the logic behind this convenient function allows us to use the algorithm implemented by this function in other libraries and environments.

We can observe that the only difference between `X_filtered` and `X_left_aligned` is on the second column of the `indices` property. To emphasize, `indices[:, 1]` skips certain integer values within each row, while `X_filtered` recreated `indices[:, 1]` so that it enumerates the array values within the row using consecutive integers. Recall in Chapter 5 that if we know how many elements there are for each row of the array, then we can use `tf.ragged.range` to create the enumerating indices. To determine the number of elements for each row, we can first replace the values of the original matrix with `1`s using `SparseTensor.with_values`, and then sum across the row using `tf.sparse.reduce:sum`. The following is the implementation that left-aligns a sparse tensor:

```
1   import tensorflow as tf
2
3   def sparse_left_align(X: tf.SparseTensor):
4       # Compute row lengths
5       row_lengths = X.with_values(tf.ones_like(X.values))
6       row_lengths = tf.sparse.reduce_sum(row_lengths, axis=1)
7       row_lengths = tf.cast(row_lengths, tf.int64)
8       # Create the flattened array enumeration index
9       array_indices = tf.ragged.range(row_lengths, dtype=tf.int64)
10      array_indices = array_indices.flat_values
11      # Recreate the SparseTensor
12      indices = tf.stack([X.indices[:, 0], array_indices])
13      indices = tf.transpose(indices)
14      y = tf.SparseTensor(
15          indices=indices,
16          values=X.values,
17          dense_shape=X.dense_shape,
18      )
19
20      return y
```

- Lines 5–7 compute the lengths of each row. An alternative or perhaps more efficient way to compute the array length of each row is by using `tf.unique_with_counts` (which is what `tft.sparse_tensor_left_align` used in the original implementation), which will be introduced in Chapter 10.
- Lines 9 and 10 create the flattened enumeration of the arrays, using consecutive integers starting from `0`.
- Lines 12–18 recreate the left-aligned sparse tensor. It replaces the second column of `indices` with the new array enumeration indices.

Recall that `tf.SparseTensor` corresponds to Scipy's `coo_matrix` format. In the later Case Study: Batchwise Set Operations, we will see another implementation of an algorithm that left-aligns a sparse tensor using Scipy's `csr_matrix`.

8.2 Index to Binary Indicator

Tensorflow has the routine `tf.sparse.to_indicator` which can convert a sparse representation of ragged lists of indices to dense binary indicators made of `True` and `False`. For instance,

```
import tensorflow as tf

X = tf.ragged.constant([
    [1, 3, 2],
    [2],
    [4, 1, 0]
]).to_sparse()

tf.sparse.to_indicator(X, vocab_size=5)

# <tf.Tensor: shape=(3, 5), dtype=bool, numpy=
# array([[False,  True,  True,  True, False],
#        [False, False,  True, False, False],
#        [ True,  True, False, False,  True]])>
```

However, if the length for each ragged row is small when compared to `vocab_size`, this dense representation can be quite memory-intensive. Alternatively, we can use a sparse tensor to indicate only the `True` (dense) entries and unfilled sparse entries to represent `False`. The following code block is an implementation of converting a set of ragged lists of indices to binary indicators using a sparse representation in Tensorflow:

```
import tensorflow as tf

def sparse_to_indicator_tf(
    X: tf.SparseTensor,
    vocab_size: int,
    out_dtype: tf.dtypes.DType = tf.float32
) -> tf.SparseTensor:
    """
```

```
9       Convert sparse tensor of indices to vectors of binary indicators
10
11      * X: tf.SparseTensor
12          Each row of the sparse matrix is the indices of the indicator
    ↪   vector
13      * vocab_size: int
14          This determines the size of the sparse indicator vector
15      * out_dtype: tf.dtypes.DType
16          DType of the values of the output tensor
17      """
18      # create indices of the output sparse tensor
19      indices = tf.stack([X.indices[:, 0], tf.cast(X.values, tf.int64)])
20      indices = tf.transpose(indices)
21      # indicator as ones
22      values = tf.ones_like(X.values)
23      values = tf.cast(values, out_dtype)
24      # (batch_size, vocab_size)
25      dense_shape = (tf.shape(X)[0], vocab_size)
26
27      # Make the sparse tensor
28      y = tf.SparseTensor(
29          indices=indices,
30          values=values,
31          dense_shape=dense_shape
32      )
33      y = tf.sparse.reorder(y)
34
35      return y
```

- Lines 19 and 20 create the indices (row and column) of the sparse matrix where the `True` (dense) values will occupy.
- Lines 22 and 23 create the dummy values of `1s` used to occupy the `True`/dense entries in the sparse matrix. Here, we can dynamically change the dtype of the output sparse matrix, which we currently default to `tf.float32`.
- Line 25 determines the shape of the sparse matrix. The `batch_size` or number of rows can be determined from the original input matrix. The number of columns is specified by `vocab_size`.
- Lines 28–32 create the binary indicator sparse matrix.
- Line 33 makes sure the indices of the new sparse tensor are sorted in ascending order, such that it can be used in subsequent operations such as `tf.sparse.to_dense`.

Figure 8.3 illustrates the process of creating the sparse representation of indicators as well. A similar implementation can be achieved using NumPy with Scipy `coo_matrix`.

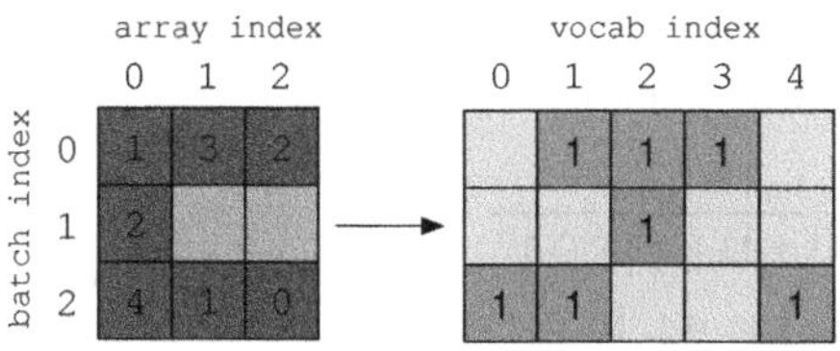

Figure 8.3 Example illustration of converting a set of variable-length lists of indices to a stack of sparse binary representations. The values in the left-aligned sparse tensor (left) represent several lists of indices for a variable that contains five categories. The same data can be represented by a sparse binary tensor (right) whose values are `1` at those indices.

```
import numpy as np
from scipy.sparse import coo_matrix

def sparse_to_indicator_scipy(X, vocab_size, out_dtype=np.float32):
    """
    Convert sparse tensor of indices to vectors of binary indicators

    * X:
        Each row of the scipy sparse matrix is the indices of the
        indicator vector
    * vocab_size:
        This determines the size of the sparse indicator vector
    * out_dtype:
        DType of the values of the output tensor
    """
    X = coo_matrix(X) # cast to coo_matrix
    # configure the sparse matrix
    row = np.array(X.row)
    col = np.array(X.data, dtype=row.dtype)
    data = np.ones_like(col)
    shape = (X.shape[0], vocab_size)
    # make the sparse matrix
    y = coo_matrix((data, (row, col)), shape=shape, dtype=out_dtype)

    return y
```

8.3 Case Study: Jaccard Similarities Using Sparse Matrix

The Jaccard Similarity metric can be useful when comparing how similar two objects are given their categorical attributes. For example, we can use the Jaccard Similarity score to compare how similar two movies are. If the two movies have many overlapping attributes such as genres and tags, then the movies are similar and should have a high similarity score; otherwise, the movies are different and should have a low similarity score. Jaccard Similarity is defined as

$$\text{Sim}_{\text{Jaccard}}(X, Y) = \frac{|X \cap Y|}{|X \cup Y|} = \frac{|X \cap Y|}{|X| + |Y| - |X \cap Y|} \tag{8.1}$$

where $|\cdot|$ represents the cardinality of the set. In the context of evaluating the performance of object detection algorithms, the Jaccard Similarity metric is also known as the Intersection-Over-Union (IOU) metric. In IOU, X and Y are coordinates of the pixels within the bounding boxes drawn around the object, either by the algorithm prediction or by the label. Jaccard Similarity metric can also be used to search for similar items. In the case where we have a library of movies and each movie's metadata (e.g. tags and genres), we can compute the Jaccard Similarity score between each pair of movies, store the results in a matrix, and then look up the scores to rank the contents whenever a user searches for similar contents to a given title. To build such a matrix, we need to use the knowledge we have gained from Chapter 6 on encoding string values to index representations, and from the previous case study on converting index representations to binary indicator representations. The following code block implements the pair-wise computation of Jaccard Similarity given a sparse binary representation of the item attributes, where we assume each item is a row of the sparse matrix, and each value along the column is an attribute of the content.

```
import numpy as np

def jaccard_similarities(mat):
    """
    Compute pairwise Jaccard similarities
    between rows of a coo sparse matrix
    """
    nnz = mat.getnnz(axis=1)
    # pair-wise binary intersection |X n Y|
    # sparse matrix multiplication
    sim = mat * mat.T
    sim = sim.astype(float)

    # for rows
    xx = np.repeat(nnz, sim.getnnz(axis=1))
    # for columns
    yy = nnz[sim.indices]

    # |X U Y| = |X| + |Y| - |X n Y|
    sim.data /= xx + yy - sim.data

    return sim
```

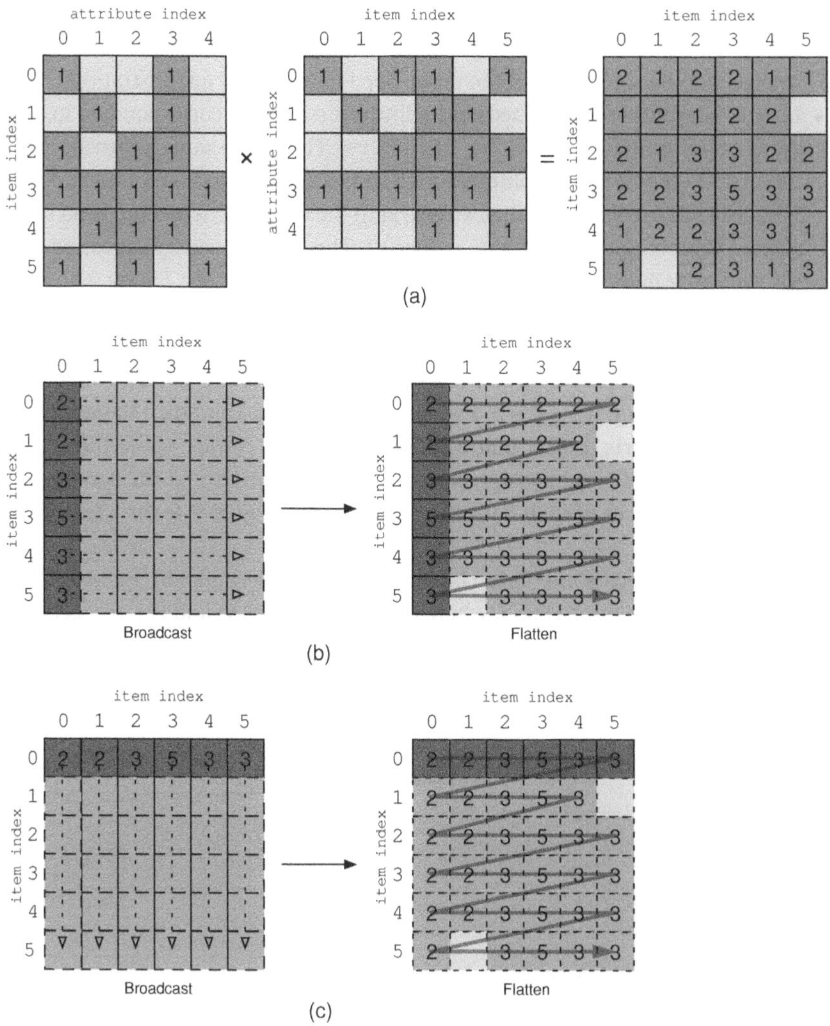

Figure 8.4 Illustration of the logic to compute pair-wise Jaccard Similarity scores between rows of a sparse matrix of attributes. (a) Pair-wise binary intersection between rows of the sparse matrix is computed by performing sparse matrix multiplication with its transpose. (b) Broadcasting the cardinality of the first set along rows by using the repeat operation. (c) Broadcasting the cardinality of the second set along columns by using the repeat operation. Once we obtained all 3 matrices, we can apply the Jaccard Similarity formula to compute the final pair-wise distances.

- Line 8 counts the number of attributes for each item.
- Lines 11 and 12 use sparse matrix multiplication to perform pair-wise binary intersection between rows of the input. See Figure 8.4 for an illustration.
- Line 15 computes the flattened cardinality of first set $|X|$ corresponding to each entry of the similarity matrix, i.e. `sim.data`. This is the same as broadcasting the `nnz` vector along the column dimension (`axis=1`) and then flattening the broadcasted matrix, except that we also skip the entries where there is no intersection (empty entries of the `sim` matrix). This is achieved by repeating `nnz` for each row by however many elements each row has. See Figure 8.4b for a schematic illustration.
- Line 17 computes the cardinality of the second set $|Y|$ corresponding to each entry of the similarity matrix, i.e. `sim.data`. This is the same as broadcasting the `nnz` vector along the row dimension (`axis=0`) and then flattening the broadcasted matrix, except that we need to skip the entries where there is no intersection. This is achieved by taking elements from `nnz` at the corresponding column indices of each data entry in `sim`. See Figure 8.4c for a schematic illustration.
- Line 20 implements the final Jaccard Similarity formula in Equation 8.1. `sim.data` is the intersection, whereas `xx` and `yy` are $|X|$ and $|Y|$, respectively.

We can test our implementation using the following example:

```
import numpy as np
from scipy.sparse import coo_matrix

mat_dense = np.array([
    [1, 0, 0, 1, 0],
    [0, 1, 0, 1, 0],
    [1, 0, 1, 1, 0],
    [1, 1, 1, 1, 1],
    [0, 1, 1, 1, 0],
    [1, 0, 1, 0, 1],
])
mat = coo_matrix(mat_dense)
# Use our implementation
sim = jaccard_similarities(mat)

# Using for loop and Python native set operations
# Convert binary to indices
mat_index = []
for row in mat_dense:
    indices = np.where(row>0)[0].tolist()
    mat_index.append(indices)
n_items = len(mat_index)

# use set operations to compute jaccard similarity
sim_loop = np.zeros((n_items, n_items))
```

```
for n in range(n_items):
    for m in range(n_items):
        if m >= n: # only compute the pair once
            continue

        x, y = set(mat_index[n]), set(mat_index[m])
        sim_loop[n, m] = len(x.intersection(y)) / len(x.union(y))

assert np.allclose(np.tril(sim.A, -1), sim_loop)
```

8.4 Case Study: Batchwise Set Operations

Set operations can be useful for efficiently computing multi-class and multi-label classification metrics. To motivate the current case study, let us consider a typical classification metric, precision, which is defined as

$$\text{Precision} = \frac{\#\text{TruePositive}}{\#\text{TruePositive} + \#\text{FalsePositive}}$$

where #TruePositive is the number of elements in the intersection between predicted positive classes and the labeled positive classes; #FalsePositive is the number of elements in the intersection between predicted positive class and negative labels. For binary classification and more generally, multi-class classification, there is only one true label class and one positive prediction for each example. In contrast, for multi-label classification tasks, there can be a variable number of labels and predictions for each example. One way to vectorize the computation of precision metric (or metrics alike) for a batch of examples is to represent the labels and predictions as one-hot (for binary and multi-class classifications) or binary vectors (for multi-label classifications). In the following code block, let us illustrate an example of computing precision in a multi-label classification task using dense binary representations

```
import numpy as np
predictions = np.array([
    [False,  True,  True, False, False],
    [ True, False, False, False, False],
    [False, False, False,  True,  True],
    [False,  True,  True, False, False],
])

labels = np.array([
    [False,  True,  True, False, False],
    [False, False,  True, False, False],
    [ True, False, False,  True, False],
```

```
    [False,   True,  False,   True,  False],
])

# Compute true positives: prediction and label both need to
    be True
tp = np.logical_and(predictions, labels)
tp = np.sum(tp, axis=1) # count
# Compute false positives: prediction is True, label is False
fp = np.logical_and(predictions, ~labels)
fp = np.sum(fp, axis=1) # count

# Compute precision
precision = tp / (tp + fp)
# array([1. , 0. , 0.5, 0.5])
```

The simple example contains only five classes. However, if the number of classes is large (say 5000), while the number of positive or label classes is small (e.g. 2 or 3), a dense representation such that may not be efficient. Alternatively, we can represent the predictions and labels as only the index of positive classes. Using the same example as above, we can represent the `predictions` and `labels` as

```
predictions = [[1, 2], [0], [3, 4], [1, 2]]
labels = [[1, 2], [2], [0, 3], [1, 3]]
```

It can be seen that `predictions` and `labels` are ragged arrays. Then, to compute precision, we can iterate one example at a time and use Python's `set.intersection` operation to compute true positives and `set.difference` to compute false positives, i.e.

```
precision = []
for pred, lab in zip(predictions, labels):
    # true positives
    tp = set(pred).intersection(set(lab))
    # false positives
    fp = set(pred).difference(set(lab))
    # precision
    p = len(tp) / (len(tp) + len(fp))
    precision.append(p)
# [1.0, 0.0, 0.5, 0.5]
```

But, as we have known since Chapter 1, this can be slow if there are more examples to compute the metric on. How can we vectorize this and compute precision without for-loops? Tensorflow implements the `tf.sets` module for set operations, where it uses either a dense `tf.Tensor` or a `tf.SparseTensor` to represent the set. We can vectorize the above as

```python
import tensorflow as tf
# Create the left-aligned ragged sparse tensors
predictions = tf.ragged.constant([[1, 2], [0], [3, 4], [1, 2]]).to_sparse()
labels = tf.ragged.constant([[1, 2], [2], [0, 3], [1, 3]]).to_sparse()
# Compute true positives
tp = tf.sets.intersection(predictions, labels)
tp = tf.sets.size(tp) # count each row
# Compute false positives
fp = tf.sets.difference(predictions, labels)
fp = tf.sets.size(fp) # count each row
# Compute precisions
precision = tp / (tp + fp)
# <tf.Tensor: shape=(4,), dtype=float64, numpy=array([1. , 0. , 0.5, 0.5])>
```

To the best of my knowledge, at the time of writing, Tensorflow is the only library that implements batch vectorized set operations. Although PyTorch and Scipy (NumPy) have sparse tensor representations, they do not implement equivalent vectorized set operations like Tensorflow. Operations under the `tf.sets` module can not only deal with numerical data but also string data, which makes the use of these operations quite convenient when dealing with the evaluation of texts. The following code block illustrates how we can implement batchwise set operations using NumPy and `scipy.sparse`.

```python
import numpy as np
from scipy.sparse import csr_matrix


def set_operation(x, y, pad=None, operation="intersection", returns="count"):
    """
    Batch-wise set operations.

    Inputs:
        * x, y: dense 2D np.ndarrays with/without padding values.
        * pad: padding values. Default to None which is no padding.
        * operation: set operation to perform. Valid values are
            ["intersection" (default), "union", "difference"].
        * returns: type of returned outputs, either "count" (default) which
            counts the cardinality of the resulting set operation for each
            row or "matrix" which returns the sparse matrix output from
            the set operations, all values are left aligned.
    """
    # Input shapes
```

```
    n_d, m_x = x.shape
    n_d, m_y = y.shape

    # Use np.unique to convert from data -> indices
    # This can appropriately handle all data types, including strings
    unique, indices = np.unique(np.hstack((x, y)), return_inverse=True)
    n_unique = len(unique)

    # From flattened index -> original shape
    indices = indices.reshape(n_d, -1)
    indices_x = indices[:, :m_x]
    indices_y = indices[:, m_x:]

    # which index from unique is the padding of the ragged representation
    pad_index = np.where(unique == pad)[0]
    if len(pad_index) > 0: # found the padding
        pad_index = pad_index[0]
    else:
        pad_index = -1

    # Use csr format to create to create binary indicator matrices
    # e.g. index = [1, 3], n_unique = 5 -> [0, 1, 0, 1, 0]
    def _create_csr_indicator(idx, m):
        # create csr
        indptr = np.repeat([m], n_d).cumsum()
        indptr = np.concatenate([[0], indptr])
        indices = idx.ravel() # flatten
        data = np.ones_like(indices, dtype=int)
        data[indices==pad_index] = 0 # filter out pad index
        sparse_matrix = csr_matrix(
            (data, indices, indptr), shape=(n_d, n_unique), dtype=int
        )
        # eliminate padding if any
        sparse_matrix.eliminate_zeros()
        return sparse_matrix

    x_hat = _create_csr_indicator(indices_x, m_x)
    y_hat = _create_csr_indicator(indices_y, m_y)

    # set operations using binary arithmetic operation
    if operation == "intersection":
        res = x_hat.multiply(y_hat)
    elif operation == "union":
        res = x_hat + y_hat
        res.data = np.minimum(res.data, 1)
    elif operation == "difference":
        res = x_hat - y_hat
        res.data = np.maximum(res.data, 0)
    else:
        raise(ValueError(f"Unrecognized operation {operation}"))

    if returns == "count": # return cardinality of set
        return res.sum(axis=1).A.ravel()
    else: # return the actual sparse matrix
```

```
74            # keep only entries of 1s
75            res.eliminate_zeros()
76            # replace the indicator back to actual intersected values
77            res.data = np.take(unique, res.indices)
78            # left align
79            l = np.diff(res.indptr) # array lengths
80            flat_indices = np.arange(len(res.data))
81            offsets = np.repeat(res.indptr[:-1], l)
82            res.indices = flat_indices - offsets
83
84            return res
```

- Lines 20 and 21 store the shapes of the inputs which will be used later.
- Line 25 uses `np.unique` to create the `unique` vector, which is the set of vocabularies used by both inputs `x` and `y`. Recall in Chapter 6 when we used `np.unqiue` in a case study to transform string values to integer values, this vector can be seen as the mapping between the original value and the integer representation of the values (corresponding to the indices `0, 1, 2, ...`). On the other hand, `indices` is the set of indices that can be used to recover the original (but flattened) inputs when used together with `unique`, i.e. `unique[indices] = <flattened_inputs>`. Note that the use of `np.unique` allows us to handle both integer and string inputs.
- Line 26 finds the overall vocabulary size of the input values from both `x` and `y`, which will be later on by the function `_create_csr_indicator`.
- Lines 29–31 first reshapes the flattened indices of values from `x` and `y`, so that it recovers the shape of the original horizontally stacked matrix `hstack((x, y))`. Then, it splits the horizontally stacked matrix to recover the respective reverse indices which can recover the original inputs.
- Lines 34–38 attempt to find the index of the padding values from the `unique` vocabulary vector, using `np.where`. If we cannot find the padding value from the `unique` vocabulary vector, then we assume there is no padding. `pad_index` will be set to `-1` which would bypass the padding value filtering steps when it is being used in the subsequent helper function `_create_csr_indicator`.
- Lines 42–54 is a helper function under the main function to create sparse binary indicator representations of the original input matrices of indices. We can do this efficiently since we assume that we are working with indices of a dense input matrix. Note that it reuses several parameters under the scope of the main function including `n_d` (batch size), `n_unique` (vocabulary size of the input value), and `pad_index`. Figure 8.5 illustrates the process of creating the sparse `csr_matrix`.
  - Lines 44 and 45 create the `indptr` of the `csr_matrix`. Since we are creating `indptr` for a dense matrix, each row has an identical number

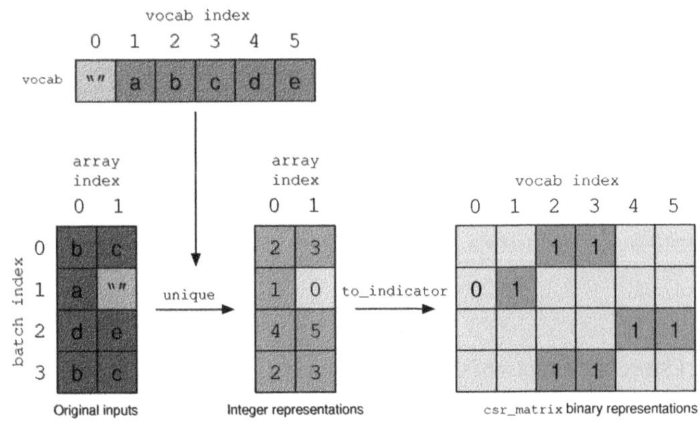

Figure 8.5 Creating a `csr_matrix` binary indicator representation using indices of a dense matrix. The empty string `""` entry in the `vocab` is the padding value in the current example.

of items. Therefore, we simply repeat the column size in a vector and then use `cumsum` to compute the cumulative length of rows (i.e. pointers for row index offset). Lastly, we prepend a `0` to follow the convention for `indptr`.

- Line 46 creates the column indices of each row. This is achieved by flattening the pre-determined `indices` matrix, which stores the integer (ordinal) representations of each item.
- Lines 47 and 48 create an array of dummy indicator data, using an array of ones. In Line 48, if the `pad` value is found in `unique`, then the indicator data for the padding values will be set to `0`, i.e. ignoring the value during subsequent set operations. In contrast, if `pad_index = -1`, then no values in the `data` would be set to `0`.
- Lines 49–51 create the `csr_matrix`.
- Line 53 eliminates any zero entries. The zero entries can only arise from Line 48 when setting the dummy indicator data to `0` for padding entries.

• Lines 56 and 57 apply the above helper function to each input's indices and create the binary indicator representations of the two input sets using `csr_matrix`. By mapping each original item to an index and then to an indicator, we can then perform subsequent set operations using the equivalent arithmetics of boolean operations.

• Lines 60–69 perform the set operations based on the user's choice:

- `"intersection"`: This is equivalent to performing `"AND"` operation (i.e. `True`/exists in set 1 AND `True`/exists in set 2) between two boolean vectors at the corresponding elements. Recall in Chapter 5 that performing

the `"AND"` operation between two boolean values (`True` or `False`) is the same as multiplying the two corresponding binary (`1`s or `0`s) values together.
  - `"union"`: This is equivalent to performing the `"OR"` operation (i.e. exists either in either set 1 `OR` set 2) between two boolean vectors at the corresponding elements. Recall in Chapter 5 that performing `the "OR"` operation between two boolean values is the same as summing the two corresponding binary values together, and then clipping the resultant values between 0 and 1.
  - `"difference"`: This is similar to performing `"XOR"` operation on two boolean vectors, except that it is directional (i.e. exists in set 1, but does not exist in set 2, not vice versa). Recall in Chapter 5 when performing the arithmetic version of `"XOR"` operations, we subtract the two binary values together and take the absolute value, in which case, we convert both `-1` (`0 - 1` or `False` vs. `True`) and `+1` (`1 - 0` or `True` vs. `False`) results as `+1` (`True`). However, by definition, `set.difference` keeps the element if it is `True` (exists) in set 1 and `False` (not exists) in set 2, while elements that correspond to `False` in set 1 and `True` in set 2 are not kept (unlike `XOR` which would keep elements in both cases). Therefore, in the arithmetic version of the set difference, we clip the resultant value to a minimum of `0` instead.
- Lines 71 and 72: After the set operation, we can choose to count the cardinality of the set for each row. This is done by simply adding together the indicators (a series of `1`s) along each row of the sparse matrix.
- Lines 73–77: Instead of returning the count, we return the resulting sparse matrix after the set operation in this case, with all entries left aligned. We first eliminate all the zero entries and keep only the positive indicators. We then replace the indicator value (`1`s) with the original value. Here, `res.indices` is the same as the indices of the original values in `unique`.
- Lines 79–82: Left align the sparse matrix after replacing the indicator with the original values of the inputs. Since we are operating on a `csr_matrix`, we only need to recompute and replace the column indices as `0, 1, ...`, which enumerates the contents of each row. Recall in a case study in Chapter 5, we used masking to implement the functionality of `tf.ragged.range` operation with NumPy functions. However, there is a better way to implement this functionality using `np.repeat` without masking, which is what is implemented here. See Figure 8.6 for a walkthrough of this algorithm.
  - Line 79 first computes the lengths of each array, i.e. each row of the `csr_matrix`.
  - Line 80 computes the `flat_indices` array according to the algorithm.
  - Line 81 creates the offset and broadcasts the offset values to each block via `np.repeat`. Here, we do not need to recompute the `offsets` by computing

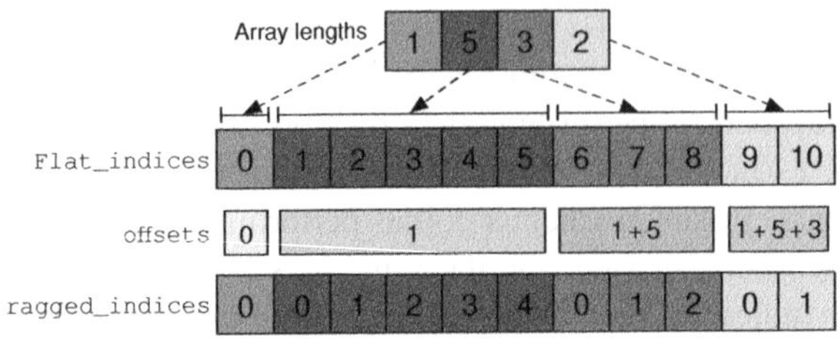

Figure 8.6 Illustration of an alternative way of computing flattened ragged indices using `np.repeat` which is used as the `indices` of the left-aligned sparse `csr_matrix`. Given array lengths `[1, 5, 3, 2]`, we can compute the ragged indices from the flat indices that enumerate the valid entries of the sparse matrix. We can color the blocks of the flat indices based on the specified array lengths. To help discover the pattern, we can first create the expected flattened ragged indices. Then, we can observe the following: for the first block of indices, the `flat_indices` and `ragged_indices` have no offset (offset 0) between each other; for the second block, the `flat_indices` and `ragged_indices` have offset `1`, corresponds to the length of the first block; for the third block, the two index arrays have an offset of `6 = 1 + 5` which is the sum of lengths of the first two blocks; lastly, for the fourth block, the two index arrays have an offset of `9 = 1 + 5 + 3` which is the sum of lengths of the first three blocks. By induction, we can conclude that the offsets for each block are computed as `cumsum([0, 1, 5, 3]) = [0, 1, 6, 9]`. Then, the offset can be subtracted from each block (as if broadcasting) of the `flat_indices` to obtain the `ragged_indices`. To broadcast, we can simply repeat the offsets by each array length, i.e. `0` repeated `one` time, `1` repeated `five` times, `6` repeated `three` times, and `9` repeated `two` times.

  the cumulative sum of the array lengths, since `indptr[:-1]` is already the offset for each row.
- Line 82 subtracts the `offsets` from the `flat_indices` and obtains the `ragged_indices` (indices that enumerate each row of the `csr matrix` starting from zero).

Let us first use the function we just implemented above to perform our example precision metric calculation:

```
predictions = np.array([[1, 2], [0, -1], [3, 4], [1, 2]])
labels = np.array([[1, 2], [2, -1], [0, 3], [1, 3]])
tp = set_operation(predictions, labels, -1, operation="intersection")
fp = set_operation(predictions, labels, -1, operation="difference")
precision = tp / (tp + fp)
# array([1. , 0. , 0.5, 0.5])
```

Let us also measure the performance of the function we implemented above against `tf.sets` and Python for-loops.

```
import numpy as np
import tensorflow as tf

# Use IPython magic function to measure performance
x = np.array([
    [1, 2, 3, 4, 5], [2, 3, 4, 5, 6], [3, 5, 1, 0, 0]
]*1024)
y = np.array([
    [5, 6, 7, 8, 9], [2, 3, 5, 7, 8], [3, 1, 0, 0, 0]
]*1024)

# Our NumPy implementation
%timeit res_np = set_operation(x, y, operation="intersection", pad=0)
# 1.01 ms ± 21.7 µs per loop (mean ± std. dev. of 7 runs, 1,000 loops
↪   each)

# Python For Loop
def batch_wise_intersection(x, y, pad=None):
    batch_size = x.shape[0]
    res = [[]] * batch_size
    for ii in range(batch_size):
        # take intersection
        r = set(x[ii, :]).intersection(set(y[ii, :]))
        if pad is not None:
            # remove padding
            r = r.difference(set([pad]))
        res[ii] = len(r)

    return res

res_python = batch_wise_intersection(x, y, pad=0)

%timeit res_python = batch_wise_intersection(x, y, pad=0)
# 4.27 ms ± 30.9 µs per loop (mean ± std. dev. of 7 runs, 100 loops each)

# Use Tensorflow tf.sets.intersection
%timeit res_tf = tf.sets.size(tf.sets.intersection(x, y))
# 1.18 ms ± 20.7 µs per loop (mean ± std. dev. of 7 runs, 1,000 loops
↪   each)
```

8.5 Case Study: Autoencoders with Sparse Inputs

An autoencoder is a shallow neural network that learns to compress the input using a latent (or hidden) representation and recovers the input from it as the output (Figure 8.7).

Mathematically, the autoencoder can be represented as

$$z = g(f(x))$$

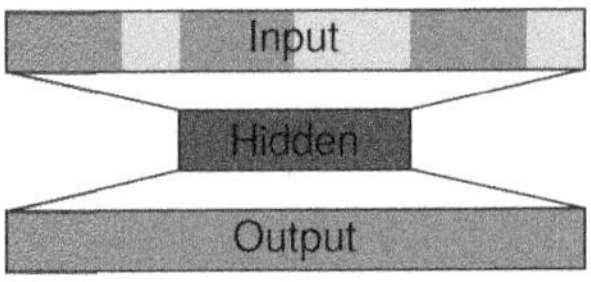

Figure 8.7 Autoencoder with categorical inputs represented by a sparse vector.

where $f(\cdot)$ is the encoder function which takes input x and produces the hidden layer y, and $g(\cdot)$ is the decoder function which takes the hidden layer representation and produces the output z. The hidden layer of the autoencoder can be represented as

$$y = \sigma(xW + b)$$

where

- x is the input vector. If x is a set of categorical values, then it is a binary vector, e.g. $x = [0, 1, 1, 0, 0]$. x is of size `(batch_size, vocab_size)`.
- W is the weight matrix, of size, `(vocab_size, hidden_size)`.
- b is the bias column vector, of size `(vocab_size, 1)`.
- σ is the activation function, e.g. sigmoid or ReLU activation.

8.5.1 Embedding Lookup on Sparse Inputs

$xW + b$ is the linear model, or the dense layer of a feedforward neural network before activation. The linear multiplication term xW results a `[batch_size, vocab_size]` × `[vocab_size, hidden_size]` = `[batch_size, hidden_size]` matrix. For example, let

$$W = \begin{bmatrix} 1 & 2 & 3 \\ 4 & 5 & 6 \\ 7 & 8 & 9 \\ 1 & 3 & 5 \\ 2 & 4 & 6 \end{bmatrix}$$

Then

$$\begin{aligned} \tilde{y} = xW &= \begin{bmatrix} 0 & 1 & 1 & 0 & 0 \end{bmatrix} \begin{bmatrix} 1 & 2 & 3 \\ 4 & 5 & 6 \\ 7 & 8 & 9 \\ 1 & 3 & 5 \\ 2 & 4 & 6 \end{bmatrix} \\ &= \begin{bmatrix} 4+7 & 5+8 & 6+9 \end{bmatrix} \\ &= \begin{bmatrix} 11 & 13 & 15 \end{bmatrix} \end{aligned}$$

Notice that this is a sparse (x) to dense matrix (W) multiplication. In addition to using dense binary representation of x, we can also use a sparse tensor to represent the binary indicator. Therefore, this operation can be achieved by `tf.sparse.sparse_dense_matmul` in Tensorflow or `torch.sparse.mm` in PyTorch, i.e.

```
import tensorflow as tf
x = tf.sparse.from_dense([[0, 1, 1, 0, 0]])
W = tf.constant([
    [1, 2, 3],
    [4, 5, 6],
    [7, 8, 9],
    [1, 3, 5],
    [2, 4, 6],
])
y_tilde = tf.sparse.sparse_dense_matmul(x, W)
# <tf.Tensor: shape=(1, 3), dtype=int32, numpy=array([[11,
   13, 15]], dtype=int32)>
```

and in PyTorch,

```
import torch
x = torch.tensor([[0, 1, 1, 0, 0]]).to_sparse()
W = torch.tensor([
    [1, 2, 3],
    [4, 5, 6],
    [7, 8, 9],
    [1, 3, 5],
    [2, 4, 6],
])
y_tilde = torch.sparse.mm(x, W)
# tensor([[11, 13, 15]])
```

If we make W the embedding matrix W_e, and let $x \mapsto s$ where s is the set of indices where the binary vector $x_i = 1$, e.g. $x = [0, 1, 1, 0, 0] \mapsto s = [1, 2]$. Then, the lookup operation on the embedding matrix will return each row of the embedding vector stacked vertically, e.g.

$$E = [W_e] \cdot \text{lookup}(s) = \begin{bmatrix} 4 & 5 & 6 \\ 7 & 8 & 9 \end{bmatrix}$$

Then, to get xW, we simply sum across the rows of the resulting matrix,

$$\tilde{y}_k = \sum_i E_{i,k}$$

Or

$$\tilde{y} = \begin{bmatrix} 4+7 & 5+8 & 6+9 \end{bmatrix} = \begin{bmatrix} 11 & 13 & 15 \end{bmatrix}$$

Embedding lookups are usually implemented with a hash map which has a time complexity of ($O(1)$). An advantage of using the embedding lookup is that, we no longer need to compute the multiplication with 0 terms. This is especially helpful when the data is very sparse (containing predominantly `0`s). Both Tensorflow and PyTorch implement a version of the above embedding lookup and summation operation. In Tensorflow, i.e.

```
import tensorflow as tf

s = tf.ragged.constant([[1, 2], [3, 0, 4]]).to_sparse()
W = tf.constant([
    [1, 2, 3],
    [4, 5, 6],
    [7, 8, 9],
    [1, 3, 5],
    [2, 4, 6],
])
weights = s.with_values(tf.ones_like(s.values)) # equal weights
y_tilde = tf.nn.embedding_lookup_sparse(
    W,
    s,
    weights,
    combiner="sum",
)
# <tf.Tensor: shape=(2, 3), dtype=int32, numpy=
# array([[11, 13, 15],
#        [ 4,  9, 14]], dtype=int32)>
```

In PyTorch, i.e.

```
import torch

# 1-based indexing; 0 is used as padding index
s = torch.tensor([[2, 3, 0, 0, 0], [4, 1, 5, 0, 0]])
W = torch.tensor([
    [0, 0, 0], # padding entry
    [1, 2, 3],
    [4, 5, 6],
    [7, 8, 9],
    [1, 3, 5],
    [2, 4, 6],
])

y_tilde = torch.nn.functional.embedding(
    s,
    W,
    padding_idx=0,
).sum(dim=1)
```

```
# tensor([[11, 13, 15],
#         [ 4,  9, 14]])
```

Notice that in Tensorflow's implementations, we need to specify `combiner="sum"` to sum together each embedding along the row. In contrast, `torch.nn.functional.embedding` returns a `(batch_size, seq_len, hidden_size)` tensor, where the padding entries are part of the output. We then need to sum along `dim=1` to combine the embeddings together.

To complete the dense layer with the bias term, we can further initialize a single vector b whose length is equal to the embedding size. Then, $\bar{y} = xW + b$ is hence represented as

$$\bar{y}_k = b_k + \sum_{i \in s} E_{i,k} \tag{8.2}$$

where i is the index of the item along each row (where a row may contain a variable number of items), and k is the index of the embedding vector.

8.5.2 Inputs with Weights

If each item in the input vector is also weighted by a weight w, then we can modify the above dense layer implementation as

$$y_k = b_k + \sum_{i \in s} w_i \cdot E_{i,k} \tag{8.3}$$

Following the above example, suppose items 1 and 2 are weighted by 1.3 and 2.7, respectively, then the input x is modified to become

$$x = \begin{bmatrix} 0 & 1.3 & 2.7 & 0.0 & 0.0 \end{bmatrix}$$

instead of a simple binary vector that contains only 0s and 1s. When converting to the indicator notations, we need two tensors created, one that stores the indices of the items, and one that stores the weights at the corresponding entry, i.e.

$$s = [1, 2]$$

$$v = [1.3, 2.7]$$

Let us realize Equation 8.3 in Tensorflow,

```
import tensorflow as tf

s = tf.ragged.constant([[1, 2], [3, 0, 4]]).to_sparse()
v = tf.ragged.constant([[1.3, 2.7], [2.1, 4.6, 0.8]]).to_sparse()
W = tf.constant([
    [1, 2, 3],
    [4, 5, 6],
```

```
    [7, 8, 9],
    [1, 3, 5],
    [2, 4, 6],
], dtype=tf.float32)
y_tilde = tf.nn.embedding_lookup_sparse(
    W,
    s,
    v,
    combiner="sum",
)

# <tf.Tensor: shape=(2, 3), dtype=float32, numpy=
# array([[24.099998, 28.1    , 32.100002],
#        [ 8.3     , 18.7    , 29.099998]], dtype=float32)>
```

And in PyTorch,

```
import torch

# 1-based indexing; 0 is used as padding index
s = torch.tensor([[2, 3, 0, 0, 0], [4, 1, 5, 0, 0]])
# using dense representations with paddings
v = torch.tensor([[1.3, 2.7, 0.0, 0.0, 0.0], [2.1, 4.6, 0.8, 0.0, 0.0]])
W = torch.tensor([
    [0, 0, 0], # padding entry
    [1, 2, 3],
    [4, 5, 6],
    [7, 8, 9],
    [1, 3, 5],
    [2, 4, 6],
], dtype=torch.float32)

# look up embeddings
y_tilde = torch.nn.functional.embedding(
    s,
    W,
    padding_idx=0,
)
# apply weights
y_tilde *= v[:, :, None]
# Combine
y_tilde = y_tilde.sum(dim=1)

# tensor([[24.1000, 28.1000, 32.1000],
#         [ 8.3000, 18.7000, 29.1000]])
```

In practice, the embedding lookup table is learned during a certain training task or can also be reloaded from a pre-trained model. Let us implement the autoencoder model which takes sparse inputs using Tensorflow/Keras and PyTorch.

Tensorflow/Keras implementation is below

```
import tensorflow as tf
from keras import Model, Input
from keras.layers import Dense, Activation


class Autoencoder(Model):
    """Autoencoder with sparse inputs."""

    def __init__(
        self,
        num_items,
        emb_size,
        hidden_activation="relu",
        output_activation="sigmoid",
    ):
        super(Autoencoder, self).__init__()
        self.num_items = num_items
        self.emb_size = emb_size
        self.hidden_activation = Activation(
            hidden_activation, name="hidden_activation"
        )
        self.output_layer = Dense(num_items, name="output_layer")
        self.output_activation = Activation(
            output_activation, name="output_activation"
        )

    def build(self, input_shape):
        self.embedding = self.add_weight(
            name="embedding",
            shape=(self.num_items, self.emb_size),
            trainable=True,
        )
        self.bias = self.add_weight(
            name="bias",
            shape=(1, self.emb_size),
            trainable=True,
        )

    def call(self, inputs, training=False):
        items = inputs["items"]
        weights = inputs.get("weights", None)
        if weights is None:
            # equal weights
            weights = items.with_value(tf.ones_like(items, tf.float32))

        # Implement the sparse dense layer (batch_size, emb_size)
        embed = tf.nn.embedding_lookup_sparse(
            self.embedding, items, weights, combiner="sum",
             ↪ name="hidden_embed"
        )

        # Add bias
        y = embed + self.bias

```

```
54         # Hidden layer activation
55         y = self.hidden_activation(y)
56 
57         # Decode to produce output
58         z = self.output_layer(y)
59         z = self.output_activation(z)
60 
61         return z
62 
63     @property
64     def model(self):
65         inputs = {
66             "items": Input((None,), sparse=True, dtype=tf.int64),
67             "weights": Input(
68                 (None,), sparse=True, dtype=tf.float32,
69             ),
70         }
71         model = Model(inputs, self.call(inputs))
72 
73         return model
```

- `__init__` **(Lines 9–25)**: This initializes the model class. It saves several model configuration parameters and initializes the activation and output dense layers.
- `build` **(Lines 27–37)**: This method creates the two weights, `embedding`, which is the embedding lookup table, and `bias`, which is the bias term added to the dense layer. Note that both weights are trainable (`trainable=True`). The use of the `build` method is the preferred way to initialize additional trainable weights in Keras.
- `call` **(Lines 39–61)**: This method implements the main logic of the autoencoder model.
  - Lines 40–44 retrieve the item indices (integer representation of the items) and the corresponding weights. If `weights` is not available from the `inputs` dictionary, then we assume the weights are all equal to 1.
  - Lines 47–49 perform the embedding lookup using `tf.nn.embedding_lookup_sparse`. This produces the hidden layer outputs. Here, we assumed that the inputs we retrieved from the previous steps are left-aligned sparse tensors. Alternatively, if we choose to use dense representations of the item indices with paddings (which necessarily needs to be 1-based indexing since 0 will be used as padding by default), then we can also use `keras.layers.Embedding` layer and set `mask_zero=True` upon initialization. This would be how we would implement the embedding lookup in PyTorch as well.
  - Line 52 adds the bias term.

- Lines 55–59 finish up the implementation of the autoencoder by applying an activation function and passing the results of the hidden layer to an output layer, which will then be activated by another activation function.

- `model` **property (Lines 63–73)**: This property is used to check if our implementation can be compiled as a graph. We can print the model architecture using `print(model.model.summary())` which gives us the following:

```
Model: "model_5"
__________________________________________________________________________
Layer (type)        Output Shape     Param #  Connected to
==========================================================================
items               [(None, None)]   0        []
(InputLayer)

weights             [(None, None)]   0        []
(InputLayer)

tf.compat.v1        (None, 128)      0        ['items[0][0]', lookup_sparse_5
.nn.embedding_                                  'weights[0][0]']
(TFOpLambda)

tf.__operators__    (None, 128)      0        ['tf.compat.v1.nn.embedding_lo
.add_5                                          okup_sparse_5[0][0]']
(TFOpLambda)

hidden_activation (None, 128)        0        ['tf.__operators__.add_5[0][0]']
(Activation)

output_layer        (None, 1000)     129000   ['hidden_activation[0][0]']
(Dense)
output_activation (None, 1000)       0        ['output_layer[0][0]']
(Activation)

==========================================================================
Total params: 129000 (503.91 KB)
Trainable params: 129000 (503.91 KB)
Non-trainable params: 0 (0.00 Byte)
__________________________________________________________________________
None
```

We can test our implementation as the following

```
inputs = {
    "items": tf.ragged.constant([[1, 2], [3, 0, 4]]).to_sparse(),
    "weights": tf.ragged.constant([[1.3,2.7], [2.1,4.6,0.8]]).to_sparse()
}

model = Autoencoder(num_items=1000, emb_size=128)
z = model(inputs) # use the input to produce output, (2, 1000)
```

PyTorch implementation is below

```
1  import torch
2
3  class Autoencoder(torch.nn.Module):
4      def __init__(self, num_items, emb_size):
5          super().__init__()
6          self.embed = torch.nn.Embedding(
7              num_items, emb_size, padding_idx=0
8          )
9          self.bias = torch.nn.Parameter(torch.zeros([1, emb_size]))
10         self.decoder = torch.nn.Sequential(
11             torch.nn.ReLU(),
12             torch.nn.Linear(emb_size, num_items),
13             torch.nn.Sigmoid()
14         )
15
16     def forward(self, items, weights=None):
17         if weights is None:
18             # equal weights
19             weights = torch.ones([items.shape[0], 1], dtype=torch.float32)
20
21         # Implement the sparse dense layer (batch_size, emb_size)
22         embed = self.embed(items)
23         # Apply weights
24         embed *= weights[:, :, None]
25         # Sum together embeddings
26         embed = embed.sum(dim=1)
27
28         # Add bias
29         y = embed + self.bias
30
31         # Hidden layer activation
32         z = self.decoder(y)
33
34         return z
```

- `__init__` **(Lines 4–14)**: This method initializes the model class. Unlike the Keras model, we initialize the embedding lookup using `torch.nn.Embedding` which takes in a dense tensor with 0 paddings (`pad_idx=0`). The bias term is also specified in this method instead of in a separate method as in Keras. The rest of the layers after the hidden layer are combined together into a single `Sequential` layer (Keras also has this container layer class, so one can potentially choose to group these layers together as well in the Keras implementation).
- `forward` **(Lines 16–34)**: This method implements the main logic of the autoencoder. Instead of using sparse inputs, this method assumes both `items` and `weights` use dense tensors with 0 paddings to represent left-aligned jagged tensors. In Lines 22–26, we first pass the list of items through the embedding

lookup, then apply the weights along the first two dimensions (batch and sequence), and finally pool the sequence dimension with summation.

Again, we can test our implementation with some faked inputs. Notice that comparing Keras, `items` are using 1-based indexing, with 0 as paddings. Both `items` and `weights` are dense tensors of the same shape.

```
inputs = {
    "items": torch.tensor([[2, 3, 0], [4, 1, 5]]),
    "weights": torch.tensor([[1.3, 2.7, 0.0], [2.1, 4.6, 0.8]]),
}

model = Autoencoder(num_items=1000, emb_size=128)
z = model(**inputs)
```

Printing the model with `print(model)` gives us

```
Autoencoder(
  (embed): Embedding(1000, 128, padding_idx=0)
  (decoder): Sequential(
    (0): ReLU()
    (1): Linear(in_features=128, out_features=1000, bias=True)
    (2): Sigmoid()
  )
)
```

Bibliography

Martın Abadi, Ashish Agarwal, Paul Barham, Eugene Brevdo, Zhifeng Chen, Craig Citro, Greg S. Corrado, Andy Davis, Jeffrey Dean, Matthieu Devin, Sanjay Ghemawat, Ian Goodfellow, Andrew Harp, Geoffrey Irving, Michael Isard, Yangqing Jia, Rafal Jozefowicz, Lukasz Kaiser, Manjunath Kudlur, Josh Levenberg, Dan Mane, Rajat Monga, Sherry Moore, Derek Murray, Chris Olah, Mike Schuster, Jonathon Shlens, Benoit Steiner, Ilya Sutskever, Kunal Talwar, Paul Tucker, Vincent Vanhoucke, Vijay Vasudevan, Fernanda Viegas, Oriol Vinyals, Pete Warden, Martin Wattenberg, Martin Wicke, Yuan Yu, and Xiaoqiang Zheng. Tensorflow RaggedTensor. https://www.tensorflow.org/api_docs/python/tf/RaggedTensor, 2023.

Denis Baylor, Eric Breck, Heng-Tze Cheng, Noah Fiedel, Chuan Yu Foo, Zakaria Haque, Salem Haykal, Mustafa Ispir, Vihan Jain, Levent Koc, Chiu Yuen Koo, Lukasz Lew, Clemens Mewald, Akshay Naresh Modi, Neoklis Polyzotis, Sukriti Ramesh, Sudip Roy, Steven Euijong Whang, Martin Wicke, Jarek Wilkiewicz, Xin Zhang, and Martin Zinkevich. Tensorflow Transform Sparse Left Align. https://

github.com/tensorflow/transform/blob/a6cca798e550c4eb7ffca90ad76288-e28fbaae88/tensorflow_transform/mappers.py#L213-L239, 2023.

Jim Pivarski, Ianna Osborne, Ioana Ifrim, Henry Schreiner, Angus Hollands, Anish Biswas, Pratyush Das, Roy Choudhury Santam, Nicholas Smith, and Manasvi Goyal. Awkward Array, October 2018.

9

Groupby, Apply, and Aggregate

Consider the following tabular dataset (Table 9.1):

Table 9.1 Example user-movie viewing hours dataset.

User	Movie	Viewing Hours
1	3	0.6
3	2	1.3
2	3	0.6
1	2	0.7
2	1	0.1
3	3	0.5
1	1	0.9

We can ask several questions about the behaviors of the users and the items from this dataset:

1. How many viewing hours each user has on an average?
2. How many movies does each user watch on an average?
3. How many hours does each user spend on each movie?

To answer these questions, we first need to group the feature values from the same users or items together and then compute summary statistics on the features within the groups. For instance, for the question of how many viewing hours each user has on average, we need to first identify how many total viewing hours user 1 has, then user 2 has, then user 3 has, … etc. In Pandas, this is often done by using the `.groupby(...).agg()` pattern applied on a `pandas.DataFrame`. We then can take the average across all the users to obtain our answer. The feature we use to identify the group is called the group(by) **key**. The feature we used to

Vectorization: A Practical Guide to Efficient Implementations of Machine Learning Algorithms,
First Edition. Edward DongBo Cui.

compute summary statistics is called the group **value**. In this particular example, the groupby key is `user`, and the group value feature is the `viewing hours` feature, where we need to add all the values together within the same user.

9.1 Pandas Groupwise Operations

Let us attempt to answer those three questions we posted above using Pandas

```
import pandas as pd

df = pd.DataFrame(
    {
        "user": [1, 3, 2, 1, 2, 3, 1],
        "movie": [3, 2, 3, 2, 1, 3, 1],
        "viewing hours": [0.6, 1.3, 0.6, 0.7, 0.1, 0.5, 0.9],
    }
)

# Question 1: How many viewing hours each user has on average?
## Compute toal viewing hours of each user
df_total = df.groupby(by="user", as_index=False).sum()
#    user  movie  viewing hours
# 0     1      6            2.2
# 1     2      4            0.7
# 2     3      5            1.8

## Compute average
df_total["viewing hours"].mean()
# 1.567

# Question 2: How many items does each user watch on average?
## Compute total number of contents each user watch
df_watch = df.groupby(by="user", as_index=False)["movie"].nunique()
#    user  movie
# 0     1      3
# 1     2      2
# 2     3      2

## Compute average
df_watch["movie"].mean()
# 2.333

# Question 3: How many hours does each user spend on each content?
## Compute total viewing hours for each item
df_user_hrs = df.groupby(by="movie", as_index=False).agg(
    {
        "viewing hours": "sum",  # compute total hours
        "user": "nunique",  # count unique users
    }
)
#    movie  viewing hours  user
```

```
# 0        1            1.0     2
# 1        2            2.0     2
# 2        3            1.7     3

## Compute average hours for each content
df_user_hrs["viewing hours"] / df_user_hrs["user"]
# 0      0.500000
# 1      1.000000
# 2      0.566667
# dtype: float64
```

- For question 1, we apply the generic groupby(by="key").<agg_func> pattern, where <agg_func> can be, for instance, sum, mean, min, max, etc. Notice that we ended up with an extraneous column movie after the groupby operation, which we ignored in our subsequent calculation.
- For question 2, we apply the aggregation on a specific group value, rather than on all the features except user. After the groupby operation, we only end up with the key user and the grouped value feature movie, which we apply the aggregation nunique to count the size of the unique set of movie ids.
- For question 3, to compute average viewing per user, we need to apply two different aggregation functions on two group values. For the viewing hours feature, we need to sum together all the numbers within the same movie id and for the user feature, we need to count the distinct number of user ids. This is, therefore, expressed using a dict to map the feature name to the aggregation function. We subsequently use the two summarized features to compute the per-user average hours of each content.

The above three examples illustrated Pandas' "groupby-aggregate" pattern. Another pattern that leverages groupby is the "groupby-apply" pattern: instead of reducing the grouped values into a single summary statistics, it performs a transformation on the existing value, retaining the size of the input. Let us illustrate this pattern by answering the following question: What percentage of time does each user spend on each movie? We assume that in the above data, each user only watches each item only once. Otherwise, we need to use the groupby-aggregate pattern first to sum together the watch hours by each user by each item first.

```
# Question 4: what percentage of time does each user spend on each movie
percentage = df.groupby(by="user")["viewing hours"].apply(
    lambda x: x / x.sum() * 100,
).droplevel(0)
percentage = percentage.rename("percentage")
df_perc = df.join(percentage, how="left")
df_perc = df_perc.sort_values(by=["user"]).reset_index(drop=True)
```

```
#     user  movie  viewing hours  percentage
# 0      1      3            0.6   27.272727
# 1      1      2            0.7   31.818182
# 2      1      1            0.9   40.909091
# 3      2      3            0.6   85.714286
# 4      2      1            0.1   14.285714
# 5      3      2            1.3   72.222222
# 6      3      3            0.5   27.777778
```

The groupby-apply pattern in the current example produces a Pandas `Series` that is sorted according to the `user` id after grouping, which is not the same as the original order in the input DataFrame. We use `DataFrame.join` to combine the new `percentage` feature back to the input DataFrame.

Note that when we were introducing vectorization and parallel computing in Chapter 1, we also illustrated an example of using the groupby-apply pattern with a custom apply function which can use even more complicated logic than the simple lambda function we illustrated in this example. Both groupby-aggregate and groupby-apply patterns are useful in either exploratory data analysis to understand data properties or data preprocessing for machine learning feature engineering.

9.2 Reshaping and Windowing of Dense Tensors

In Chapter 2, we introduced reduction operation, in which we apply an arithmetic operation along one or more dimension(s) or axis (axes) of a tensor. When combined with a reshaping operation, a reduction operation can be seen as applying the groupby-aggergate pattern. For instance, suppose we would like to downsample a time series by taking the average over chunks of 10 elements.

```
import numpy as np
import matplotlib.pyplot as plt

# Simulate time series
x = np.linspace(0, 10, 500)
y = np.sin(x) + np.random.rand(len(x))

# Downsample by averaging groups of 10 elements
x_downsample = x[::10]
y_reshaped = y.reshape((-1, 10))
y_downsample = y_reshaped.mean(axis=1)

# Plot out the results
fig, ax = plt.subplots(1, 1, figsize=(8, 4))
ax.plot(x, y, label="original", zorder=0)
ax.plot(x_downsample, y_downsample, label="downsampled", c="r", zorder=1)
```

```
17 ax.scatter(x_downsample, y_downsample, c="r", zorder=2)
18 ax.set_xlabel("Time")
19 ax.set_ylabel("Value")
20 ax.legend()
```

Here, in Line 10, we reshaped `y` from a shape of `(500, )` array to a 2D `(50, 10)` matrix. The reshaping operation collects every 10 elements in the second dimension (`axis=1`) and stacks each of the collections along the first dimension (`axis=0`). Therefore, this is as if using the groupby pattern. Then, we apply the `mean` reduction operation along `axis=1` in Line 11, which takes the average within the grouped values. Together, this reshape–reduce operation combination completes the implementation of a groupby-aggregate pattern. See Figure 9.1 for the results of the plot. See also Figure 9.2 for a schematic illustration of the concept of combining reshaping and reduction operations to implement the groupby-aggregate pattern. Note that to use the reshape–reduce operation, we need to assume that the length of the sequence is divisible by the number of groups we would like to create via reshape. Otherwise, we should pad the sequence so that it can be grouped/reshaped properly.

Similarly, in Chapter 2, we also introduced the concept of broadcasting, where we apply a lower-dimensional tensor to a higher-dimensional tensor to perform arithmetic operations. Another way to look at broadcasting is from the perspective of groupby-apply. Again, using the above time series example, let us now suppose we would like to remove the mean of the time series for every chunk of 10 data points.

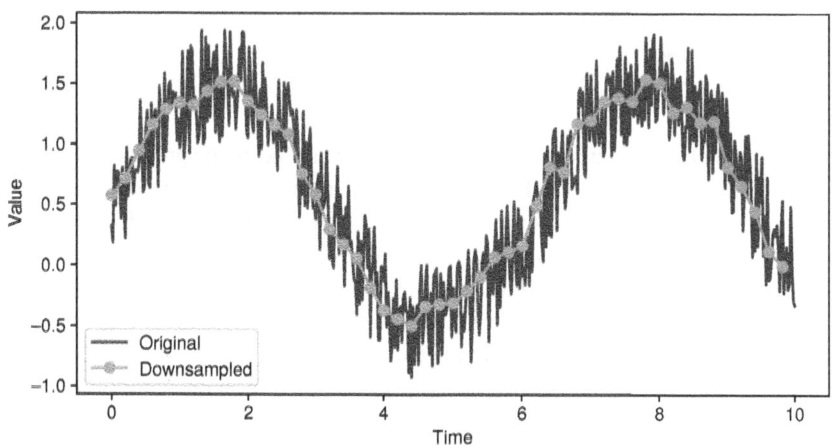

Figure 9.1 Downsampled time series from the code example.

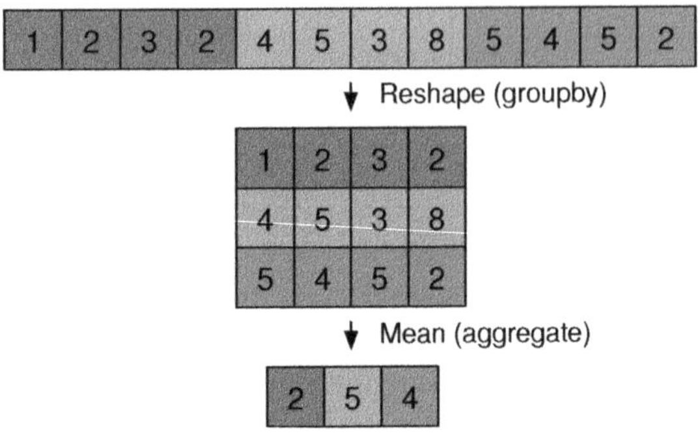

Figure 9.2 Schematic illustration of reshape and reduce operations used to implement groupby-aggregate.

```
1  import numpy as np
2  import matplotlib.pyplot as plt
3
4  # Simulate time series
5  x = np.linspace(0, 10, 500)
6  y = np.sin(x) + np.random.rand(len(x))
7
8  # Downsample by averaging groups of 10 elements
9  x_downsample = x[::10]
10 y_reshaped = y.reshape((-1, 10))
11 y_downsample = y_reshaped.mean(axis=1)
12 y_mean_removed = y_reshaped - y_downsample[:, None]
13 y_mean_removed = y_mean_removed.reshape((-1, ))
14
15 # Plot out the results
16 fig, ax = plt.subplots(1, 1, figsize=(8, 4))
17 ax.plot(x, y, label="original", zorder=0)
18 ax.plot(x, y_mean_removed, label="mean-removed", zorder=1)
19 ax.set_xlabel("Time")
20 ax.set_ylabel("Value")
21 ax.legend()
```

Up to Line 11, we compute the groupwise mean just like in the previous example. Then, in Line 12, we remove the mean by broadcasting the `y_downsample` array to `y_reshaped` along the second dimension (i.e. the dimension that contains each group). Line 13 flattens the grouped values back to a 1D time series vector. See Figure 9.3 for the results of the plot. See also Figure 9.4 for a schematic illustration of the concept of combining reshaping, reduction, and broadcasting operations to implement the groupby-apply pattern.

Groupwise operation on dense tensor inputs is seen in many computer vision models including You Only Look Once (YOLO) [Terven and Cordova-Esparza, 2023], Visual Transformer(ViT) [Dosovitskiy et al., 2021], and Masked AutoEncoders (MAE) [He et al., 2021]. In the subsequent case study, we will implement ViT from scratch to illustrate how this idea is being used in an application.

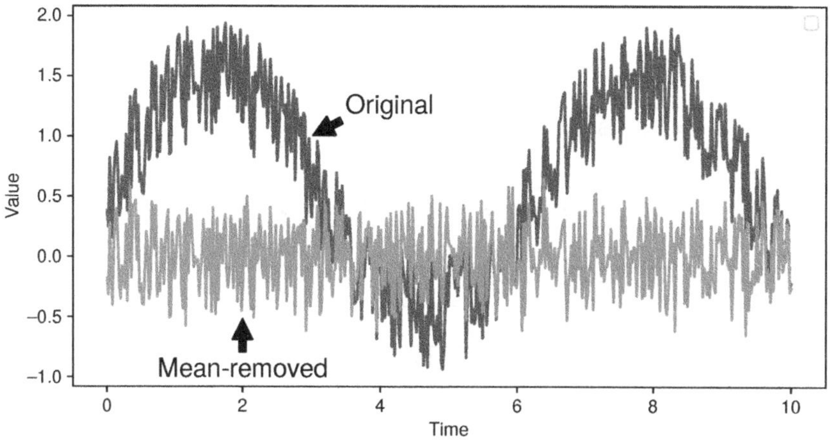

Figure 9.3 Mean-removed time series from the code example.

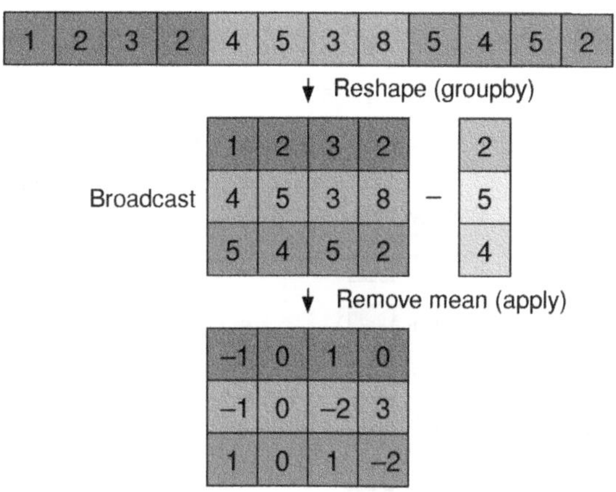

Figure 9.4 Schematic illustration of broadcasting used to implement groupby-apply.

A more general operation than the reshape operation is the sliding window operation: instead of discretely separating each group of values into chunks, a sliding window operation allows the groups to overlap with each other between the start and end of the sequence of value groups. See Figure 9.5 for an illustration.

NumPy creates sliding window views via `np.lib.stride_tricks.sliding_window_view`, while PyTorch implements equivalent operations `torch.Tensor.unfold` and `torch.nn.functional.unfold`. To the best of my knowledge, Tensorflow does not have a ready-to-use sliding window operation, but one can achieve the same effect using `tf.gather` (see Figure 9.6 for an illustration).

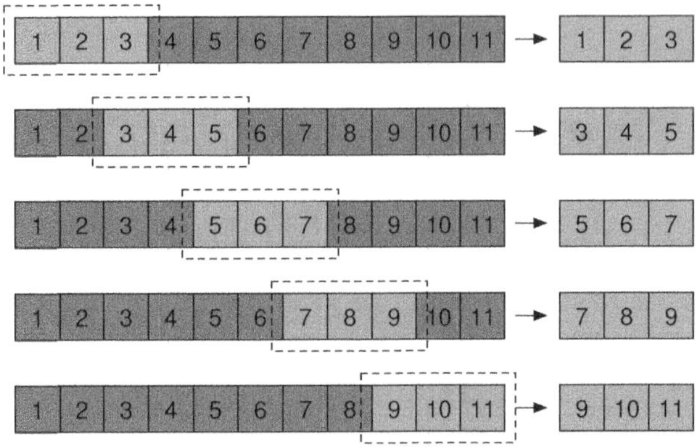

Figure 9.5 Sliding window of size 3 and stride 2.

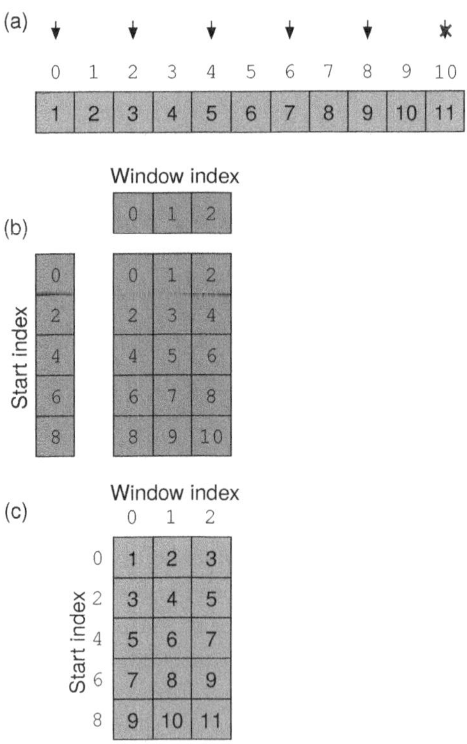

Figure 9.6 Illustration of implementation logic of `sliding_window_tf` using Tensorflow's `tf.gather`. (a) We first need to determine the starting indices based on the valid indices of the input, stride, and window size. Notice that we eliminated the last index since the window it will create will exceed the valid indices of the input. (b) Then, we gather indices based on the start index offset and within-window index by summing the two indices together via broadcasting. (c) Finally, we use `tf.gather` to take elements from the original input, based on the indices created from (b).

```
1  import tensorflow as tf
2
3  def sliding_window_tf(X, window_size, stride, axis=0):
4      """Sliding window."""
5      start_indices=tf.range(0, tf.shape(X)[axis] - window_size + 1, stride)
6      start_indices = tf.reshape(start_indices, (-1, 1))
7      slicing_indices = tf.range(window_size)
8      slicing_indices = start_indices + slicing_indices
9      # use tf.gather to extract slices
10     windowed = tf.gather(X, slicing_indices, axis=axis)
11     # Change axis so that it follows the np conventions
12     windowed = tf.experimental.numpy.moveaxis(windowed, axis+1, -1)
13
14     return windowed
```

- Line 5 creates the expected starting index of each window. The valid window should start and end within valid indices of the input. Therefore, the last start index that would create a valid window only covering the valid indices would end at `len(input) - window_size + 1`.
- Line 6 reshapes `start_indices` to a column vector for broadcasting later.
- Lines 7 and 8 create the slicing indices. Within each slice, we will take the `window_size` number of values. The starting index of each window is offset by the values in `start_indices`.
- Line 10 gathers the values based on the window-slicing indices.
- Line 12 fixes the order of axes, making it equivalent to the order of NumPy and PyTorch's sliding window outputs.

The following code block illustrates how to use these operations to create groups of sliding windowed values in NumPy, PyTorch, and Tensorflow.

```
import numpy as np
import tensorflow as tf
import torch

X = np.arange(1, 12)
# Sliding window of size 3, and stride/step 2
windowed_np = np.lib.stride_tricks.sliding_window_view(
    X, window_shape=(3, ), axis=0
)[::2]
windowed_torch = torch.tensor(X).unfold(dimension=0, size=3, step=2)
windowed_tf = sliding_window_tf(X, window_size=3, stride=2, axis=0)
assert np.allclose(windowed_np, windowed_torch.numpy())
assert np.allclose(windowed_np, windowed_tf.numpy())
```

An example application of creating a sliding window is to compute the moving average over a time series.

```
1  import numpy as np
2  import pandas as pd
3  import matplotlib.pyplot as plt
4
5  # Simulate time series
6  x = np.linspace(0, 10, 500)
7  y = np.sin(x) + np.random.rand(len(x))
8
9  # Create sliding window
10 x_windowed = np.lib.stride_tricks.sliding_window_view(x, (25, ), 0)[::2]
11 y_windowed = np.lib.stride_tricks.sliding_window_view(y, (25, ), 0)[::2]
12 # Compute moving average
13 x_moving_avg = x_windowed.mean(axis=1)
14 y_moving_avg = y_windowed.mean(axis=1)
15
16 # Equivalently in Pandas
17 y_moving_avg_pd = pd.Series(y).rolling(window=25, step=2).mean()
18 y_moving_avg_pd.dropna(inplace=True)
19
20 # Plot out the results
21 fig, ax = plt.subplots(1, 1, figsize=(8, 4))
22 ax.plot(x, y, label="original", zorder=0)
23 ax.plot(x_moving_avg, y_moving_avg, label="moving-average", zorder=1,
   ↪ c="r")
24 ax.set_xlabel("Time")
25 ax.set_ylabel("Value")
26 ax.legend()
```

Note that in Lines 17 and 18, I have also included an example of using `pandas.Series.rolling(...).mean()` to compute moving average as well.

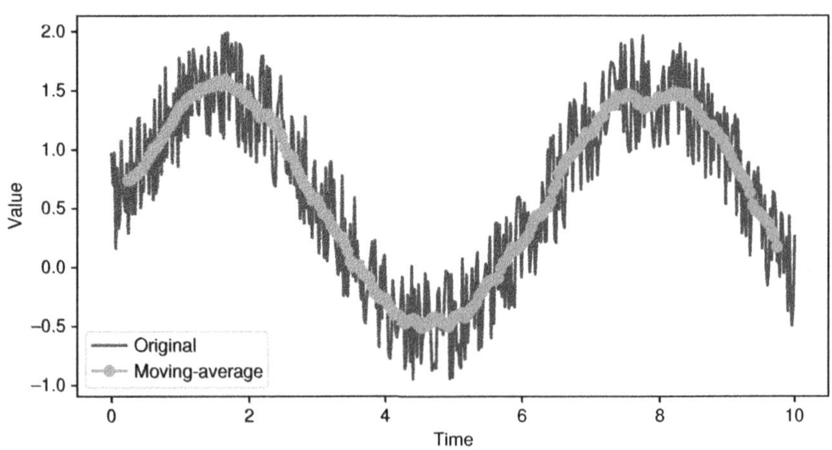

Figure 9.7 Moving average of the time series from the code example.

`pandas.Series.rolling` and `pandas.DataFrame.rolling` are Pandas' implementation of sliding window operation. See Figure 9.7 for the result of the plot.

9.3 Case Study: Vision Transformer (ViT)

The transformer [Vaswani et al., 2017] model is a powerful sequence representation model that is responsible for the subsequent advancements in natural language processing, computer vision, audio processing, and time series forecasting. The transformer model was originally developed around the context of natural language processing, focusing on the task of sequence-to-sequence neural machine translation. However, in the current case study, we will look at a research paper on how to apply a transformer model in computer vision tasks. Specifically, we will look at the use of Vision Transformer, or ViT [Dosovitskiy et al., 2021], to classify images. Figure 9.9 illustrates how ViT works. Conceptually, a transformer model can be seen as a backbone architecture to represent and extract information from a sequence of embeddings of discrete tokens. It takes an input of shape `(sequence:length, embed_size)` and then outputs a tensor of identical shape. Since image data typically come in the shape of `(num_channels, height, width)`, we need to come up with a strategy to reshape the image inputs to sequences of embeddings. (Here, I am using PyTorch's convention for input image shape; for Tensorflow/Keras convention, it would typically be `(width, height, num_channels)`, but this should not prevent us from generalizing our implementation in the current study using PyTorch to Tensorflow or other libraries.) ViT converts images into sequences by first breaking the 2D images into smaller, fixed-size patches, which are then flattened along the channel, height, and width dimensions to form a sequence

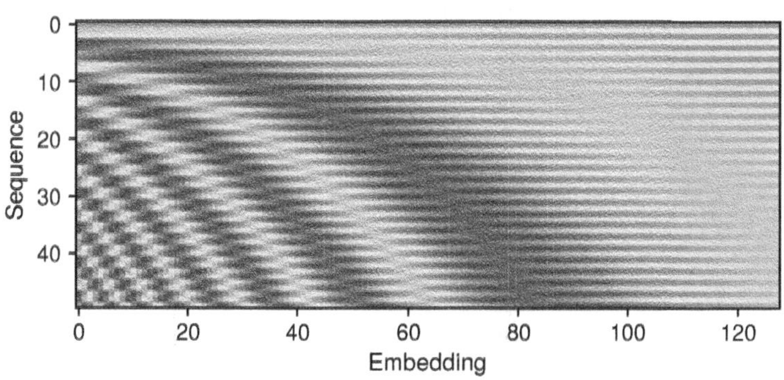

Figure 9.8 Sinusoidal positional encoding of the transformer architecture.

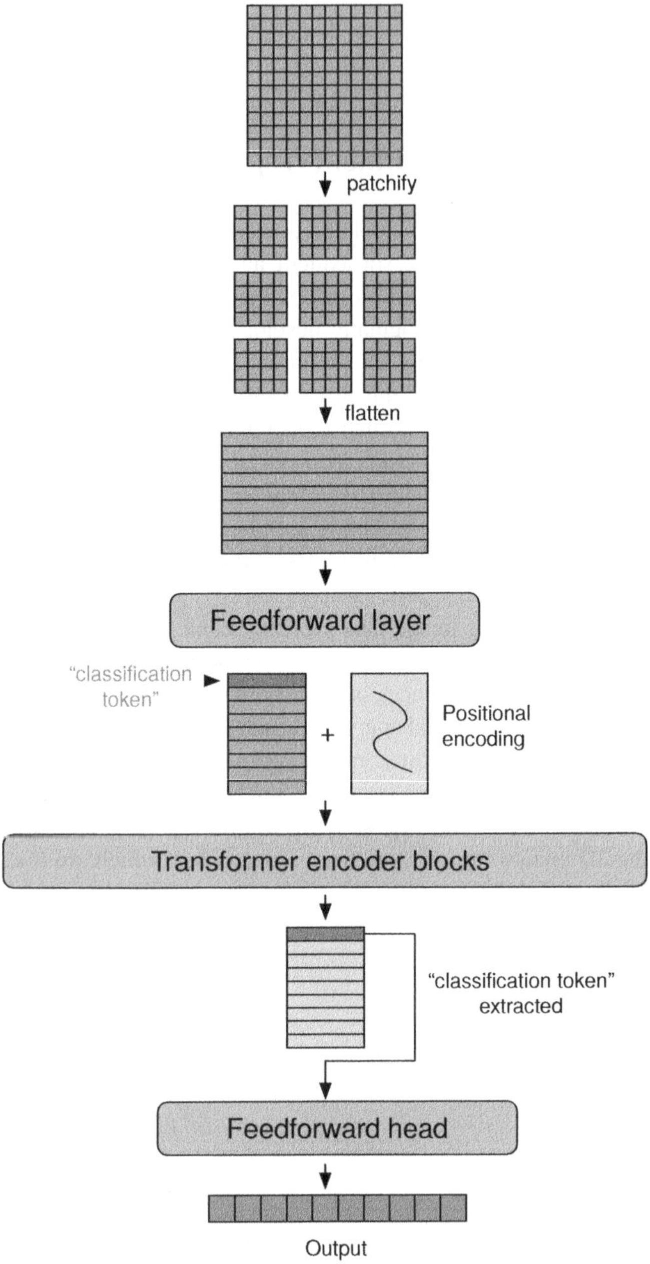

Figure 9.9 ViT architecture overview.

of 1D vectors. These vectors are first fed into a linear/dense layer to reduce the dimensionality of the flattened vectors, which now form the embedding vector representations of the patches. Then, a helper "classification token" embedding is added to the beginning of the sequence, which is dynamically learned, and later used to make the final prediction of the class. This tensor of shape `(num_patches + 1, emb_size)` forms the input to the transformer encoder model. The input tensor is added with a sinusoidal positional encoding matrix as per the original transformer paper [Vaswani et al., 2017]. The sinusoidal positional encoding of the transformer encoder model inputs is defined as

$$\begin{aligned} \mathrm{PE}(\mathrm{pos}, 2i) &= \sin(\mathrm{pos}/10000^{2i/d_{\mathrm{model}}}) \\ \mathrm{PE}(\mathrm{pos}, 2i+1) &= \cos(\mathrm{pos}/10000^{2i/d_{\mathrm{model}}}) \end{aligned} \tag{9.1}$$

where

- pos is the position along the sequence dimension.
- i is the index of the embedding dimensions. For every even index of the embedding, we apply a sine function; for every odd index, we apply a cosine function.
- d_{model} is the embedding size of the tokens.

This forms a 2D matrix of `(sequence:length, embed_size)`, which is illustrated in Figure 9.8. The positionally encoded inputs are then fed into multiple blocks of transformer encoders (see Figure 9.10). ViT uses a Pre-LayerNormalization or Pre-LN transformer encoder architecture which is more common nowadays instead of the architecture from the original transformer paper, which is now often referred as a Post-LayerNormalization or Post-LN architecture. See [Xiong et al., 2020] for a more detailed discussion. Finally, we extract the first "classification token" embedding from the output sequence and use this vector to make the final multi-class classification prediction by feeding it into a dense layer that outputs the probability of each class (Figure 9.9).

As mentioned in Section 9.2, we will use ViT to illustrate how reshaping and grouping are used in practical applications. The following code block implements a ViT for the task of image classification using PyTorch.

```
import torch
from torch import nn
from torchsummary import summary # !pip install torch-summary==1.4.5


class TransformerEncoderBlock(nn.Module):
    def __init__(self, embed_dim, ffn_size, num_heads=2, dropout=0.1):
        super(TransformerEncoderBlock, self).__init__()
        self.mhattn = nn.MultiheadAttention(
            embed_dim=embed_dim,
            num_heads=num_heads,
```

```
            dropout=dropout,
            batch_first=True,
        )
        self.ffn = nn.Sequential(
            nn.Linear(embed_dim, ffn_size),
            nn.GELU(),
            nn.Dropout(dropout),
            nn.Linear(ffn_size, embed_dim),
        )
        self.input_ln = nn.LayerNorm(embed_dim)
        self.ffn_ln = nn.LayerNorm(embed_dim)

    def forward(self, x):
        y = self.input_ln(x)
        y, _ = self.mhattn(y, y, y)
        y += x

        z = self.ffn_ln(y)
        z = self.ffn(z)
        z += y

        return z


class VisionTransformer(nn.Module):
    def __init__(
        self,
        image_dim=(1, 28, 28),  # channel, height, width
        num_patches=(7, 7),  # along each dimension
        num_classes=10,
        embed_dim=8,
        ffn_size=32,
        num_heads=2,
        dropout=0.1,
        num_blocks=2,
    ):
        super(VisionTransformer, self).__init__()

        self.image_dim = image_dim
        self.num_patches = num_patches

        # Compute patch sizes
        err_msg = "Cannot evenly divide into patches"
        assert image_dim[1] % num_patches[0] == 0, err_msg
        assert image_dim[2] % num_patches[1] == 0, err_msg
        px_size = image_dim[1] // num_patches[0]
        py_size = image_dim[2] // num_patches[1]
        self.patch_size = px_size, py_size

        # Linear embedding of each patch
        self.patch_embed = nn.Linear(px_size * py_size, embed_dim)

        # Class token
        self.class_token = nn.Parameter(torch.rand(1, embed_dim))
```

```

        # Sinusoidal Positional encoding
        seq_len = num_patches[0] * num_patches[1] + 1
        positions = torch.arange(seq_len)
        timescales = torch.pow(
            1 / seq_len, 2 * torch.arange(embed_dim) // 2 / embed_dim
        )
        angles = positions[:, None] * timescales[None, :]
        pos_encoding = torch.zeros((seq_len, embed_dim))
        pos_encoding[0::2, :] = torch.sin(angles[0::2, :])
        pos_encoding[1::2, :] = torch.cos(angles[1::2, :])
        self.register_buffer("pos_encoding", pos_encoding)

        # Transformer encoder blocks
        self.blocks = nn.ModuleList(
            TransformerEncoderBlock(embed_dim, ffn_size, num_heads,
            dropout)for _ in range(num_blocks)
        )
        # final layer norm
        self.blocks.append(nn.LayerNorm(embed_dim))

        # Prediction head
        self.prediction_head = nn.Linear(embed_dim, num_classes)


    def patchify(self, x: torch.Tensor):
        # (batch_size, channel, patch_x, height, patch_y, width)
        patches = x.view(
            x.shape[0],
            x.shape[1],
            self.num_patches[0],
            self.patch_size[0],
            self.num_patches[1],
            self.patch_size[1],
       )
        # (batch_size, patch_x, patch_y, channel, height, width)
        patches = patches.transpose(3, 4).moveaxis(1, 3)
        # Flatten the patch dims
        # (batch_size, num_patches, channel, height, width)
        patches = patches.flatten(1, 2)
        # Faltten the patch
        # (batch_size, num_patches, num_pixels)
        patches = patches.flatten(2, 4)

        return patches

    def forward(self, x: torch.Tensor):
        # Create patches
        # (batch_size, num_patches, num_pixels)
        patches = self.patchify(x)

        # Pass through linear embedding
        # (batch_size, seq_len, emb_size)
        patches = self.patch_embed(patches)
```

```
120
121         # Adding in the class embedding
122         patches = torch.cat(
123             [
124                 self.class_token.expand(patches.size(0), 1, -1),
125                 patches,
126             ],
127             dim=1,
128         )
129
130         # Add positional encoding
131         hidden = patches + self.pos_encoding[None, :, :]
132
133         # Pass through transformer encoder layer
134         for block in self.blocks:
135             hidden = block(hidden)
136
137         # Get the class token embedding only
138         output = hidden[:, 0, :]
139
140         # Output
141         output = self.prediction_head(output)
142
143         return output
```

- `TransformerEncoderBlock`: This is the implementation of the transformer encoder block using Pre-LN architecture.
  - `__init__` (**Lines 7–22**): This initializes all the necessary parameters and layers in the encoder block.
  - `forward` (**Lines 24–33**): This implements a single block of transformer encoder layer using the Pre-LN architecture. See Figure 9.10 for an illustration when reviewing the code.
- `VisionTransformer`: This is the main module of the ViT model.
  - `__init__` (**Lines 37–88**): This initializes the parameters and layers in the ViT model.
    * Lines 50–51 store some parameters to be used later on.
    * Lines 54–59 compute the patch size based on the input image size and the number of desired patches. Notice that the image width and height need to be divisible by the number of desired patches along each dimension. Otherwise, an assertion error will be thrown.
    * Line 62 initializes a feedforward layer which will be used to reduce the dimensionality of the patches.
    * Line 65 initializes a model parameter for "classification token" embedding. This embedding vector will be added at the beginning of the input sequence before feeding into the transformer encoder blocks.

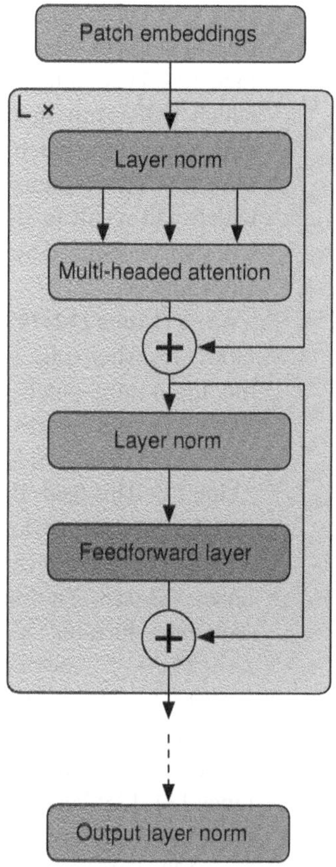

Figure 9.10 Transformer pre-LN architecture. After L blocks of encoder layers, the outputs will be additionally passed to a normalization layer before outputting the final token sequence representation.

   * Lines 68–77 implement the sinusoidal encoding based on Equation 9.1. See Figure 9.8 for an illustration of the resulting positional encoding added to the patches. The `self.register_buffer` method keeps the precomputed sinusoidal positional embeddings as a static copy in the class attributes which will then be later referenced by the `forward` method.
   * Lines 80–83 create a series of transformer encoder blocks.
   * Line 85 adds a final LayerNormalization layer after the series of transformer encoder blocks, as designed by the Pre-LN architecture (see Figure 9.10).
   * Line 88 creates the prediction head, which outputs the logit value (i.e. before softmax activation) of the probabilities of each class.
- `patchify` (**Lines 91–110**): This method converts the 2D images into flattened 1D vectors of each patch.
   * Lines 93–100 reshape the images in the height and width dimensions, creating a tensor of shape `(batch_size, channel, patch_x, height, patch_y, width)`.

  * Line 102: These operations rearrange the dimensions of the tensor so that we can perform subsequent flattening.
  * Line 105: This flattens the `patch_x` and `patch_y` dimensions together, which will result in the final `num_patches` or `sequence_length` dimension of the patch embeddings.
  * Line 108: This flattens channel, patch height, and patch width together, which will result in the final `num_pixels` dimension. This dimension will then be reduced by the subsequent feedforward layer to form the patch embeddings.
- `forward` (**Lines 112–143**): The main logic of the ViT is implemented here.
  * Line 115: This calls the `patchify` method on the input to reshape the input into patch vector representations, resulting in a tensor of shape `(batch_size, sequence_length, num_pixels)`, where `sequence:length` is the number of patches from the image.
    * Line 119: This feeds the output from the above step to a feedforward layer to reduce the dimensionality of the vector of each patch, forming a tensor of shape `(batch_size, sequence_length, embed_size)`.
  * Lines 122–128: This adds the "classification token" embedding to the sequence, forming a tensor of shape `(batch_size, sequence_length + 1, embed_size)`.
  * Line 131: Adds the precomputed sinusoidal positional encoding to the image patch embeddings, in preparation for feeding it to the subsequent blocks of transformer encoder layers.
  * Lines 134–135: Feed the inputs to the transformer encoder layers, then to the final LayerNormalization layer.
  * Line 138: Extracts the "classification token" embedding from the transformer encoder's output. This is the latent representation of the final class to be predicted.
  * Line 141: Decodes the class from the latent embedding vector by outputting the logit values of the probabilities of each class.

Let us train the model using the MNIST dataset to classify the handwritten digits in the images.

```
from tqdm import tqdm
import torch.nn.functional as F
from torch.utils.data import DataLoader
# !pip install torchvision==0.16.0
from torchvision.datasets.mnist import MNIST
from torchvision.transforms import ToTensor
from torchsummary import summary

train_set = MNIST(
```

```
    root="./mnist", train=True, download=True, transform=ToTensor()
)
eval_set = MNIST(
    root="./mnist", train=False, download=True, transform=ToTensor()
)

train_loader = DataLoader(train_set, shuffle=True, batch_size=128)
eval_loader = DataLoader(eval_set, shuffle=False, batch_size=128)
train_data_size = len(train_loader)
eval_data_size = len(eval_loader)

if torch.cuda.is_available():  # NVIDIA GPU
    device = torch.device("cuda")
elif torch.backends.mps.is_available():  # Apple Silicon
    device = torch.device("mps")
else:
    device = torch.device("cpu")

# Training loop
model = VisionTransformer().to(device)
optimizer = torch.optim.Adam(model.parameters(), lr=0.005)
num_epochs = 10

def train_epoch(model, device, train_loader, optimizer):
    model.train()
    loss_train = 0.0
    n_examples = 0
    for X, y in tqdm(train_loader):
        # Ship data to training device
        X, y = X.to(device), y.to(device)
        n_examples += len(y)
        # Forward pass
        optimizer.zero_grad()
        y_pred = model(X)
        loss = F.cross_entropy(y_pred, y, reduction="sum")
        loss_train += loss # accumulating
        loss = loss / len(y) # average
        loss.backward()
        optimizer.step()

    # Gather losses
    avg_loss = loss_train / n_examples

    return avg_loss


def eval_epoch(model, device, eval_loader):
    loss_eval = 0.0
    accuracy = 0.0
    n_examples = 0
    with torch.no_grad():
        for X, y in tqdm(eval_loader):
            X, y = X.to(device), y.to(device)
            n_examples += len(y)
```

```
                y_pred = model(X)
                loss = F.cross_entropy(y_pred, y, reduction="sum")
                loss_eval += loss.item()

                # Compute accuracy
                accuracy += (y_pred.argmax(dim=1) == y).sum()

        # Gather losses
        avg_loss = loss_eval / n_examples
        avg_acc = accuracy / n_examples
        return avg_loss, avg_acc


    for epoch in range(num_epochs):
        train_loss = train_epoch(model, device, train_loader, optimizer)
        eval_loss, avg_acc = eval_epoch(model, device, eval_loader)
        print(
            f"Epoch {epoch+1}/{num_epochs}:",
            f"train_loss - {train_loss:.2f},",
            f"eval_loss - {eval_loss:.2f},",
            f"eval_acc - {avg_acc:.2f}",
        )
```

The training results are the following:

```
100%|| 469/469 [00:18<00:00, 25.20it/s]
100%|| 79/79 [00:02<00:00, 37.86it/s]
Epoch 1/10: train_loss - 1.54, eval_loss - 1.05, eval_acc - 0.63
100%|| 469/469 [00:13<00:00, 34.59it/s]
100%|| 79/79 [00:01<00:00, 51.05it/s]
Epoch 2/10: train_loss - 0.88, eval_loss - 0.77, eval_acc - 0.75
100%|| 469/469 [00:13<00:00, 34.10it/s]
100%|| 79/79 [00:01<00:00, 51.78it/s]
Epoch 3/10: train_loss - 0.68, eval_loss - 0.62, eval_acc - 0.80
100%|| 469/469 [00:13<00:00, 34.70it/s]
100%|| 79/79 [00:01<00:00, 51.54it/s]
Epoch 4/10: train_loss - 0.59, eval_loss - 0.54, eval_acc - 0.83
100%|| 469/469 [00:13<00:00, 34.69it/s]
100%|| 79/79 [00:01<00:00, 50.84it/s]
Epoch 5/10: train_loss - 0.54, eval_loss - 0.51, eval_acc - 0.84
100%|| 469/469 [00:13<00:00, 34.52it/s]
100%|| 79/79 [00:01<00:00, 51.75it/s]
Epoch 6/10: train_loss - 0.52, eval_loss - 0.51, eval_acc - 0.84
100%|| 469/469 [00:13<00:00, 34.30it/s]
100%|| 79/79 [00:01<00:00, 51.49it/s]
Epoch 7/10: train_loss - 0.50, eval_loss - 0.48, eval_acc - 0.84
100%|| 469/469 [00:13<00:00, 34.44it/s]
100%|| 79/79 [00:01<00:00, 51.06it/s]
Epoch 8/10: train_loss - 0.48, eval_loss - 0.50, eval_acc - 0.84
100%|| 469/469 [00:13<00:00, 34.11it/s]
100%|| 79/79 [00:01<00:00, 50.31it/s]
Epoch 9/10: train_loss - 0.47, eval_loss - 0.43, eval_acc - 0.86
```

```
100%|| 469/469 [00:13<00:00, 33.88it/s]
100%|| 79/79 [00:01<00:00, 50.92it/s]
Epoch 10/10: train_loss - 0.46, eval_loss - 0.43, eval_acc - 0.87
```

9.4 Bucketizing Values

Suppose we have a continuous variable in a given range, and we would like to divide the range into several smaller ranges, which we call buckets (a.k.a bins) so that we could assign an integer index to indicate which bucket each value falls into. This is called **Binning** or **Bucketization**. Hence, the continuous variable becomes a categorical or ordinal variable. For example, given the continuous feature, $x = [0, 1, 2, 3, 4, 5, 6, 7, 8]$, we would like to assign integer values according to the following range buckets:

$$x \in \begin{cases} [0,3) \rightarrow 0 \\ [3,6) \rightarrow 1 \\ [6,9) \rightarrow 2 \end{cases}$$

Then, the bucketized values are indices $= [0, 0, 0, 1, 1, 1, 2, 2, 2]$.

There are several ways we can bucketize values using vectorization. NumPy's `np.digitize`, Tensorflow's `tf.raw_ops.Bucketize`, and PyTorch's `torch.bucketize` conveniently serve this purpose, though the default formulation of these libraries can be different. Given a list of continuous values and the buckets specified as a monotonically increasing list $[b_0, b_1, b_2, b_3, \ldots, b_N]$, where $b_0 < b_1 < b_2 < \cdots < b_N$, and the buckets are arranged as $(-\infty, b_0), [b_0, b_1), [b_1, b_2) \ldots, [b_{N-1}, b_N), [b_N, +\infty)$ (i.e. left inclusive, right exclusive, or $b_{i-1} \leq x < b_i$ for the ith bucket). This totals $N + 2$ buckets since these functions add two special buckets that account for values beyond the specified range. Let us denote this as the "left-inclusive" formulation. Alternatively, the buckets can also be formulated as $(-\infty, b_0], (b_0, b_1], \ldots, (b_{N-1}, b_N], (b_N, +\infty)$ (i.e. left exclusive, right inclusive, or $b_{i-1} < x \leq b_i$ for the ith bucket). Let us call this the "left-exclusive" formulation. Interestingly, `np.digitize` (`numpy==1.26.1`) uses the "left-inclusive" formulation when it is called with the default `right=False` setting, while setting `right=True` switches it to use the "left-exclusive" formulation; Tensorflow's `tf.raw_ops.Bucketize` also uses and can only use "left-inclusive" formulation. Interestingly, the default behavior of `torch.bucketize` is the opposite of the default behavior of `np.digitize`. `torch.bucketize` uses "left-exclusive" formulation by default when setting `right=False`, while setting `right=True` switches its behavior to "left-inclusive."

To illustrate,

```
import numpy as np
import torch
x = np.arange(10)
buckets = [0, 3, 6, 9]

# Left-inclusive [1, 1, 1, 2, 2, 2, 3, 3, 3, 4]
indices_np_false = np.digitize(x, buckets, right=False)
indices_torch_true = torch.bucketize(
    torch.tensor(x), torch.tensor(buckets), right=True
)
assert np.allclose(indices_np_false, indices_torch_true)

# Left-exclusive [0, 1, 1, 1, 2, 2, 2, 3, 3, 3]
indices_np_true = np.digitize(x, buckets, right=True)
indices_torch_false = torch.bucketize(
    torch.tensor(x), torch.tensor(buckets), right=False
)
assert np.allclose(indices_np_true, indices_torch_false)
```

An alternative operation is `searchsorted` implemented by all three libraries. The following illustrates that these operations are equivalent:

```
import numpy as np
import tensorflow as tf
import torch

x = np.arange(10)
buckets = np.array([0, 3, 6, 9])
x_t = torch.tensor(x)
buckets_t = torch.tensor(buckets)

# Left inclusive [1, 1, 1, 2, 2, 2, 3, 3, 3, 4]
res = [
    ## using digitize/bucketize
    np.digitize(x, buckets, right=False),
    torch.bucketize(x_t, buckets_t, right=True),
    ## using searchsorted
    np.searchsorted(buckets, x, side="right"),
    torch.searchsorted(buckets_t, x_t, side="right"),
    tf.searchsorted(buckets, x, side="right")
]
# Pair-wise comparison
assert all([np.allclose(res[ii], res[ii+1]) for ii in range(len(res)-1)])


# Left exclusive [0, 1, 1, 1, 2, 2, 2, 3, 3, 3]
res = [
```

```
        ## using digitize/bucketize
        np.digitize(x, buckets, right=True),
        torch.bucketize(x_t, buckets_t, right=False),
        ## using searchsorted
        np.searchsorted(buckets, x, side="left"),
        torch.searchsorted(buckets_t, x_t, side="left"),
        tf.searchsorted(buckets, x, side="left")
    ]
    # Pair-wise comparison
    assert all([np.allclose(res[ii], res[ii+1]) for ii in range(len(res)-1)])
```

We can also implement bucketization from scratch using masking, though it may not be as efficient as using either `digitize` or `searchsorted`. See Figure 9.11 for an illustration of the masking process.

```
1   import numpy as np
2
3   def bucketize_via_masking(x, buckets, side="left"):
4       """Bucketize values based via broadcasting and
         ↪ masking."""
5       # keep original shape in case we need to deal
6       # with multi-dimensional arrays
7       shape = x.shape
8       x = x.flatten() # flatten
9       # convert to a column vector for subsequent broadcasting
10      buckets = buckets[:, None]
11      if side == "left":
12          mask = x > buckets # left-exclusive
13      else:
14          mask = x >= buckets # left-inclusive
15
16      # Count the indices
17      indices = mask.sum(axis=0)
18
19      # reshape back to the shape of inputs
20      indices = indices.reshape(shape)
21
22      return indices
```

- Line 7: Stores the shape of the original input. This is because we will be performing broadcasting on the flattened inputs, but for the mapped bucket indices, we would still like to return the same shape as that of the input.
- Line 8: Flattens the input.
- Line 10: Converts `buckets` into a column vector.
- Lines 11–14: If we follow the `searchsorted` argument `side`, and when `side="left"`, we formulate as left-exclusive, then input values should be

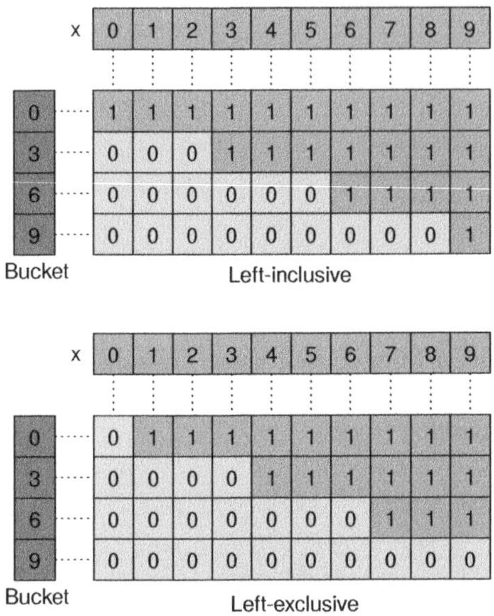

Figure 9.11 Example illustration of left-inclusive and left-exclusive masking results.

strictly greater than the bucket values. Otherwise, to be left-inclusive, the input values are greater than or equal to the bucket values. See Figure 9.11 for illustrations of the resulting masks under the two formulations.

- Line 17: Here we are counting how many indicators satisfy the formulation condition. Using `x=3` as an example (Figure 9.11), if we are formulating as "left-inclusive" (`side="right"`), then `3 >= 0 -> True -> 1`, `3 >= 3 -> True -> 1`, `3 >=6 -> False -> 0`, and `3 >=9 -> False -> 0`. The value `3` has indicators `[1, 1, 0, 0]` along the buckets. Adding the indicators together, we obtain an index value of `2`, which corresponds to the bucket range $[3, 6)$ (with index `1` being $[0, 3)$, and index `0` being $(-\infty, 0)$). By contrast, if we are formulating as "left-exclusive" (`side="left"`), then `3 > 0 -> True -> 1`, `3 > 3 -> False -> 0`, `3 > 6 -> False -> 0`, and `3 > 9 -> False -> 0`. Now, the value `3` has accumulated indicators as `[1, 0, 0, 0]`. Adding the indicators together, we obtain an index value of `1`, which corresponds to the bucket range $(0, 3]$ (recall the index `0` corresponds to $(-\infty, 0]$).
- Line 20: Reshapes the flattened, mapped indices back to the shape of the input.

There are several applications that need value bucketization. For example, we can engineer a continuous feature as a categorical feature since the model may benefit from learning the features as a group rather than a fine-grained continuous value that may not represent any important aspects of the feature. Consider the user age feature which is typically stored as numerical values in the data. However, when predicting if a user will watch mature content, it may be more beneficial to use a binary feature indicating whether or not the user is an adult, rather than their exact age. Thus, bucketization is a powerful feature engineering technique that helps data scientists and machine learning engineers build more efficient models. In another example, if we would like to plot a histogram given a list of values, we can first assign the continuous values in a predefined list of buckets, and then use `np.unique` to count the number of occurrences of each bin. We can also use bucketization operations to downsample float values into lower-precision integer values to compress the size of the model weights, for example. This is known as **quantization** and is a key technique to speed up model inference by sacrificing output quality due to reduced precision. Both Tensorflow [Abadi et al., 2023] and PyTorch [Paszke et al., 2023] have dedicated operations and quantization integer representations (`qint8`, `qint32`, etc.) to help the parameter quantization and model compression processes.

9.5 Segment-wise Aggregation

Although Pandas is the best-known library for the groupby-aggregate-apply pattern, more basic numerical libraries including NumPy as well as deep-learning-oriented ones like Tensorflow and PyTorch also implement operations to facilitate the use of this pattern for data transformation. More specifically, these are called segment operations, where, in general,

- Sequences of elements of variable lengths are arranged along a single dimension.
- An array of integers labels the group identity of the segments.
- Elements within the group are aggregated together by a reduction operation, such as summation.

See Figure 9.12 for an illustration of an example of segment-wise summation operations. Table 9.2 provides a list of segment operations in NumPy, Tensorflow, and PyTorch.

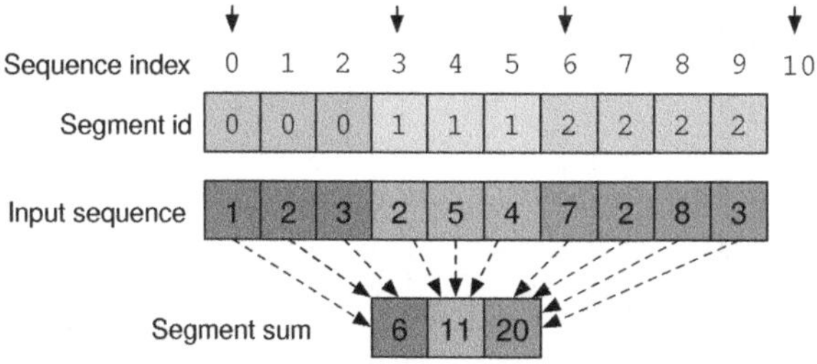

Figure 9.12 Illustration of segment-wise summation operation, with each color indicating a segment/group of values. There are two ways to denote the segment, either through sequence index or segment id.

Table 9.2 List of groupwise aggregation operations in NumPy, Tensorflow, and PyTorch.

Operation	NumPy	Tensorflow	PyTorch
Groupwise sum[a)]	`np.add.reduceat`	`tf.math.segment_sum`, `tf.math.unsorted_segment_sum`, `tf.sparse.segment_sum`	`torch.scatter_add`, `torch.scatter_reduce(..., reduce="sum")`, `torch.segment_reduce(arr, "sum", offsets)`

Notes:

a) There are in general two ways to label the segments (see also Figure 9.12):

- **Start–end indices**: This is reminiscent of `csr_matrix`'s `indptr` property from Scipy, where an array of ascending integer indices indicates the starts and ends of each segment along a flattened array. This also assumes the elements of the same group are adjacent to each other. When working with NumPy's groupwise operations `np.ufunc.reduceat`, the value to the `indices` argument is an array of starting indices of each segment, signifying the slices to aggregate (this means, it does not have to be strictly monotonically increasing). It assumes the last segment starts from the last element of `indices` to the last element of the array, i.e. `data[indices[-1]:]`. `indices` here is also equivalent to `indptr[:-1]` which excludes the last element that denotes the end of the segment. Example usage:

```
import numpy as np
data = np.array([1, 2, 3, 2, 5, 4, 7, 2, 8, 3])
indices = np.array([0, 3, 6])
res_np = np.add.reduceat(data, indices)
# sum(data[0:3]), sum(data[3:6]), sum(data[6:])
# array([ 6, 11, 20])

# non-monotonically increasing indices
indices = np.array([0, 3, 2, 6])
res_np = np.add.reduceat(data, indices)
# sum(data[0:3]), sum(data[3]), sum(data[2:6]), sum(data[6:])
# array([ 6,  2, 14, 20])
```

In the case of non-monotonically increasing indices, the second slice `[3:2]` is interpreted as `[3]`, which then takes the element at index 3 only. Therefore, there is no good way to create segments with zero elements. On the other hand, PyTorch's `torch.segment_reduce` operation has two arguments to indicate the start and end indices:

- `offsets`: Unlike NumPy, the value to this argument needs to include the end index of the last segment, just like the `indptr` property of `csr_matrix`. Unlike the `indices` argument of `np.ufunc.reduceat`, the `offsets` property needs to be strictly monotonically increasing, since `offsets` is being used to compute the length of the segments. Note that it is possible to specify zero-length segments, e.g. (continue on next page)

```
import torch
data = torch.tensor([1, 2, 3, 2, 5, 4, 7, 2, 8, 3], dtype=float)
offsets = torch.tensor([0, 3, 6, 6, 10])
res_torch = torch.segment_reduce(data, "sum", offsets=offsets)
# tensor([ 6., 11.,  0., 20.], dtype=torch.float64)
```

- `lengths`: Instead of specifying the start and end indices, specify the length of each segment in the sequence. The `lengths` of the segments needs to be nonnegative. So, it is possible to create segments with zero lengths, e.g.

```
import torch
data = torch.tensor([1, 2, 3, 2, 5, 4, 7, 2, 8, 3], dtype=float)
lengths = torch.tensor([3, 3, 0, 4])
res_torch = torch.segment_reduce(data, "sum", lengths=lengths)
# tensor([ 6., 11.,  0., 20.], dtype=torch.float64)
```

At the time of writing, PyTorch's `torch.segment_reduce` operation is only implemented for float dtype arrays.

- **Segment id**: Tensorflow's segment operations as well as PyTorch's scattering operations use this strategy. Instead of explicitly inferring the segments based on the intervals between the start and end indices, segment ids specifically label which group a segment belongs to. Here, the `segment_id` input is treated as the group key. This can be advantageous because we can then use operations that do not necessarily have all elements of the same group arranged next to each other before aggregation. Tensorflow implements three variants of segment aggregation operations, which assume the first axis (`axis=0`) is the segment dimension to be aggregated.
  - `tf.math.segment_sum`: This operation still assumes that the values of the same segments are adjacent to each other. It requires the segment ids to be sorted. The integer values can be skipped to create 0 entries in between segments with lengths greater than 0; it cannot create the 0-length entries as the last segments. Negative segment ids will throw an error.
  - `tf.math.unsorted_segment_sum`: This operation does not assume the values of the same segments are adjacent to each other. That is, the elements of each segment can be scattered around. The integer values can be skipped to create 0 entries in between the segments with lengths greater than 0; it can also create 0-length entries as the last segments by specifying the `num_segments` argument. Negative segment ids will be ignored. PyTorch's `torch.scatter_reduce` operation can be viewed as the equivalent operation to `tf.math.unsorted_segment_*`.
  - `tf.sparse.segment_sum`: The previous two operations require the number of segment ids to be equal to the number of elements in the first dimension of the data input. This operation loosens that requirement. It requires an additional argument `indices`, which is paired with `segment_ids` to extract the elements from the data input by the group. Suppose, we have `data = [[1], [2], [3], [4], [5]]` and grouping ids `[0, 0, 1, 1, 1]`. If we specify `indices = [0, 1, 2, 3, 4]`, then this is equivalent to using `tf.math.segment_sum` or `tf.math.unsorted_segment_sum`, obtaining `[3, 12]`, which is from `[data[0]+data[1], data[2]+data[3]+data[4]]`, or `[1+2, 3+4+5]`. But, if we specify `indices = [0, 1, 2, 3, 0]`, then we will obtain `[3, 8]` which is from `[data[0]+data[1], data[2]+data[3]+data[0]]`, or `[1+2, 3+4+1]`. Similar to `tf.math.segment_sum`, negative segment ids will throw an error.

(continued)

Table 9.2 (Continued)

Operation	NumPy	Tensorflow	PyTorch
Groupwise mean[b)]	N/A	`tf.math.segment_mean`, `tf.math.unsorted_segment_mean`, `tf.sparse.segment_mean`	`torch.scatter_reduce(..., reduce="mean")`, `torch.segment_reduce(arr, "mean", offsets)`
Groupwise product[c)]	`np.multiply.reduceat`	`tf.math.segment_prod`, `tf.math.unsorted_segment_prod`	`torch.scatter_reduce(..., reduce="prod")`, `torch.segment_reduce(arr, "prod", offsets)`

b) NumPy does not implement a groupwise mean function. This is probably because the mean operation is technically not an element-wise reduction operation, as it not only needs to track the sum but also the total number of elements within each group. However, an easy workaround is to utilize `np.add.reduceat` to implement the equivalent function, where we compute the groupwise sum first, and then count the number of elements within the group by summing together an indicator array of 1s. The implementation is as follows:

```
import numpy as np
def np_mean_reduceat(arr, indices, **kwargs):
    # Compute groupwise sum
    gsum = np.add.reduceat(arr, indices, **kwargs)
    # Compute groupwise count
    indicator = np.ones_like(arr)
    gcount = np.add.reduceat(indicator, indices, **kwargs)
    # Compute mean
    return gsum / gcount
```

c) Tensorflow does not implement the `tf.sparse` version of the segment multiplication operation. But, it is possible to work around this by performing segment addition of log values, then raising back to the exponent (with some sacrifice on numerical precision), i.e.

```python
import tensorflow as tf

def tf_sparse_segment_prod(data, indices, segment_ids, num_segments=None):
    # take log values
    data_out = tf.math.log(tf.cast(data, tf.float32))
    # Segment sum of log
    data_out = tf.sparse.segment_sum(
        data_out, indices, segment_ids, num_segments
    )
    # Raise back to exponents
    data_out = tf.exp(data_out)
    if data.dtype in (tf.int32, tf.int64):
        data_out = tf.round(data_out)
    data_out = tf.cast(data_out, data.dtype)
    return data_out

data = tf.constant([1, 2, 3, 2, 3, 6, 3, 4])
indices = tf.constant([0, 1, 2, 3, 4, 5, 6, 7])
segment_ids = tf.constant([0, 0, 0, 1, 1, 1, 2, 2])
data_out = tf_sparse_segment_prod(data, indices, segment_ids)
# <tf.Tensor: shape=(3,), dtype=int32, numpy=array([ 6, 36, 12], dtype=int32)>
```

(continued)

Table 9.2 (Continued)

Operation	NumPy	Tensorflow	PyTorch
Groupwise sum divided by sqrt[d)]	N/A	`tf.math.unsorted_segment_sqrt_n`, `tf.sparse.segment_sqrt_n`	N/A
Groupwise maximum[e)]	`np.maximum.reduceat`	`tf.math.segment_max`, `tf.math.unsorted_segment_max`	`torch.scatter_reduce(..., reduce="amax")`, `torch.segment_reduce(arr, "max", offsets)`
Groupwise minimum[e)]	`np.minimum.reduceat`	`tf.math.segment_min`, `tf.math.unsorted_segment_min`	`torch.scatter_reduce(..., reduce="amin")`, `torch.segment_reduce(arr, "min", offsets)`

d) `segment_sqrt_n` is a special function that was motivated by previous natural language modeling studies that attempted to summarize embeddings of tokens within a sentence, i.e. representing a sentence as a bag of words (e.g. pretrained feed-forward Neural-Net Language Model, or NNLM [Bengio et al., 2003], from TFHub [Google, 2023]). These studies found that taking the square root of the sum of the embeddings is more effective in representing sentences than averaging the embeddings of the tokens together. Similar to the groupwise mean function, this function can also be implemented using `np.add.reduceat` in NumPy or `torch.scatter_add` in PyTorch. We illustrate an implementation in the following code block using PyTorch:

```
import torch

def torch_scatter_reduce_sqrt_n(arr, segment_ids, **kwargs):
    # Initialize accumulators
    numerator = torch.zeros(segment_ids.max() + 1)
    denom = torch.zeros_like(numerator)
    # Compute groupwise sum
    gsum = numerator.scatter_add_(0, segment_ids, arr)
    # Compute groupwise count
    indicator = torch.ones_like(arr)
    gcount = denom.scatter_add_(0, segment_ids, indicator)
    # Compute the sum / sqrt_n
    return gsum / torch.sqrt(gcount)
```

e) These operations compute the segment-wise minimum and maximum values. A caveat with the Tensorflow operations using unsorted segment ids, i.e. `tf.math.unsorted_segment_max` and `tf.math.unsorted_segment_min`, is that if we skip over some integer ids with the intention to create zero entries, we will find that these skipped entries will have $-\infty$ (e.g. -3.4028235e+38) or $+\infty$ (e.g. +3.4028235e−38) filled, respectively, instead of having zeros. So, care needs to be taken when using these operations.

9.6 Case Study: EmbeddingBag

PyTorch has a convenient neural network layer class `torch.nn.EmbeddingBag` EmbeddingBag which can summarize the embeddings from the same group of values, e.g. a sequence of embeddings for events occurring on the same date. In Chapter 8, we looked at a case study that leveraged the idea of sparse matrix multiplication and used sparse tensor inputs to look up and add the embeddings together. In this case study, we will generalize the embedding lookup and summarization process in Tensorflow. We will implement four summarization modes, `"mean"`, `"max"`, `"sum"`, and `"sqrt_n"`, to combine the embeddings from the same group (i.e. values from the same row of a batch of variable-length categorical values). The following code block implements the `EmbeddingBag` layer using Tensorflow and Keras.

```
import tensorflow as tf
from keras import Model, Input
from keras.layers import Layer, Embedding

class EmbeddingBag(Layer):
    """
    Keras EmbeddingBag class that summarizes embeddings of a sparse input.

    Inputs:
        * num_items: number of embeddings to create.
        * emb_size: size of embeddings.
        * mode: embedding summarization mode,
            one of ["mean", "max", "sum", "sqrt_n"].
        * init_weights: NumPy array of embedding values. To keep using the
            initialized values, set 'trainable=False' to prevent the
            embedding values being updated during training.
        * **kwargs: additional keyword arguments to pass to the Keras
            Embedding layer.
    """
    def __init__(
            self,
            num_items,
            emb_size,
            mode="mean",
            init_weights=None,
            name="embedding_bag",
            **kwargs
        ):
        super(EmbeddingBag, self).__init__(name=name)
        assert mode in ("mean", "max", "sum", "sqrt_n")
        self.mode = mode
        self.emb_size = emb_size
        self.embedding = Embedding(
            num_items, emb_size, **kwargs
        )
```

```
        # load pretrained weights
        if init_weights is not None:
            self.embedding.build([]) # initialize
            self.embedding.set_weights([init_weights])

    def call(self,
         x: tf.SparseTensor,
         weights: tf.SparseTensor = None,
         training: bool = False
    ):
        if weights is None:
            # equal weights
            weights = x.with_values(tf.ones_like(x.values, tf.float32))

        # flattened (batch_size x seq_len, emb_size)
        emb = self.embedding(x.values)
        emb *= weights.values[:, None]

        # Group-wise aggregation
        # Assuming x is already left aligned and indices sorted
        if self.mode == "mean":
            emb = tf.sparse.segment_mean(
                emb,
                indices=tf.range(tf.shape(emb)[0]),
                segment_ids=x.indices[:, 0],
                num_segments=tf.shape(x)[0],
            )
        elif self.mode == "sum":
            emb = tf.sparse.segment_sum(
                emb,
                indices=tf.range(tf.shape(emb)[0]),
                segment_ids=x.indices[:, 0],
                num_segments=tf.shape(x)[0],
            )
        elif self.mode == "sqrt_n":
            emb = tf.sparse.segment_sqrt_n(
                emb,
                indices=tf.range(tf.shape(emb)[0]),
                segment_ids=x.indices[:, 0],
                num_segments=tf.shape(x)[0],
            )
        elif self.mode == "max":
            emb = tf.math.segment_max(emb, x.indices[:, 0])
            # Appending the potentially missing rows where the ids are
             ↪ empty
            emb = tf.pad(
                emb, [(0, tf.shape(x)[0] - tf.shape(emb)[0]), (0, 0)]
           )

        return emb

    @property
    def model(self):
        inputs = {
```

```
89                "x": Input((None,), sparse=True, dtype=tf.int64, name="items"),
90                "weights": Input(
91                    (None,), sparse=True, dtype=tf.float32, name="weights"
92                ),
93            }
94            model = Model(inputs=inputs, outputs=self.call(**inputs))
95
96            return model
```

- __init__ (**Lines 20–39**): This method initializes the EmbeddingBag layer class. It leverages Keras' built-in Embedding layer to create the embedding lookup table (Lines 33–35). Additionally, we allow the embedding values to be created from a set of pretrained weights, by specifying and loading the weights of the Embedding layer. To keep using the same embedding values during training without further updates, users can set trainable=True when initializing the layer.
- call (**Lines 41–84**): This method takes two data inputs, x which is a sparse tensor of the integer index representation of categorical values, and weights which is a sparse tensor of weights for each category, with shape and number of elements identical to that of x.
  - Lines 46–48 if weights is not specified, then we default all embeddings having identical weights of 1.
  - Lines 51–52 perform the embedding lookup and apply the weights of the category.
  - Lines 56–82 apply the groupby and aggregate operations of the embedding vectors using the mode specific groupby and aggregate operations. Note that for "mean", "sum", and "sqrt_n", we use tf.sparse.segment_<mean/sum/sqrt_n> operations. This is because when x is a sparse input, there may be rows within the batch that have empty entries, and we would like to represent the sequence with empty entries with a vector of all zeros. By specifying the num_segments=tf.shape(x)[0], we are guaranteed to return a dense tensor of shape (batch_size, emb_size). If we do not specify the num_segments argument, then if an empty row is the last row of the batch, we will be missing its representation. For "max" mode, we use tf.math.segment_max since there is no corresponding operation under the tf.sparse module. For the same reason, when implementing the "max" mode, we need to append zero vectors to the batch such that the output summarized embedding tensor has an expected shape of (batch_size, emb_size). To illustrate:

```
import tensorflow as tf
# 6 valid values total
x = tf.ragged.constant([[1, 3], [2], [], [6, 8, 9], []]).to_sparse()
# After embedding lookup, we have a matrix of (7, 8)
emb = tf.random.uniform((len(x.values), 8))
# Summarized correctly: (5, 8)
emb_sum_correct = tf.sparse.segment_sum(
    emb, x.indices[:, 1], x.indices[:, 0], num_segments=tf.shape(x)[0]
)
# Summarized incorrect: (4, 8)
emb_sum_wrong = tf.sparse.segment_sum(
    emb, x.indices[:, 1], x.indices[:, 0]
)
# Summarized wrong using tf.math.segment_max: (4, 8)
emb_max_wrong = tf.math.segment_max(
    emb, x.indices[:, 0]
)
# Need to pad the empty entries occurring at the end
pad_size = tf.shape(x)[0] - tf.shape(emb_max_wrong)[0]
emb_max_correct = tf.pad(emb_max_wrong, [(0, pad_size), (0, 0)])
```

- `model` property (**Lines 86–96**): This property is used to check if our implementation can be compiled as a graph. We can print the model architecture using `print(layer.model.summary())` which gives us the following:

```
Model: "model_3"
____________________________________________________________________
Layer (type)          Output Shape      Param #   Connected to
====================================================================
items                 [(None, None)]    0         []
(InputLayer)

weights               [(None, None)]    0         []
(InputLayer)

input.values_6        (None,)           0         ['items[0][0]']
(InstanceProperty)

input.values_7        (None,)           0         ['weights[0][0]']
(InstanceProperty)

embedding_16          (None, 4)         4000      ['input.values_6[0][0]']
(Embedding)

tf.__operators__      (None, 1)         0         ['input.values_7[0][0]']
.getitem_12
(SlicingOpLambda)

input.indices_3       (None, 2)         0         ['items[0][0]']
(InstanceProperty)
```

```
tf.math.multiply_3  (None, 4)         0          ['embedding_16[0][0]',
(TFOpLambda)                                      'tf.__operators__.getitem_12
                                                  [0][0]']

tf.__operators__    (None,)           0          ['input.indices_3[0][0]']
.getitem_13
(SlicingOpLambda)

tf.math.segment_    (None, 4)         0          ['tf.math.multiply_3[0][0]',
max_3 (TFOpLambda)                               'tf.__operators__.getitem_13
                                                 [0][0]']

tf.compat.v1        (2,)              0          ['items[0][0]']
.shape_6
(TFOpLambda)

tf.compat.v1        (2,)              0          ['tf.math.segment_max_3
.shape_7                                          [0][0]']
(TFOpLambda)

tf.__operators__    ()                0          ['tf.compat.v1.shape_6[0][0]']
.getitem_14
(SlicingOpLambda)

tf.__operators__    ()                0          ['tf.compat.v1.shape_7[0][0]']
.getitem_15
(SlicingOpLambda)

tf.math.subtract_3  ()                0          ['tf.__operators__.getitem_14
(TFOpLambda)                                      [0][0]','tf.__operators__
                                                  .getitem_15[0][0]']

tf.compat.v1.pad_3  (None, 4)         0          ['tf.math.segment_max_3[0][0]'
(TFOpLambda)                                     , 'tf.math.subtract_3[0][0]']

=====================================================================================
Total params: 4000 (15.62 KB)
Trainable params: 4000 (15.62 KB)
Non-trainable params: 0 (0.00 Byte)
_____________________________________________________________________________________
```

We can test our implementation as the following

```
layer = self = EmbeddingBag(1000, 4, mode="max")
x = tf.ragged.constant([[1, 3], [2], [], [6, 8, 9], []]).to_sparse()
emb = layer(x)

# Slice one by one
emb2 = [
    self.embedding.weights[0].numpy()[[1, 3], :].max(axis=0),
    self.embedding.weights[0].numpy()[2, :],
```

```
    tf.zeros((4,), dtype=tf.float32),
    self.embedding.weights[0].numpy()[[6, 8, 9], :].max(axis=0),
    tf.zeros((4,), dtype=tf.float32),
]

assert np.allclose(emb, emb2)
```

9.7 Case Study: Vocal Duration Constrained by Note Duration

Singing voice synthesis (SVS) is an extension of the text-to-speech (TTS) problem, where lyrics and notes are converted from texts to artificial singing voices. The notes input dictates the tempo (i.e. speed as well as timing) of the generated vocal singing and provides the SVS model with contextual information to conform the synthesized singing to the musicality of the vocal piece. In one of the SVS projects that I have done previously, my model predicted the duration of each vocal unit at the phoneme level (i.e. one syllable at a time). For example, "I saw you smiling at me. Was it real or just my fantasy" becomes ["I," "saw," "you," "smi," "ling," "at," "me," "was," "it," "real," "or," "just," "my," "fan," "ta," "sy"] ("Eyes on Me" from the main theme of Final Fantasy VIII, by Faye Wong; Figure 9.13), where the duration of each syllable is predicted by the model output. However, due to the nature of music, the singer's vocal duration and timing need to align with the tempo of the sheet music or sometimes the background music accompaniment. That is to say, if the tempo denoted by the music sheet requires the vocalist to finish the singing of a sentence within 16 beats, then the total duration of all the syllables sung by the vocalist cannot exceed or be less than 16 beats (give or take with some expressiveness allowed). Therefore, there is a temporal constraint that needs to be put on the predicted duration of the phoneme. A simple approach to solve this problem is to predetermine the

Figure 9.13 "Eyes On Me" sheet music from my previous SVS project.

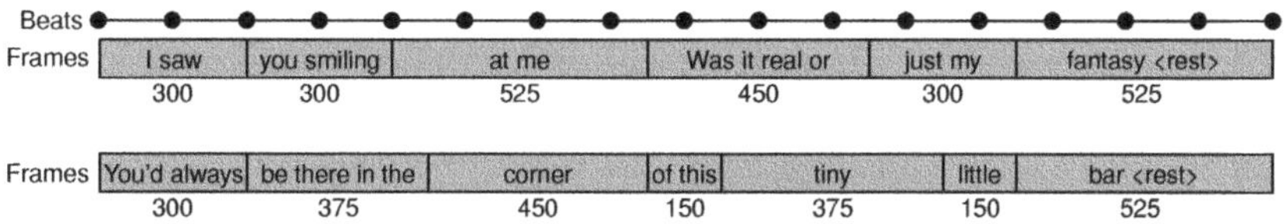

Figure 9.14 Example note duration constraints by group.

total note duration of a measure (a.k.a bar, where it denotes a segment of the music that contains a fixed number of beats, say, four beats per measure). For the example of Eyes on Me, let us assume there are four beats per measure, and the tempo is 80 beats per minute. This means for the sentence/lyric above that covers a total of four measures, we need to make sure the full sentence is sung within (4 beats/measure) × (4 measures)/(80 beats/minute) = 12 seconds. In addition, in TTS and SVS modeling tasks, it is more common to represent temporal units of the time series as "*frames*," where each frame denotes a fixed time duration, say 5 ms. This is because it is more typical to model the audio signal as a spectrogram in the time-frequency domain, where each frame corresponds to a temporal window used to perform the Fourier transformation. Therefore, 12 seconds corresponds to 2400 frames or 150 frames per beat. Lastly, we can define even finer-grained temporal constraints by groups of words, based on the duration of the notes. In our particular example, let us define the following six groupings (also illustrated in Figure 9.14 top):

1. "I saw," which corresponds to a total of 2 beats or 300 frames.
2. "you smiling," which corresponds to 2 beats or 300 frames.
3. "at me," which corresponds to 3.5 beats or 525 frames.
4. "Was it real or," which corresponds to 3 beats or 450 frames.
5. "just my," which corresponds to 2 beats or 300 frames.
6. "fantasy <rest>," which corresponds to 3.5 beats or 525 frames.

The following code block implements a function that can adjust the predicted duration of each syllable/word given the predicted number of frames and the total duration constraints within the grouping:

```
import tensorflow as tf

def scatter_batch_2d(index, inputs, out=None, mode="add"):
    """Batch-wise scattering over a 2D matrix."""
    # Determines the length of each row for the output
    row_length = tf.cast(tf.reduce_max(index), tf.int32) + 1
    row_length = tf.maximum(tf.shape(inputs)[1], row_length)
    # start and end index of each row
    # row 0: inc[0]:inc[1], row 1: inc[1]:inc[2], ...
```

```
    inc = row_length * tf.range(tf.shape(inputs)[0]+1)
    # Compute flat indices for each element
    indices = tf.cast(index, inc.dtype) + tf.reshape(inc[:-1], (-1, 1))
    # Flatten to column vector
    indices = tf.reshape(indices, (-1, 1))
    # Flatten updates
    updates = tf.reshape(inputs, (-1,))
    if out is None: # create from zeros
        shape = (inc[-1], )
        out = tf.scatter_nd(indices, updates, shape)
    else:
        out = tf.reshape(out, (-1, ))
        if mode == "add":
            out = tf.tensor_scatter_nd_add(out, indices, updates)
        elif mode == "sub":
            out = tf.tensor_scatter_nd_sub(out, indices, updates)
        elif mode == "min":
            out = tf.tensor_scatter_nd_min(out, indices, updates)
        elif mode == "max":
            out = tf.tensor_scatter_nd_max(out, indices, updates)
        elif mode == "update":
            out = tf.tensor_scatter_nd_update(out, indices, updates)
    # reshape: (batch_size, row_length)
    out = tf.reshape(out, (tf.shape(index)[0], -1))

    return out


def apply_duration_constraints(dur_pred, dur_true, index, rounding):
    """
    Apply duration constraints for each phoneme.

    Inputs:
    * dur_pred: predicted duration of each phoneme.
    * dur_true: total duration within the grouping.
    * index: index to indicate which group each phoneme belongs to.
    """
    # Sum together total predicted duration within the group
    pred = scatter_batch_2d(index, dur_pred)
    # Broadcast the total values across all phonemes within the group
    pred = tf.gather(pred, index, axis=1, batch_dims=1)
    label = tf.gather(dur_true, index, axis=1, batch_dims=1)
    # Compute ratio: constraints / total
    ratio = tf.math.divide_no_nan(
        tf.cast(label, tf.float32), tf.cast(pred, tf.float32)
    )
    # ratio = label / pred  = dur_adj / dur_pred
    # dur_adj = dur_pred * (label / pred)
    dur_adj = ratio * tf.cast(dur_pred, tf.float32)
    # Rounding: will be off by at most 1 frame
    dur_adj = tf.cast(tf.round(dur_adj), dur_true.dtype)

    # Costly! Correct rounding error,
    # if 1 frame fewer, add 1 frame to the smallest duration
```

```
64        # if 1 frame more, subtract 1 frame from the largest duration
65        if rounding == "strict":
66            # Compute groupwise statistics and broadcast to phoneme level
67            min_dur = tf.ones_like(dur_adj) * tf.reduce_max(dur_adj)
68            min_dur = scatter_batch_2d(index, dur_adj, out=min_dur, mode="min")
69            min_dur = tf.gather(min_dur, index, axis=1, batch_dims=1)
70            max_dur = tf.zeros_like(dur_adj)
71            max_dur = scatter_batch_2d(index, dur_adj, out=max_dur, mode="max")
72            max_dur = tf.gather(max_dur, index, axis=1, batch_dims=1)
73            total_adj = scatter_batch_2d(index, dur_adj)
74            total_adj = tf.gather(total_adj, index, axis=1, batch_dims=1)
75
76            # Adjust
77            adjustment = label - total_adj
78            cond_fewer = tf.math.logical_and(
79                adjustment > 0, # missing 1 frame
80                dur_adj == min_dur # need to adjust on min_dur
81            )
82            cond_more = tf.math.logical_and(
83                adjustment < 0, # extra 1 frame
84                dur_adj == max_dur # need to adjust on max_dur
85            )
86            # combine two conditions
87            cond = tf.math.logical_or(cond_fewer, cond_more)
88            dur_adj = tf.where(cond, dur_adj + adjustment, dur_adj)
89
90        return dur_adj
```

- Lines 3–35 implement the function `scatter_batch_2d`. This is a modification of a function that we implemented in a case study in Chapter 3. In addition to scattering using `tf.scatter_nd`, we added additional modes of accumulation using `tf.tensor_scatter_nd_*` including `"add"`, `"sub"`, `"min"`, `"max"`, and `"update"`.
- Line 48: This accumulates (summing) together the predicted durations within the same grouping. For example, given grouping `index = [0, 0, 1, 1, 1]` and `dur_pred = [1, 2, 3, 4, 5]`, this results `pred = [1+2, 3+4+5] = [3, 12]`.
- Line 50: This broadcasts the within-group duration sum back to the individual syllables. Following the example in the previous step, `pred = [3, 12]` is now broadcasted back to each syllable based on the grouping `index = [0, 0, 1, 1, 1]`, becoming `[3, 3, 12, 12, 12]`.
- Line 51: This broadcasts the total duration constraints of each group, `dur_true` to each individual syllable as well. `pred` and `label` should have identical lengths.
- Lines 53–58: Compute the ratio and then adjust the predicted duration. Since we assume the ratio between groupwise predicted duration and duration

constraints translates to durations of each individual syllable, i.e. as noted in the comment, `ratio = label / pred = dur_adj / dur_pred`, we can solve for the constraints-adjusted duration by rearranging `dur_adj = dur_pred * (label / pred)`. Note that we are performing the computation in float dtypes.

- Line 60: Rounds the adjusted duration so that it still represents the number of frames using integers. But, this comes with rounding errors, where the within-group total duration is either one frame more or one frame fewer than the constraints. In practice, such rounding errors occurred rarely; it also resulted in the output audio duration being 5 ms longer or shorter, which is barely noticeable to the human perception. So, I typically do not strictly enforce the groupwise total duration constraints.
- Lines 65–88: Nonetheless, it is still interesting to come up with a strategy to correct the rounding error that occurred above. The strategy I had was that if the adjusted within-group total duration is one frame fewer than the constraints, then we add this missing frame back to the syllable that has the shortest duration in the group; on the other hand, if it is one frame more than the duration specified by the constants, then we subtract the extra frame from the syllable that has the longest duration in the group. To apply this strategy, we first need to compute three groupwise metrics: max, min, and sum. This can be costly compute-wise, so we need to carefully consider the trade-off between practicality and rigor.
  - Lines 67–69 compute the within-group minimum duration and broadcast this value back to the syllables.
  - Lines 70–72 compute the within-group maximum duration and broadcast this value back to the syllables.
  - Lines 73–74 recompute the within-group total duration after the adjustment and broadcast this value back to the syllables.
  - Line 77 this computes the number of frames needed to be added or subtracted so that the adjusted durations can further be corrected to strictly meet the duration constraints.
  - Lines 78–87 these compute a series of boolean conditions to determine which syllable needs to be corrected based on the strategy we set above.
  - Line 88 applies the final correction using `tf.where` based on the conditions we determined above.

Let us apply the function we implemented above to adjust the predicted durations. We compare the results when setting `rounding="loose"` which allows rounding errors, and `rounding="strict"` which strictly enforces the duration constraints.

```
phonemes = tf.constant([
    [
        "I", "saw", "you", "smi", "ling", "at", "me",
        "was", "it", "real", "or", "just", "my", "fan", "ta", "sy",
        "<rest>", "", "", ""
    ],
    [
        "You'd", "al", "ways", "be", "there", "in", "the", "cor", "ner",
        "of", "this", "ti", "<rest>", "ny", "lit", "tle", "bar",
        "<rest>", "", ""
    ],
])
index = tf.constant([
    [0, 0, 1, 1, 1, 2, 2, 3, 3, 3, 3, 4, 4, 5, 5, 5, 5, 6, 6, 6],
    [0, 0, 0, 1, 1, 1, 1, 2, 2, 3, 3, 4, 4, 4, 5, 5, 6, 6, 7, 7]
])

dur_true = tf.constant([
    [300, 300, 525, 450, 300, 525,   0, 0, 0, 0],
    [300, 375, 450, 150, 375, 150, 525, 0, 0, 0],
])

dur_pred = tf.constant([
    [
        71, 244, 69, 138, 74, 163, 374,
        82, 71, 231, 70, 154, 158, 81, 76, 289,
        79, 0, 0, 0
    ],
    [
        80, 143, 81, 77, 152, 79, 68, 82, 362,
        69, 71, 311, 79, 78, 93, 95, 443,
        73, 0, 0
    ]
])

# Adjusted durations

# When setting rounding = "loose" to allow rounding errors of +/- 1 frame
dur_adj = apply_duration_constraints(
    dur_pred, dur_true, index, rounding="loose"
)
# <tf.Tensor: shape=(2, 20), dtype=int32, numpy=
# array([[ 68, 232,  74, 147,  79, 159, 366,  81,  70, 229,  69, 148, 152,
#           81,  76, 289,  79,   0,   0,   0],
#         [ 79, 141,  80,  77, 152,  79,  68,  83, 367,  74,  76, 249,  63,
#           63,  74,  76, 451,  74,   0,   0]], dtype=int32)>
dur_adj_group = scatter_batch_2d(index, dur_adj)
# tf.Tensor(
# [[300 300 525 449 300 525   0   0   0   0   0   0   0   0   0   0   0   0
#     0   0]
#  [300 376 450 150 375 150 525   0   0   0   0   0   0   0   0   0   0   0
#     0   0]], shape=(2, 20), dtype=int32)

# When setting rounding = "strict" to enforce total duration
```

```
# to conform to the label
dur_adj = apply_duration_constraints(
    dur_pred, dur_true, index, rounding="strict"
)
# <tf.Tensor: shape=(2, 20), dtype=int32, numpy=
# array([[ 68, 232,  74, 147,  79, 159, 366,  81,  70, 229,  70, 148, 152,
#          81,  76, 289,  79,   0,   0,   0],
#        [ 79, 141,  80,  77, 151,  79,  68,  83, 367,  74,  76, 249,  63,
#          63,  74,  76, 451,  74,   0,   0]], dtype=int32)>

dur_adj_group = scatter_batch_2d(index, dur_adj)
# <tf.Tensor: shape=(2, 20), dtype=int32, numpy=
# array([[300, 300, 525, 450, 300, 525,   0,   0,   0,   0,   0,   0,   0,
#           0,   0,   0,   0,   0,   0,   0],
#        [300, 375, 450, 150, 375, 150, 525,   0,   0,   0,   0,   0,   0,
#           0,   0,   0,   0,   0,   0,   0]], dtype=int32)>
```

Of course, if we arrange the data inputs as a Pandas DataFrame and use its `DataFrame.groupby.apply` pattern, we can also solve the problem of adding duration constraints. Nonetheless, this case study illustrates that the "groupby-apply" pattern can be implemented using libraries like Tensorflow and PyTorch from scratch when needed.

9.8 Case Study: Filling of Missing Values in a Sequence

Pandas implement a well-known `.fillna` method for its `pandas.DataFrame` and `pandas.Series` classes to fill in missing values represented by `NaN`s. In addition to filling the missing values with a constant value like 0, it can also do sequence-based filling based on the value either before or after it, corresponding to the `ffill` and `bfill` methods, respectively. In many neural network implementations, we may sometimes need to fill in missing entries such that a set of features all have exactly the same lengths. Consider the following example with a pair of features `event_type` and `event_timestamp`:

```
import tensorflow as tf

event_type = tf.constant(
    [
        [
            "view",
            "click",
            "click",
            "add to cart",
            "buy",
```

```
        ],
        ["view", "view", "click", "view", "click"],
    ]
)
event_timestamp = tf.constant(
    [
        [
            "2022-03-15 06:23:24",
            "2022-03-15 06:23:38",
            "",
            "2022-03-15 06:24:58",
            "",
        ],
        ["", "", "2022-03-18 06:24:37", "", ""],
    ]
)
```

Each entry of the `event_timestamp` feature corresponds to a specific time of occurrence of the event in the `event_type` feature. Notice, we are missing multiple entries in the `event_timestamp` feature. One strategy to guess the time of the event occurrence is to look for the nearest timestamp, either before or after the missing value. See Figure 9.15 for the detailed strategy for `ffill` that we implemented below using Tensorflow, along with the line-by-line explanation of the code below.

```
import tensorflow as tf

def ffill(x):
    """Forward filling the values of a 2D matrix row"""
    # Track indices for all values
    r, c = tf.cast(tf.shape(x), dtype=tf.int64)
    flat_indices = tf.range(r * c)
    indices = tf.stack([
        flat_indices // c,
        flat_indices % c
      ], axis=0)
    indices = tf.transpose(indices)

    # Indices of consecutive integers
    empty_indices = tf.where(x == "")
    f = (empty_indices[1:, 1] - empty_indices[:-1, 1]) == 1
    f = tf.concat(([False], f), 0) != tf.concat((f, [False]), 0)
    f = tf.where(f)
    f = tf.reshape(f, (-1, 2))

    # Compute backtracking offset values
    offset_indices = tf.ragged.range(f[:, 0], f[:, 1] + 1).flat_values
    offset_values = tf.ragged.range(f[:, 1] - f[:, 0] + 1).flat_values
```

```
24        offset_values = -(offset_values+1)
25
26        offsets = -tf.ones_like(empty_indices[:, 0])
27        offsets = tf.tensor_scatter_nd_update(
28            offsets, offset_indices[:, None], offset_values
29        )
30
31        # Replace the empty indices with offsets
32        flat_empty_indices = empty_indices[:, 0] * c + empty_indices[:, 1]
33        update_indices = tf.stack(
34            [flat_empty_indices, tf.ones_like(flat_empty_indices)], axis=0
35        )
36        update_indices = tf.transpose(update_indices)
37        indices_offset = tf.tensor_scatter_nd_add(
38            indices, update_indices, offsets
39        )
40
41        # Address any remaining -1s. These are empty values starting at
42        # the beginning that cannot be filed. So use original indices,
43        # which is index 0
44        indices_offset = tf.maximum(indices_offset, 0)
45
46        # Finally, use tf.gather_nd to create the ffilled matrix
47        y = tf.gather_nd(x, indices_offset) # flat
48        y = tf.reshape(y, (r, c))
49
50        return y
```

- Line 6 determines the number of rows (`batch_size`) and number of columns of the 2D matrix.
- Line 7 creates the flat indices of the matrix.
- Lines 8–12 create the multi-index of each value of the matrix, in the form of `[[r0, c0], [r1, c1], ...]`.
- Line 15 finds the multi-index where the values are missing/empty. The function can also be generalized to fill integer and float values by looking for corresponding padding values such as `NaN`.
- Lines 16–19 find the start and end indices (e.g. `2, 3, 4, 5 -> [2, 5]`) of the segment of empty entries. Recall we implemented the same logic in a case study in Chapter 3. We cluster/group a segment of empty entries together since all the empty values within the current group need to be filled by the same previous value.
- Lines 22–24 compute the backtracking offset values for empty segments with more than 1 value. For example, if indexes `3` and `4` are empty, then we need to use the value from index `2` to fill these two values. This means, we need to backtrack `-1` and `-2`, respectively. See Figure 9.15b for an illustration. For example, for the first occurrence of the empty value at multi-index `(0, 2)`, we need to backtrack 1 index, to indicate that in the filled matrix, we need to

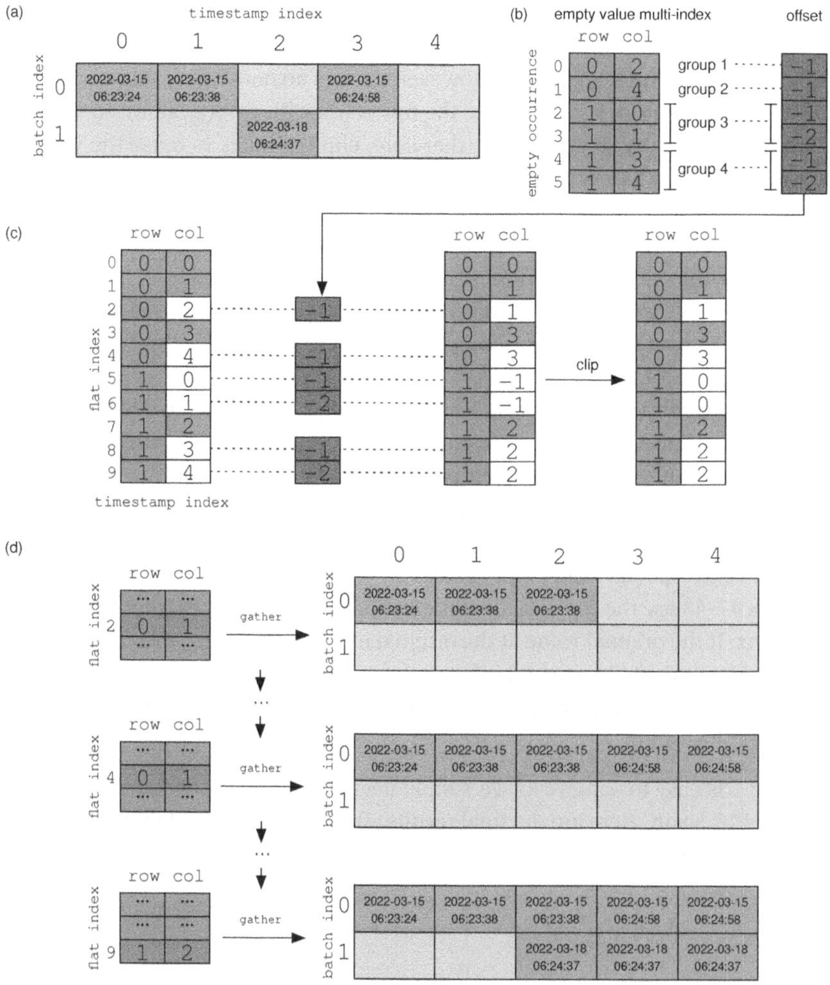

Figure 9.15 Implementation logic for `ffill`. (a) Illustration of the input data with missing values; (b) creation of the backtracking offset value logic. In the current example, there are four groups of empty entries. Groups 1 and 2 have only single entries of empty values, while Groups 3 and 4 have two empty entries. Therefore, for Groups 1 and 2, the offset only needs to backtrack 1 value (looking back 1), while for Groups 3 and 4, each subsequent index within the group needs to backtrack 1 more, i.e. `-1, -2`, ..., etc. (c) Given the set of all indices of the matrix, we apply the backtracking logic at each empty entry. (d) Iterative illustration of using gather to recreate a matrix with empty value filled. Notice that the beginning two values of the second row cannot be filled using the `ffill` method.

use the non-empty value at `(0, 1)`. Using the same logic, for the empty value occurring at multi-index `(1, 1)`, we need to backtrack two indices to use the non-empty value at `(1, -1)`. However, there is no non-empty value occurring before the value at `(1, 1)` since the row starts with empty values, so we need to keep this entry as empty. We, therefore, clip the index to reuse the value at index `0`, which happens to be an empty entry.

- Lines 26–29 create the full backtracking offset array. We first create an array of `-1`s. We then replace backtrack offset values for the empty segments. See Figure 9.15b for an illustration.
- Lines 32–39 update the matrix value indices, replacing any indices of the empty values with a corresponding index used to replace the empty values. See Figure 9.15c for an illustration.
- Line 44 handles any remaining `-1`s. The algorithm attempts to use the nearest previous value to fill the current empty values, but there is not really a previous value that can be used if the empty value starts at the beginning of the row. Therefore, these empty entries remain empty after `ffill`. We clip the index matrix to have a minimum of `0`, which replaces the remaining `-1`s as `0` indices, which corresponds to an empty value.
- Lines 47–48 use the `gather` operation to resample the values from the original matrix. If the original value at the original index is not empty, then they remain as is. However, if the original value at the original index is an empty entry, then we gather from the nearest index where the value is not empty. See Figure 9.15d for an illustration.

`bfill` is easy to achieve as we simply need to reverse the sequence and then call `ffill` again. Here are the final results after first `ffill` and then `bfill` the missing values.

```
event_timestamp = tf.constant( # reproduced from above
    [
        [
            "2022-03-15 06:23:24",
            "2022-03-15 06:23:38",
            "",
            "2022-03-15 06:24:58",
            "",
        ],
        ["", "", "2022-03-18 06:24:37", "", ""],
    ]
)

# ffill
event_timestamp_filled = ffill(event_timestamp)
# bfill
event_timestamp_filled = ffill(event_timestamp_filled[:, ::-1])
```

```
# <tf.Tensor: shape=(2, 5), dtype=string, numpy=
# array([[b'2022-03-15 06:24:58', b'2022-03-15 06:24:58',
#          b'2022-03-15 06:23:38', b'2022-03-15 06:23:38',
#          b'2022-03-15 06:23:24'],
#         [b'2022-03-18 06:24:37', b'2022-03-18 06:24:37',
#          b'2022-03-18 06:24:37', b'2022-03-18 06:24:37',
#          b'2022-03-18 06:24:37']], dtype=object)>
```

Bibliography

Martín Abadi, Ashish Agarwal, Paul Barham, Eugene Brevdo, Zhifeng Chen, Craig Citro, Greg S. Corrado, Andy Davis, Jeffrey Dean, Matthieu Devin, Sanjay Ghemawat, Ian Goodfellow, Andrew Harp, Geoffrey Irving, Michael Isard, Yangqing Jia, Rafal Jozefowicz, Lukasz Kaiser, Manjunath Kudlur, Josh Levenberg, Dan Mane, Rajat Monga, Sherry Moore, Derek Murray, Chris Olah, Mike Schuster, Jonathon Shlens, Benoit Steiner, Ilya Sutskever, Kunal Talwar, Paul Tucker, Vincent Vanhoucke, Vijay Vasudevan, Fernanda Viegas, Oriol Vinyals, Pete Warden, Martin Wattenberg, Martin Wicke, Yuan Yu, and Xiaoqiang Zheng. Tensorflow quantization. https://www.tensorflow.org/model_optimization/guide/quantization/post_training, 2023.

Yoshua Bengio, Réjean Ducharme, Pascal Vincent, and Christian Jauvin. A neural probabilistic language model, 2003.

Alexey Dosovitskiy, Lucas Beyer, Alexander Kolesnikov, Dirk Weissenborn, Xiaohua Zhai, Thomas Unterthiner, Mostafa Dehghani, Matthias Minderer, Georg Heigold, Sylvain Gelly, Jakob Uszkoreit, and Neil Houlsby. An image is worth 16x16 words: Transformers for image recognition at scale, June 2021.

Google. TFHub NNLM Kaggle Models. https://www.kaggle.com/models/google/nnlm, 2023.

Kaiming He, Xinlei Chen, Saining Xie, Yanghao Li, Piotr Dollár, and Ross Girshick. Masked autoencoders are scalable vision learners, December 2021.

Adam Paszke, Sam Gross, Francisco Massa, Adam Lerer, James Bradbury, Gregory Chanan, Trevor Killeen, Zeming Lin, Natalia Gimelshein, Luca Antiga, Alban Desmaison, Andreas Köpf, Edward Yang, Zach DeVito, Martin Raison, Alykhan Tejani, Sasank Chilamkurthy, Benoit Steiner, Lu Fang, Junjie Bai, and Soumith Chintala. PyTorch quantization. https://pytorch.org/docs/stable/quantization.html#module-torch.ao.quantization, 2023.

Juan Terven and Diana Cordova-Esparza. A comprehensive review of YOLO: From YOLOv1 and beyond, October 2023.

Ashish Vaswani, Noam Shazeer, Niki Parmar, Jakob Uszkoreit, Llion Jones, Aidan N. Gomez, Lukasz Kaiser, and Illia Polosukhin. Attention is all you need. *Advances in Neural Information Processing Systems*, 2017-Decem (Nips), pages 5999–6009, 2017. ISSN 10495258.

Ruibin Xiong, Yunchang Yang, Di He, Kai Zheng, Shuxin Zheng, Chen Xing, Huishuai Zhang, Yanyan Lan, Liwei Wang, and Tie-Yan Liu. On layer normalization in the transformer architecture, June 2020.

10

Sorting and Reordering

Whether it is processing of continuous time series or analysis of other forms of sequentially represented discrete data, it is usually more efficient to process the sequence data in a batch rather than using for-loops to apply the processing logic one sequence at a time. Sequence data are often represented as dense tensors if all sequences are having the same length. However, not all sequences have the same length, and therefore, sequence data are often represented as a dense tensor with paddings (whose data entries can be determined by a padding mask), or left-aligned ragged or sparse tensors. In this chapter, we will focus on several case studies on various ways to sort and reorder a batch of variable-length sequence data. We will use all the knowledge we have gained so far on vectorization to implement solutions to these problems.

Note

Typically, Tensorflow represents a batch of sequence data as a tensor of shape `(batch_size, seq_len, <embed_size> ...)`, while PyTorch represents it with shape `(batch_size, <emb_size> ..., seq_len)`. This is a small detail on the different conventions used by the libraries. In this chapter, I will use the Tensorflow convention in our sample code examples, but implementing the PyTorch convention can be achieved relatively easily by transposing the tensors properly.

10.1 Sorting Operations

Sorting is a basic operation in computer science that rearranges the elements within an array according to certain ordering rules. Table 10.1 provides a list of equivalent sorting operations implemented in NumPy, Tensorflow, and PyTorch.

Vectorization: A Practical Guide to Efficient Implementations of Machine Learning Algorithms, First Edition. Edward DongBo Cui.

Table 10.1 List of sorting operations in NumPy, Tensorflow, and PyTorch.

Operation	NumPy	Tensorflow	PyTorch
Sorting[a)]	`np.sort`	`tf.sort`	`torch.sort`
Indices to sort an array[a)]	`np.argsort`	`tf.argsort`	`torch.argsort`
Lexicographical sorting[b)]	`np.lexsort`	N/A	N/A
Unique values/indices[c)]	`np.unique`	`tf.unique`, `tf.unique_with_counts`	`torch.unique`
Max value[d)]	`np.max`	`tf.reduce:max`	`torch.max`
Index of the max value[d)]	`np.argmax`	`tf.argmax`	`torch.argmax`
Min value[d)]	`np.min`	`tf.reduce:min`	`torch.min`
Index of the min value[d)]	`np.argmin`	`tf.argmin`	`torch.argmin`
Top K min/max values[e)]	N/A	`tf.math.top_k`	`torch.topk`
Partitioning[f)]	`np.partition`	N/A	N/A
Indices to partition an array[f)]	`np.argpartition`	N/A	N/A
Insertion index of sorted array[g)]	`np.searchsorted`	`tf.searchsorted`	`torch.searchsorted`

Notes:

a) The `.sort` operation sorts the array in ascending order (from smallest to largest) by default. In Tensorflow and PyTorch's implementations, however, users can specify arguments to sort descendingly (from largest to smallest). For Tensorflow, this is `tf.sort(direction="DESCENDING)"`, and for PyTorch, this is `torch.sort(descending=True)`. For NumPy, we can reverse the sorted order with `[::-1]`. The `.argsort` operation returns the indices that can be used to sort the original array.

b) `np.lexsort` is unique to NumPy. It sorts a set of arrays together based on the keys specified. In one of the following case studies, we will see how we can use `np.lexsort` to sort rows of a matrix.

c) The `.unique` operation returns the unique set of contents. It can also return:
- indices that can be used to convert the original tensor to the unique set. In NumPy, this is to set `return_index=True`; in Tensorflow, this is to use `tf.unique` and use the second returned output. PyTorch does not seem to implement this functionality at the time of writing this chapter.
- indices to recover the original tensor from the unique set. In NumPy and PyTorch, this is to set `return_inverse=True`. In Tensorflow, this is to call `tf.unique_with_counts` and use the second returned output.

- counts of the number of occurrences of the item in the unique set. In NumPy and PyTorch, this is to set `return_counts=True`. In Tensorflow, this is to call `tf.unique_with_counts` and use the third returned output. In the case study where we implement a function to sort rows of a matrix, we will use `np.unique` to sort the rows lexicographically.

d) These operations get the maximum or minimum values (or the index of it) of an array. They correspond to the mathematical notations max, min, argmax, and argmin.

e) The top-k operation returns the top-k values from an array along with the indices of the top-k values. Although NumPy does not have a ready-to-use function to extract top-k values, as we will see in one of the studies in Section 10.2, we can use `np.argpartition` to implement the top-k operation.

f) `np.partition` is unique to NumPy. It partitions an array into two parts, based on the *k*th value of the final sorted array (but not actually doing the sorting on the whole array). The values in the left partition are all smaller, while the values in the right partition are all larger than this value. See Figure 10.1 in the following case study section for an illustration. `np.argpartition` gets the indices to partition and reorder the array.

g) Given a sorted array, the `.searchsorted` operation finds the index where a new value can be inserted while the sorting order of the new array is still maintained (i.e. still from smallest to largest). If the value is equal to one of the values already existing in the array, then it is either inserted to the left or right of the original number, depending on the argument set for `side="left"` or `side="right"`, respectively. The implementations of the `.searchsorted` method by NumPy, Tensorflow, and PyTorch vectorize the search for multiple insertion values simultaneously. To illustrate,

```
import numpy as np
X = np.array([1, 3, 6, 8, 10])
insertion_values = np.array([2, 5, 9, 10, 11])
insertion_indices = np.searchsorted(X, insertion_values)
# array([1, 2, 4, 4, 5])
# Insert the values based on the search sorted indices
y = np.insert(X, insertion_indices, insertion_values)
# array([ 1,  2,  3,  5,  6,  8,  9, 10, 10, 11])
```

10.2 Case Study: Top-k Using `argsort` and `argpartition`

Both Tensorflow (`tf.math.top_k`) and PyTorch (`torch.topk`) implement a method that returns the top-k largest elements along with the indices from an array. For Tensorflow,

```
import tensorflow as tf
X = tf.constant([1, 3, 2, 5, 9, 8, 4])
top_values, top_indices = tf.math.top_k(X, k=3)
# top_values: <tf.Tensor: shape=(3,), dtype=int64, numpy=array([9, 8, 5])>
# top_indices: <tf.Tensor: shape=(3,), dtype=int32, numpy=array([4, 5, 3], dtype=int32)>
```

and for PyTorch

```
import torch
X = torch.tensor([1, 3, 2, 5, 9, 8, 4])
top_values, top_indices = torch.topk(X, k=3)
# top_values: tensor([9, 8, 5])
# top_inidces: tensor([4, 5, 3])
```

At the time of writing this chapter, NumPy does not have a ready-to-use function to retrieve the top-k values and indices. However, it is possible to use sorting to achieve the same goal. A naive approach is to use `np.argsort` to determine the indices of the top values, then use `np.take` to retrieve the top values.

```
1  import numpy as np
2
3  def top_k_argsort(X, k=3):
4      """Get top-k values and indices along the last dimension."""
5      top_indices = np.argsort(X)[..., -k:][..., ::-1]
6      top_values = np.take_along_axis(X, top_indices, axis=-1)
7      return top_values, top_indices
8
9  X = np.array([1, 3, 2, 5, 9, 8, 4])
10 top_values, top_indices = top_k_argsort(X)
11 # top_values: array([9, 8, 5])
12 # top_indices: array([4, 5, 3])
```

- Line 5: `np.argsort` first gets the index used to sort each of the arrays. Since it sorts the values in ascending order (from smallest to largest), we then slice out the last *k* indices for the top-k values along the last dimension of the input, using

[..., -k:]. Then, [..., ::-1] reverses the top-k indices so that it is now arranged from largest to smallest.

- Line 6: `np.take_along_axis` applies the sorting order obtained from `top_indices` to the input, along the last dimension.

However, this would involve sorting the entire sequence. The time complexity of this approach is $O(n \log n)$, where n is the size of the sequence. In contrast, another strategy to retrieve top-k values is using `np.argpartition` to return the unordered top-k values first, and then sort the top-k values and indices using `np.argsort`. Since `np.argpartition` runs in $O(n)$ [Harris et al., 2023] time complexity, the overall time complexity is then $O(n + k \log k)$, where $k \ll n$.

```
1  import numpy as np
2
3  def top_k_argpartition(X, k=3):
4      """Get top-k values and indices along the last dimension."""
5      # Unsorted top k values and indices
6      top_indices = np.argpartition(X, -k, axis=-1)[..., -k:]
7      top_values = np.take_along_axis(X, top_indices, axis=-1)
8
9      # Sort based on top_values
10     indices = np.argsort(top_values, axis=-1)[..., ::-1]
11     top_indices = np.take_along_axis(top_indices, indices, axis=-1)
12     top_values = np.take_along_axis(top_values, indices, axis=-1)
13     return top_values, top_indices
14
15 X = np.array([1, 3, 2, 5, 9, 8, 4])
16 top_values, top_indices = top_k_argpartition(X)
17 # top_values: array([9, 8, 5])
18 # top_indices: array([4, 5, 3])
```

- Line 6 partitions the array by the kth largest value. `np.argpartition` will get the indices of the array after the partitioning. Here, we use `-k` to specify the kth largest value instead of the kth smallest value. See Figure 10.1 for an illustration

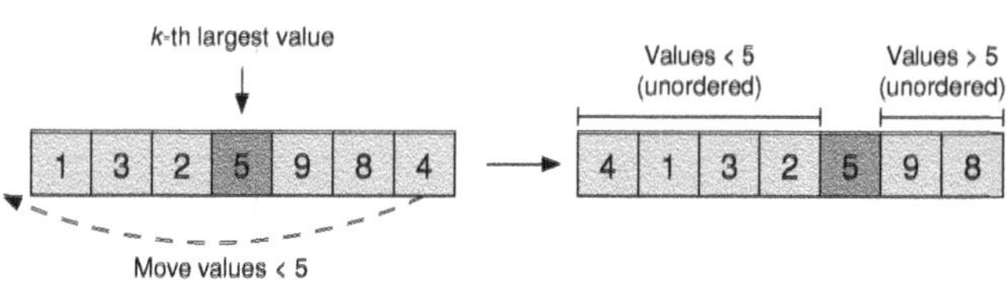

Figure 10.1 Schematic illustration of the effect of partitioning the array. The algorithm first determines the kth largest value as specified, which we denote as the "pivot" (as in quicksort). Then, it moves any value less than the pivot to the left of the pivot and moves any value greater than the pivot to the right of the pivot. As a result, all values to the left of the pivot are smaller than the pivot, and all values to the right of the pivot are greater than the pivot. Within the two partitions, however, the values are not ordered.

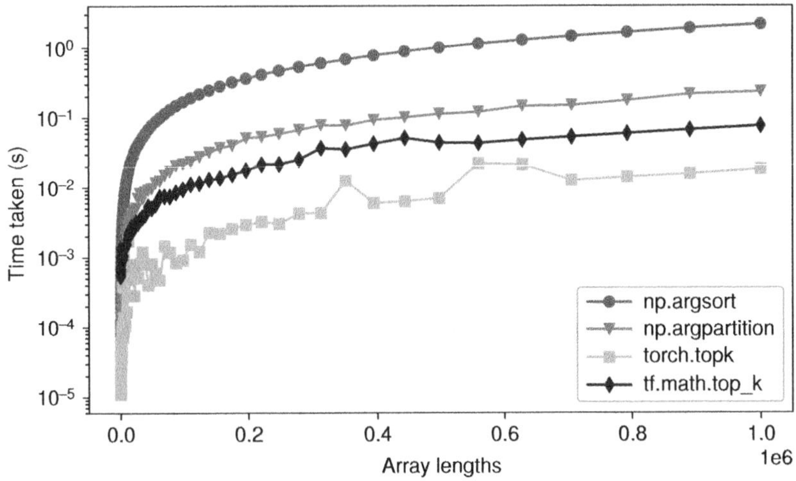

Figure 10.2 Performance of two top-k functions implemented in NumPy, Tensorflow, and PyTorch, measured on a 2021 M1 MacBook Pro.

of how `np.partition` works (`np.argpartition` gets the indices to be used for partitioning instead of the partitioned results, similar to how `np.argsort` gets the sort indices rather than the sorted results). Then, we slice out only the indices for the top-k largest values, `[..., ...k:]`.

- Line 7 takes the top-k values from the original inputs. At this point, neither `top_indices` nor `top_values` are sorted.
- Lines 10–12 sort the `top_values` and apply the sorting order to `top_indices`.

We can simulate some data and compare the performance of the two functions, `top_k_argsort` vs. `top_k_argpartition`, as well as implementations from Tensorflow `tf.math.top_k`, and PyTorch `torch.topk`. The results are plotted in Figure 10.2.

```
import numpy as np
import time
import matplotlib.pyplot as plt

batch_size = 32
max_array_length = int(1E6)
lengths = np.concatenate([
    np.geomspace(10, max_array_length, num=100).astype(int)
])

time_taken = {
    "top_k_argsort": [],
```

```
        "top_k_argpartition": [],
        "torch_top_k": [],
        "tf_top_k": [],
    }
    for ii, l in enumerate(lengths):
        X = np.random.randint(0, max_array_length+1, size=(batch_size, 1))
        # measure top_k_argsort
        tnow = time.time()
        _ = top_k_argsort(X, k=5)
        time_taken["top_k_argsort"].append(time.time() - tnow)
        # measure top_k_argpartition
        tnow = time.time()
        _ = top_k_argpartition(X, k=5)
        time_taken["top_k_argpartition"].append(time.time() - tnow)
        # measure torch
        X_torch = torch.tensor(X)
        tnow = time.time()
        _ = torch.topk(X_torch, k=5)
        time_taken["torch_top_k"].append(time.time() - tnow)
        # measure tf
        tnow = time.time()
        _ = tf.math.top_k(X, k=5)
        time_taken["tf_top_k"].append(time.time() - tnow)

    # Plot the results
    fig, ax = plt.subplots(1, 1, figsize=(7, 4))
    ax.plot(lengths, time_taken["top_k_argsort"], label="np.argsort")
    ax.plot(lengths, time_taken["top_k_argpartition"], label="np.argpartition")
    ax.plot(lengths, time_taken["torch_top_k"], label="torch.topk")
    ax.plot(lengths, time_taken["tf_top_k"], label="tf.math.top_k")
    ax.legend()
    ax.set_xlabel("Array lengths")
    ax.set_ylabel("Time Taken (sec)")
```

10.3 Case Study: Sort the Rows of a Matrix

Suppose we have a 2D matrix of integers, and we would like to sort the rows of the integers, first by column 1, then by column 2, then by column 3, … etc. For example,

```
[   [3, 2, 1],
    [3, 1, 2],
    [1, 3, 2],
]
```

becomes

```
[    [1, 3, 2],
     [3, 1, 2],
     [3, 2, 1],
]
```

We can imagine that we are sorting from `[321, 312, 132]` to `[132, 312, 321]`. However, combining the columns into a single integer number only works for integer cases, it proved to be more difficult if we are sorting integers with an arbitrary number of digits. Still, sorting the whole row as a single representation will not help us generalize to float and string data types. While `np.sort` can only sort along a single dimension, we can use `np.lexsort` to sort based on multiple columns. The following code is an implementation of sorting rows using NumPy's `np.lexsort`.

```
1  import numpy as np
2
3  def sort_rows_lex(A, col=None):
4      """Return sorted A, and index, such that A_old[index] = A_new"""
5      if col is None:
6          col = np.s_[::-1] # build a whole slice
7      else:
8          col = col[::-1]
9      # Sort by columns.
10     # lexsort sort by last columns first, and first columns last
11     index = np.lexsort(A[:, col].T)
12     # Apply the resulting sorting
13     A_sorted = A[index, :]
14
15     return A_sorted, index
```

- Lines 5–8 parse the `col` argument. If it is `None`, then we build the whole slice `np.s_[:]` object; otherwise, we will select the columns based on the array of indices provided by `col`. Here, we also reversed the `col` with `[::-1]`, because `np.lexsort` will sort based on the last columns first and the first columns last. To keep our function intuitive, we reverse the indices in this preprocessing step, such that the user of the function can still provide the order of columns to be sorted in ascending order (e.g. `col=[1, 3, 2]` to tell the function to sort column 1 first, then column 3, then column 2, and ignore column 0.)
- Line 11 performs the lexsort. We slice the input `A` based on the columns we want to sort on. We then transpose the matrix since `np.lexsort` operates by a list/tuple of columns. This returns the sorted index.
- Line 13 applies the sorted index to the original input `A` using slicing syntax.

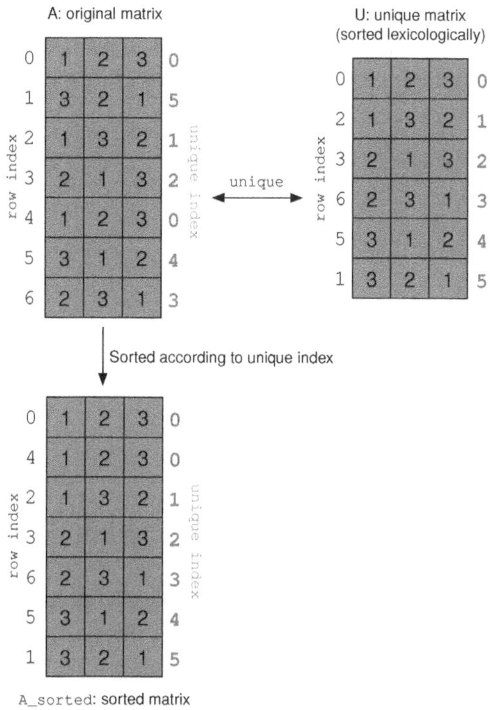

Figure 10.3 Illustration of using `np.unique` to sort rows of a matrix. In this example, we have a 7 × 3 matrix `A` where we want to sort its rows. Note that row `0` and row `4` are duplicated. After `np.unique` along `axis=0`, the rows are deduplicated to produce the matrix, and then sorted lexicologically (i.e. first by column 0, then by column 1, …, etc.), which produces unique matrix `U`. We can see that the first row of `U` came from row `0` of `A`; the second row of `U` came from row `2` of `A`; the third row of `U` came from row `3` of `A`, and so on. Hence we have a mapping between the rows in `A` and `U` (which is the `row index` on the left and the `unique index` on the right of the unique matrix `U` at corresponding rows). As we also specified `return_inverse=True`, `np.unique` returns a set of reverse indices to indicate from which indices in `U` can recover the original matrix `A`. This is illustrated as the `unique index` on the right side of the original matrix `A`. For examples, we can use row `0` from the unique matrix `U` as the first row of the original matrix `A`; we can use row `5` from `U` as the second row of `A`; we can use row `1` from `U` as the third row of `A`, and so on. The ordinal value of the `unique index` indicates the lexicological order of the rows. So if we would like to sort the original matrix `A` according to the lexicological ordering, then we should sort the rows according to the ordinal values of the reverse indices. In other words, we would like to place the rows in the order of the `unique index`. As shown in the `A_sorted` matrix, rows from the original matrix are sorted according to the ordinal values of the `unique index`.

An alternative solution is to use `np.unique` which can detect unique rows of a matrix. It can also return the indices from the unique set of rows to reconstruct the original matrix, by specifying `return_inverse=True`. This is implemented below

```
1  import numpy as np
2
3  def sort_rows_unique(A, col=None):
4      """Return sorted A, and index, such that A_old[index] = A_new"""
5      if col is None:
6          col = np.s_[:] # build a whole slice
7      # Sort each row
8      _, index = np.unique(A[:, col], axis=0, return_inverse=True)
9      # Apply the resulting sorting
10     A_sorted = A[np.argsort(index), :]
11
12     return A_sorted, index
```

- Lines 5–6 create the whole slice object if `col` is `None`, indicating that we are sorting based on all columns.
- Line 8 uses `np.unqiue` to find the reverse index which can be used to reconstruct the original inputs using the unique set (returned as _ in this particular example). The unique set is already lexicologically sorted by columns 0, 1, 2, …, etc.
- Line 10: `np.argsort` on the reverse `index` creates the correct index to be applied to the original input to create the lexicologically sorted unique set, but without dropping the duplicated entries. See Figure 10.3 for an illustration of how this works.

Since Tensorflow and PyTorch both miss the `lexsort` method, the `.unique` method is, therefore, used to implement `sort_rows` in these two libraries. To test our implementations, suppose we would like to sort the following matrix:

```
A = np.array([
      [3, 2, 1],
      [1, 3, 2],
      [2, 1, 3],
      [1, 2, 3],
      [3, 1, 2],
      [2, 3, 1],
])

A_sorted_lex, _ = sort_rows_lex(A)
A_sorted_unique, _ = sort_rows_unique(A)
assert np.allclose(A_sorted_lex, A_sorted_unique)
```

```
# array([[1, 2, 3],      # 123
#        [1, 3, 2],      # 132
#        [2, 1, 3],      # 213
#        [2, 3, 1],      # 231
#        [3, 1, 2],      # 312
#        [3, 2, 1]])     # 321
```

We can also apply the sorting to string data

```
B = np.array([
    ["a", "c", "b"],
    ["b", "a", "c"],
    ["c", "b", "a"]
])
B_sorted_lex, _ = sort_rows_lex(B)
B_sorted_unique, _ = sort_rows_unique(B)
assert np.all(B_sorted_lex == B_sorted_unique)
# array([['a', 'c', 'b'],
#        ['b', 'a', 'c'],
#        ['c', 'b', 'a']], dtype='<U1')
```

An example application of the `sort_rows` functions that we implemented above is to sort the *xy*-coordinates of a set of points that we want to fit a single-exponential curve on, using the method introduced by Jean Jacquelin (Chapter 4). In that specific case study, we have assumed that the data points are already sorted in the correct order. But, in practice, we would need to use `sort_rows` to preprocess the data before fitting the curve.

10.4 Case Study: Reverse Padded Sequence

Tensorflow has a rather magical function `tf.reverse_sequence` when I need to process my sequence backward. It reverses padded sequence batchwise. For example,

```
import tensorflow as tf
sequence = tf.constant([
    [3, 4, 7, 8, 0, 0, 0, 0],
    [2, 3, 5, 0, 0, 0, 0, 0],
    [8, 7, 6, 5, 4, 0, 0, 0],
])
seq_lengths = tf.constant([4, 6, 5])
reversed_sequence = tf.reverse_sequence(
```

```
        sequence, seq_lengths, seq_axis=1, batch_axis=0
    )
```

which gives me

```
<tf.Tensor: shape=(3, 8), dtype=int32, numpy=
array([[8, 7, 4, 3, 0, 0, 0, 0],
       [0, 0, 0, 5, 3, 2, 0, 0],
       [4, 5, 6, 7, 8, 0, 0, 0]], dtype=int32)>
```

Notice that for the second row of the example, I have three additional 0s in the front. This is because I specified the length of my second sequence to have six elements rather than three (of the nonzero) elements. Therefore, the `tf.sequence` function only assumes a left-aligned, right-ragged tensor to be reversed. PyTorch does not have such a function, so how can we implement one without resorting to `for-loop`?

```
import numpy as np

def reverse_sequence(sequence, seq_lengths, seq_axis=1, batch_axis=0):
    """
    Reverse a batch of variable-length sequences given known sequence lengths.
    Assuming the sequences are left-aligned with padding on the right.

    * sequence: input sequence tensor. Must be at least 2D.
    * seq_lengths: list of integer sequence lengths. Its values must be less
            than the size of the sequence dimension, sequence.shape[seq_axis]
    * seq_axis: axis of the sequence dimension. Default to 1.
    * batch_axis: axis of the batch dimension. Default to 0.
    """
    seq_lengths = np.asarray(seq_lengths)[:, None]
    assert sequence.shape[batch_axis] == len(
        seq_lengths
    ), "Length of seq_lengths must be equal to sequence.shape[batch_axis]"

    # creating reversing indices via masking (batch_size, seq_len)
    indices = np.arange(sequence.shape[seq_axis])
    # if within specified sequence length, create the reverse index
    # otherwise, retain the forward index
    indices = np.where(
        indices < seq_lengths, seq_lengths - 1 - indices, indices
    )

    # Add additional singleton dimensions for broadcasting
```

```
    # e.g. (1, ..., batch_size, ...., seq_size, ..., 1, ...)
    shape = [1] * sequence.ndim
    shape[batch_axis] = indices.shape[0]
    shape[seq_axis] = indices.shape[1]
    indices = indices.reshape(shape)

    # reindexing
    reversed_sequence = np.take_along_axis(sequence, indices, seq_axis)

    return reversed_sequence
```

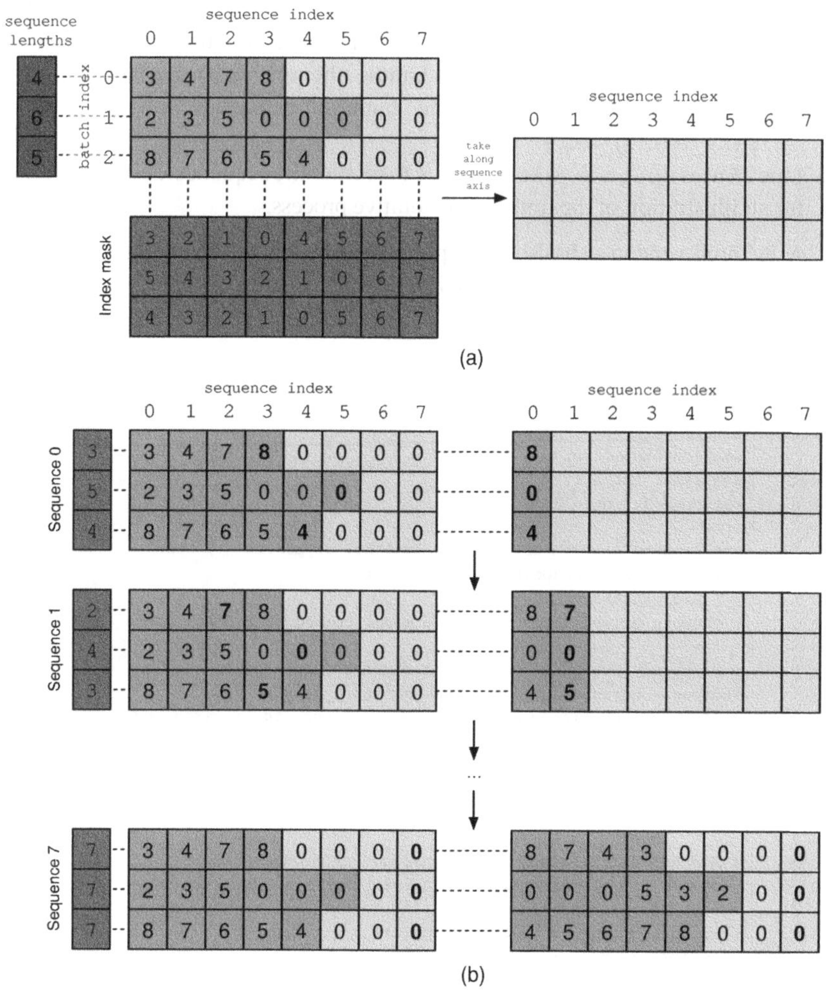

Figure 10.4 Illustration of the implementation logic to reverse a padded sequence. (a) Illustration of creating the reverse index mask within the specified sequence lengths. (b) Illustration of the step-by-step process of reordering the sequence.

- Lines 14–17 get the lengths of the sequences within the batch. It performs a check using `assert` to make sure there is one sequence length specified for each sequence in the batch.
- Lines 20–25 create an index reversing mask. We first start with a forward sequence index, `[0, 1, 2, ..., seq_len]`. Then, use `np.where` to replace the values via broadcasting: if the forward index is less than the specified sequence length, then compute the reverse sequence index as `seq_lengths - 1 - indices`; otherwise, retain the current forward index. See Figure 10.4a for an illustration (bottom matrix labeled as "index mask").
- Lines 29–32 reshape the index masking in order to handle higher-dimensional cases. The batch and sequence dimensions make up the two dimensions of the index masking matrix. To handle the rest of the dimensions, we rely on broadcasting to slice the higher-dimensional sequences using `np.take_along_axis`.
- Line 35 uses `np.take_along_axis` to reorder the sequence. See Figure 10.4b for an illustration of the underlying iterative process.

A PyTorch version can also be implemented by simply changing the NumPy function to the corresponding PyTorch functions

```
import torch

def torch_reverse_sequence(
    sequence: torch.Tensor,
    seq_lengths: list,
    seq_axis: int = 1,
    batch_axis: int = 0,
):
    seq_lengths = torch.asarray(seq_lengths, dtype=int)[:, None]
    assert sequence.shape[batch_axis] == len(
        seq_lengths
   ), "Length of seq_lengths must be equal to sequence.shape[batch_axis]"

    # creating reversing indices via masking (batch_size, seq_len)
    indices = torch.arange(sequence.shape[seq_axis])
    # if within specified sequence length, create the reverse index
    # otherwise, retain the forward index
    indices = torch.where(
        indices < seq_lengths, seq_lengths - 1 - indices, indices
   )

    # Add additional singleton dimensions for broadcasting
    # e.g. (1, ..., batch_size, ...., seq_size, ..., 1, ...)
    shape = [1] * sequence.ndim
    shape[batch_axis] = indices.shape[0]
    shape[seq_axis] = indices.shape[1]
    indices = indices.reshape(shape)
```

```

        # reindexing
        reversed_sequence = torch.take_along_dim(sequence, indices, seq_axis)

        return reversed_sequence
```

We can verify the implementation by testing a couple of examples against Tensorflow's implementation.

```
import numpy as np
import tensorflow as tf

# Create a 2D example
sequence = np.array([
    [3, 4, 7, 8, 0, 0, 0, 0],
    [2, 3, 5, 0, 0, 0, 0, 0],
    [8, 7, 6, 5, 4, 0, 0, 0],
])
batch_axis = 0
seq_axis = 1
seq_lengths = [4, 6, 5]

reversed_sequence_np = reverse_sequence(
    sequence, seq_lengths, seq_axis=seq_axis, batch_axis=batch_axis
)
reversed_sequence_tf = tf.reverse_sequence(
    sequence, seq_lengths, seq_axis=seq_axis, batch_axis=batch_axis
)

assert np.allclose(reversed_sequence_np, reversed_sequence_tf)

# Create a 3D example. The batch and sequence axes are not defaults.
# batch_axis = 1, seq_axis = 2
sequence = torch.asarray(
    np.array(
        [
            [
                [3, 4, 7, 8, 0, 0, 0, 0],
                [2, 3, 5, 0, 0, 0, 0, 0],
                [8, 7, 6, 5, 4, 0, 0, 0],
            ],
            [
                [2, 4, 6, 8, 0, 0, 0, 0],
                [4, 6, 10, 0, 0, 0, 0, 0],
                [16, 14, 12, 10, 8, 0, 0, 0],
            ],
        ]
    )
)
batch_axis = 1
```

```
seq_axis = 2
# Here padding is also not always 0 and valid sequence value is not always > 0
seq_lengths = [4, 6, 3]

reversed_sequence_np = reverse_sequence(
    sequence, seq_lengths, seq_axis=seq_axis, batch_axis=batch_axis
)
reversed_sequence_tf = tf.reverse_sequence(
    sequence, seq_lengths, seq_axis=seq_axis, batch_axis=batch_axis
)

assert np.allclose(reversed_sequence_np, reversed_sequence_tf)
```

10.5 Case Study: Gumbel-Max Sampling with Weights

When training a multi-class classification algorithm, we activate the final layer, whose output we denote as θ, with a softmax function to obtain the probabilities of each class,

$$p_i = \text{softmax}(\theta_i) = \frac{e^{\theta_i}}{\sum_j^N e^{\theta_j}}$$

where we denote p_i as the probability of the class i; N is the number of classes; θ_i as the unactivated final layer output for class i, which is often referred to as the logit value of the class. By inverting the softmax function, we can compute the logit value from the probability as

$$\theta_i = \log\left(p_i \sum_j^N e^{\theta_j}\right) = \log(p_i) + \log(C) = \log(p_i) + C$$

where C denotes some constants which we can ignore later on.

In certain areas like recommendation systems and object detection in which the problems were often formulated as multi-class or multi-label classifications, it is common to see negative classes outnumbering the positive ones. The overwhelming number of negative classes can also reduce training efficiency. A way to tackle this is by selecting a set number of negative classes via random sampling. On the other hand, the distribution of positive labels may skew toward those more popular classes. These factors can lead to class imbalance and a model bias toward the more generic classes. To mitigate class imbalance, a viable solution could be reducing the number of negative label classes in the training data by sampling. The sampling can favor more generic classes, making them appear more frequently as negative labels whenever appropriate. This strategy, in turn, would make the tail-end classes show up as negative labels less frequently, thus, motivating the model to learn to predict these less generic classes more effectively. To achieve these goals, we need to implement a sampling without replacement strategy; to achieve these goals efficiently with vectorization, we need to rely on one-pass rather than iterative algorithms for sampling. Hence, our discussion on the Gumbel-Max trick [Huijben et al., 2022] in this section.

Gumbel distribution has the probability density function (PDF) and cumulative density function (CDF) as

$$\mathrm{PDF}_{\mathrm{Gumbel}}(x;\mu,\beta) = \frac{1}{\beta}e^{-\frac{x-\mu}{\beta}-e^{-\frac{x-\mu}{\beta}}}$$

$$\mathrm{CDF}_{\mathrm{Gumbel}}(x;\mu,\beta) = e^{-e^{-\frac{x-\mu}{\beta}}}$$

The standard Gumbel distribution is obtained by setting $\mu = 0, \beta = 1$, and the PDF and CDF become

$$\mathrm{PDF}_{\mathrm{Gumbel}}(x;0,1) = e^{-x-e^{-x}}$$

$$\mathrm{CDF}_{\mathrm{Gumbel}}(x;0,1) = e^{-e^{-x}}$$

Let G denotes a sample obtained from the standard Gumbel distribution. The Gumbel-Max trick suggests that to sample from the softmax distribution whose unactivated logits are denoted by θ_i, we can first obtain the index I of the sample among the N classes using

$$I = \mathrm{argmax}_{i\in\{1,\dots,N\}}(\theta_i + G^{(i)}) \tag{10.1}$$

where the term inside the parentheses is often called *perturbed logits*. Since we can predetermine how we want to sample (e.g. popularity based or simply random uniform), we can designate the probability of each class p_i. Therefore, we can rewrite Equation 10.1 in terms of p_i,

$$\begin{aligned} I &= \mathrm{argmax}_{i\in\{1,\dots,N\}}(\theta_i + G^{(i)}) \\ &= \mathrm{argmax}_{i\in\{1,\dots,N\}}(\log(p_i) + C + G^{(i)}) \end{aligned} \tag{10.2}$$

where C is the normalizing constant. However, C does not impact the order of the perturbed logits term since it simply shifts the values for all classes by the same value. We can, therefore, drop the term, and obtain

$$I = \mathrm{argmax}_{i\in\{1,\dots,N\}}(\log(p_i) + G^{(i)}) \tag{10.3}$$

To sample $G^{(i)}$ for each class i, we use the inverse transform sampling technique. This is done by first sampling from the uniform distribution $y \sim U(0,1)$, and then using the inverse cumulative distribution function (ICDF) to compute the sample value. The ICDF of the standard Gumbel distribution is

$$\mathrm{CDF}^{-1}_{\mathrm{Gumbel}}(y) = -\log(-\log y) \tag{10.4}$$

Furthermore, to sample multiple items simultaneously without replacement, instead of obtaining the argmax of the perturbed logits, we can obtain top-k indices from the list of perturbed logits for all classes.

In the following code sample, we will implement a function that solves the negative sampling problem when training a multi-label classification model, where the positive labels may include multiple classes. For example, in object detection tasks, a single image may contain multiple objects (e.g. an apple and a cup on a table), so the label will be `["apple", "cup", "table"]`. We would like to create negative labels so that the model can learn the image does not contain certain items, such as `["cat", "cereal", "spoon", "fork"]`. Since an image may contain a variable number of objects, we therefore use a ragged sparse tensor to represent the labels (recall in Chapter 8). We also assume the labels are already converted from string to 1-based indices (recall in Chapter 6).

```
import numpy as np
import tensorflow as tf

def gumbel_max_sample_without_replacement(
        true_labels: tf.SparseTensor,
        num_neg_samples: int,
        vocab_size: int,
        vocab_freq: tf.Tensor = None,
        shuffle: bool = True,
):
    """
    Use Gumbel-Max trick to select negative samples.
    * true_labels: (batch_size, None)
        Positive / label class indices, 1-based indexing
    * num_nega_samples:
        Number of negative samples to take
    * vocab_size:
        Number of items in the categorical variable
    * vocab_freq:
        The distribution of the classes to take samples from.
        Does not have to be normalized.
        If None, assuming uniform sampling.
    * shuffle:
        Whether or not to shuffle the results so that the
        positive labels are not always at the beginning of
        the resulting sample.
    """
    # Default to uniform sampling if vocab_freq is None
    if vocab_freq is None:
        vocab_freq = tf.ones((vocab_size,), tf.float32)

    # Compute proba for the batch
    true_proba = tf.sparse.to_indicator(
        true_labels, vocab_size + 1
    )
```

```
        true_proba = true_proba[:, 1:]  # get rid of OOV bucket
        true_proba = tf.cast(true_proba, tf.float32)
        neg_proba = (1.0 - true_proba) * vocab_freq

        # Gumbel-Max trick: perturbed logits
        z = -tf.math.log(
            -tf.math.log(
                tf.random.uniform(tf.shape(neg_proba), 0, 1)
            )
        )
        z += tf.math.log(neg_proba + 1e-8)
        # set true targets as -inf, so it won't be sampled
        z = tf.where(
            tf.greater(true_proba, 0.5), -np.inf, z
        )

        # Take negative samples
        _, neg_samples = tf.math.top_k(
            z, num_neg_samples, sorted=False
        )  # (batch_size, num_neg_samples)
        # offset by 1, since we use 1-based indexing
        neg_samples = tf.cast(neg_samples, tf.int64) + 1
        # Concat to make the full sample: positive + negative
        samples = tf.concat([
            tf.sparse.to_dense(true_labels),
            -neg_samples
        ], axis=1)

        # shuffle
        if shuffle:
            # generating random ordered indices
            r = tf.random.uniform(tf.shape(samples))
            r = tf.argsort(r, axis=1)
            # Apply the shuffled indices
            samples = tf.gather(samples, r, axis=1, batch_dims=1)

        # Convert to sparse
        samples = tf.sparse.from_dense(samples)

        # Left align
        row_indices = samples.indices[:, 0]
        row_counts = tf.unique_with_counts(
            row_indices, out_idx=tf.int64
        ).count
        column_indices = tf.ragged.range(row_counts).flat_values
        indices = tf.transpose(tf.stack([row_indices, column_indices]))
        binary_labels = tf.SparseTensor(
            indices=indices,
            values=tf.where(samples.values > 0, 1.0, 0.0),
            dense_shape=samples.dense_shape,
        )
        samples = tf.SparseTensor(
            indices=indices,
            values=tf.abs(samples.values),
```

```
90            dense_shape=samples.dense_shape,
91        )
92
93        return samples, binary_labels
```

- Lines 29–30: When `vocab_freq` is `None`, we assume we want to take the negative samples uniform randomly. Hence, we initialize the `vocab_freq` to be an array of `1`s.
- Lines 33–35 convert the labels from sparse indices to the binary representation of the data, where the values at the label indices will be set to `True` and everything else is set to `False`. Since we use 1-based indexing, we need `vocab_size + 1` number of buckets.
- Line 36 removes the out-of-vocabulary (oov) bucket. During training time, we should not expect any oov items.
- Line 37 casts from boolean dtype to float dtype, since we will use the binary representations to implement the Gumbel-Max trick later on. Values at the label indices will be set to `1.0` and everything else is set to `0.0`.
- Line 38 creates an array of sampling probabilities for the negative classes. Here, the positive classes will have a sampling probability of `0.0`, while everything else has a probability proportional to the `vocab_freq`. The `neg_proba` is the unnormalized probability. We will not normalize `neg_proba` since it will not matter for our Gumbel-Max formulation, as we have seen in our derivations of Equations 10.2 and 10.3.
- Lines 41–45 take an array of random samples from the standard Gumbel distribution based on Equation 10.4.
- Line 46 completes the implementation of perturbed logits formulation, by adding the $\log(p_i)$ term. Inside the `tf.math.log` operation, we add a small smoothing factor to make sure we can take the log on the positive labels as well, which currently is set to zero.
- Lines 48–50 set the perturbed logits for the positive labels to $-\infty$. Although the perturbed logits at the positive labels were set to a small value, which after log transformation, become a negative number with large magnitude, there is still a chance that positive values will be sampled, if some of the negative samples have unnormalized probability less than our smoothing factor 10^{-8}. Therefore, these lines will make sure the positive labels will never be sampled in practice.
- Lines 53–55 take top-k negative samples based on the Gumbel-Max formulation.
- Line 57 shifts the indices by `1` so that the indices of the negative samples align with the indices of the input positive labels.
- Lines 59–62 concatenate the two tensors, one containing the positive label indices (which is first converted to a dense tensor), and one containing the negative sample indices (which are denoted as negative numbers). Since we are

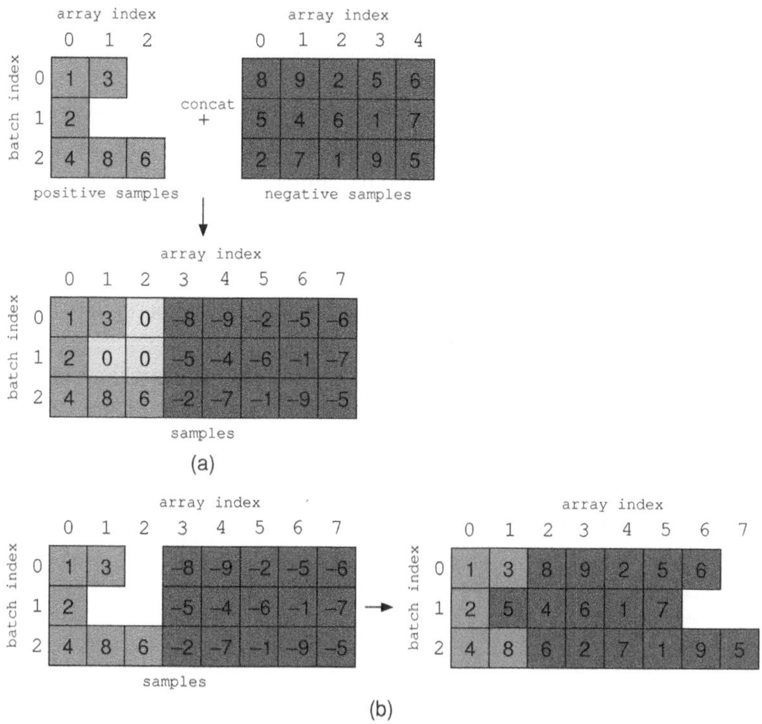

Figure 10.5 Evolution of `samples` tensor when implementing the Gumbel-Max trick for negative sampling. (a) The sparse positive label input is converted into a dense tensor (with 0 padded) and concatenated with the negative samples obtained from the Gumbel-Max sampling trick. Here, the positive samples remain as positive index numbers, while the negative samples are denoted by negative numbers. This can be seen as a ledger for us to determine which entries are from positive labels and which are from negative samples, which is helpful for us to create the `binary_labels` tensor later on. (b) After converting from a dense tensor to a sparse tensor, the `samples` tensor is left aligned to eliminate the zero-filled "gaps." At the same time, index values for the negative samples are converted to positive numbers.

using 1-based indexing, all valid entries of the resulting tensor `samples` should be nonzeros. Also, since the positive labels vary in length for each example in the batch, there will be some "gaps" between positive labels and negative samples (which will be padded as 0s. See Figure 10.5a for an illustration of the `samples` tensor), which will be corrected later on.

- Lines 65–70 implement the logic to shuffle each row independently. This can be seen as another Gumbel-Max trick implementation, with the sampling probabilities of each class being equal to each other. Therefore, we only need to take samples from a uniform distribution to sort the entries in the array, skipping the log and double negative log transformation steps.

- Line 73 converts the dense tensors (after shuffling, if needed) to sparse tensors. Still, the original concatenated dense tensors have zero entries, which become "gaps" when converted to dense tensors.
- Lines 76–81 implement indices that are used to left-align the sparse tensors, as we have seen in Chapter 8. The same indices can be applied to both `samples` and `binary_labels`, therefore, I opted to reimplement the logic rather than calling a preexisting function.
- Lines 82–86 create the `binary_labels` left-aligned sparse tensor to indicate which entries within the final `samples` tensor will be positive labels, and which ones will be negative samples. We determine the value of the label (1 or 0) based on the sign of the entries in `samples`.
- Lines 87–91 create the left-aligned `samples` tensor and convert the sampled class indices to positive values by taking the absolute value. See Figure 10.5 for an illustration.

Finally, we can verify the implementation by examining the sampling frequency of the negative samples via a simulation

```
from collections import Counter
import numpy as np

def simulate_gumbel_max(
    batch_size=1024,
    max_num_pos_samples=5,
    num_neg_samples=10,
    vocab_size=100,
    num_iter=100,
    vocab_freq=None,
):
    """Get sampling frequency of each negative class."""
    # Initialize the counter
    neg_sample_freq = Counter()
    for _ in range(num_iter):
        true_labels = []
        # Generate positive samples iteratively
        for num_pos in np.random.choice(
                np.arange(1, max_num_pos_samples+1),
                size=batch_size,
                replace=True
            ):
            pos_samples = np.random.choice(
                np.arange(1, vocab_size+1),
                size=num_pos,
                replace=False,
            )
            true_labels.append(pos_samples)

        # Convert to left aligned sparse
        true_labels = tf.ragged.constant(
            true_labels, dtype=tf.int64
        ).to_sparse()
```

```

        # Use Gumbel-Max negative sampling
        samples, binary_labels = gumbel_max_sample_without_replacement(
            true_labels,
            num_neg_samples=num_neg_samples,
            vocab_size=vocab_size,
            vocab_freq=vocab_freq,
            shuffle=False,
        )

        # Extract negative labels only
        neg_samples = samples.values.numpy() * (
            1 - binary_labels.values.numpy()
        )
        neg_samples = neg_samples[neg_samples > 0]
        neg_samples = neg_samples.astype(int).tolist()
        neg_sample_freq.update(neg_samples)

    return neg_sample_freq
```

Let us first simulate by sampling from a uniform distribution.

```
import matplotlib.pyplot as plt
# Simulate: default from uniform distribution
neg_sample_freq = simulate_gumbel_max()
# Plot
fig, ax = plt.subplots(1, 1, figsize=(7, 4))
ax.bar(list(neg_sample_freq.keys()), list(neg_sample_freq.values()))
ax.set_xlabel("Vocab Index")
ax.set_ylabel("Frequency")
```

Figure 10.6 illustrates the simulated sampling frequency, which is close to a uniform distribution.

We can also sample from a geometric-like categorical distribution. See Figure 10.7 for the result.

```
rho = 0.05
vocab_freq = (1 - rho)**(np.arange(100)) * rho * 1E6
neg_sample_freq = simulate_gumbel_max(
    vocab_freq=vocab_freq
)
fig, ax = plt.subplots(1, 1, figsize=(7, 4))
ax.bar(list(neg_sample_freq.keys()), list(neg_sample_freq.values()),
ax_twin = ax.twinx()
ax_twin.bar(np.arange(1, 100+1), vocab_freq, color="#FF7F0E", alpha=0.7)
ax.set_xlabel("Vocab Index")
ax.set_ylabel("Frequency", color="#1F77B4")
ax_twin.set_ylabel("Underlying Distribution", color="#FF7F0E")
ax.tick_params(axis='y', colors="#1F77B4")
ax_twin.tick_params(axis='y', colors="#FF7F0E")
```

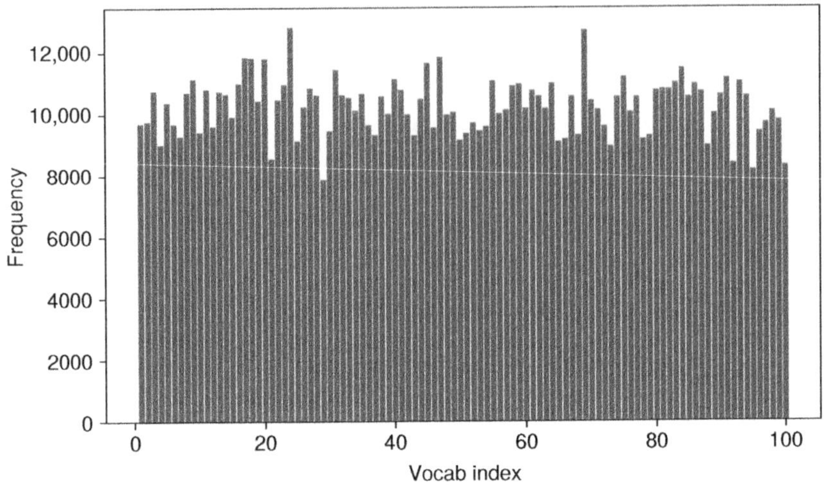

Figure 10.6 Simulated Gumbel-Max negative sampling frequency when using uniform sampling.

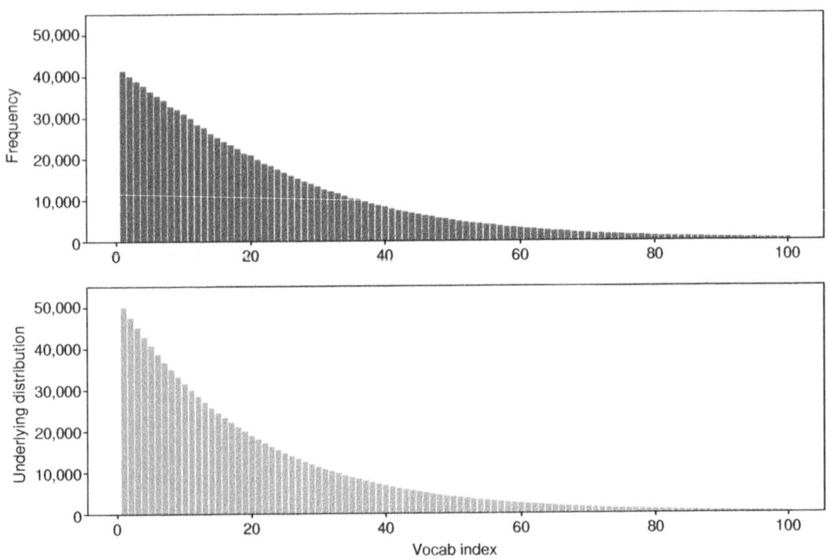

Figure 10.7 Simulated Gumbel-Max negative sampling frequency when sampling from a geometric distribution. Top: simulated frequency; bottom: underlying frequency from `vocab_freq`.

10.6 Case Study: Sorting Articles Around Anchored Advertisements

Suppose we created a website that recommends news articles to readers. To monetize our website, we would like to insert some advertisements among the news articles that we recommend. Figure 10.8 illustrates the effect we would like to achieve. Our imaginary product manager has set the following requirements for the illustrated effect:

1. Anchored ads may occur at any position. This could be an output from a separate machine learning model.
2. The number of anchored ads may differ between each user.
3. News articles need to be sorted based on the priority scores for each item.
4. Separately, ads also need to be sorted based on their own priority scores, independent of the scoring of the news.
5. Ads can only occur in the prespecified anchor positions, while the news can only occur in the rest of the free positions.
6. The final output should display up to eight contents including both ads and news, even though we may be provided more news and ads to sort. We also assume that there is always enough content to make up eight total displayable items.

The following code block implements a vectorized operation to rearrange the news and ads among each other. The general strategy is to sort news and ads separately based on their respective priority scores and then rearrange the two pieces into the same matrix, using masking and gathering.

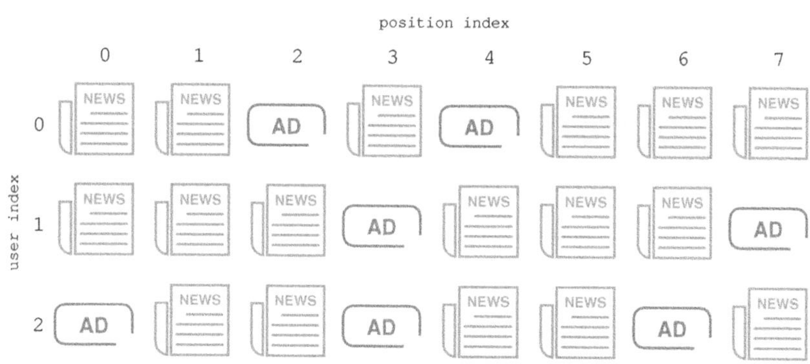

Figure 10.8 Illustration of arranging news articles around anchored ads.

```
import tensorflow as tf

def arrange_ads_among_news(
    news_scores,
    ads_scores,
    ads_positions,
    news_vocab=None,
    ads_vocab=None,
    k=8
):
    """
    Arrange ads within a series of news articles for a batch of users.

    Inputs:
        * news_scores: (batch_size, news_vocab_size) Tensor
            Scores for each news, corresponding to each news in news_vocab
        * ads_scores: (batch_size, ads_vocab_size) Tensor
            Scores for each ad, corresponding to each ad in ads_vocab
        * ads_positions: (batch_size, None) Left Aligned SparseTensor
            List of allowed ads positions for each user
        * news_vocab: (news_vocab_size,) array
            List of vocabulary for news. If None, the function returns
            the indices in the concatenated vocab space
        * ads_vocab: (ads_vocab_size,) array
            List of vocabulary for ads. If None, the function returns
            the indices in the concatenated vocab space
        * k: int
            Number of total contents to keep.
    """
    # Sort the news and ads
    # switch to 1-based indexing
    news_items = tf.argsort(news_scores, direction="DESCENDING") + 1
    # padding at the beginning of each row
    news_items = tf.pad(news_items, [(0, 0), (1, 0)])
    news_items = tf.cast(news_items, tf.int64)
    ads_items = tf.argsort(ads_scores, direction="DESCENDING") + 1
    ads_items = tf.pad(ads_items, [(0, 0), (1, 0)])
    ads_items = tf.cast(ads_items, tf.int64)

    # Compute unanchored positions mask
    anchored_indices = tf.transpose(
        tf.stack([ads_positions.indices[:, 0], ads_positions.values])
    )
    # remove any anchor positions beyond the max number of contents to keep
    anchored_indices = tf.boolean_mask(
        anchored_indices, anchored_indices[:, 1] < k, axis=0
    )
    # Use sparse tensor to inidcate the anchored positions
    anchored_mask = tf.sparse.SparseTensor(
        indices = anchored_indices,
        values = tf.ones_like(anchored_indices[:, 0]),
        dense_shape=(tf.shape(news_scores)[0], k)
```

```
53         )
54         anchored_mask = tf.sparse.to_dense(anchored_mask)
55
56         # Create the indices matrix for ads
57         anchored_positions = tf.cumsum(anchored_mask, axis=1) * anchored_mask
58         # Gather along axis=1
59         output_index_ads = tf.gather(
60             ads_items, anchored_positions, batch_dims=1, axis=1
61         )
62
63         # Create the indices matrix for news
64         free_mask = 1 - anchored_mask
65         free_positions = tf.cumsum(free_mask, axis=1) * free_mask
66         output_index_news = tf.gather(
67             news_items, free_positions, batch_dims=1, axis=1
68         )
69
70         # Combine the two sets of indices together
71         news_vocab_size = tf.cast(tf.shape(news_scores)[1], tf.int64)
72         output_index = tf.where(
73             free_mask > 0,
74             # convert back to 1 based indexing
75             output_index_news-1,
76             # index of ads in the combined vocab
77             news_vocab_size + output_index_ads - 1
78         )
79
80         if news_vocab is not None and ads_vocab is not None:
81             # Create the string outputs
82             vocab = tf.concat([news_vocab, ads_vocab], axis=0)
83             output = tf.gather(vocab, output_index)
84             return output, free_mask
85         else:
86             # return the indices in the combined space
87             return output_index, free_mask
```

- Lines 32–38 sort the news and ads separately and get the corresponding sorting indices (i.e. we use an integer representation of the items). We changed to a 1-based indexing system to prepare for subsequent masking operations. We also padded the beginning of each row by a `0` so that the subsequent `gather` operation can still pretend it is gathering using a 0-based indexing system (more details below).
- Lines 41–43 determine the multi-index of the anchored positions in the final output matrix. The row/batch indices correspond to the batch indices of the original sparse tensor; the column indices of the anchored positions correspond to the values of the original sparse tensor. See Figure 10.9 for an illustration.
- Lines 45–47 use `tf.boolean_mask` to remove any anchor positions beyond the required final output matrix. For example, if the input specifies an anchor position 9, but the function is asked to return only the first eight positions, then

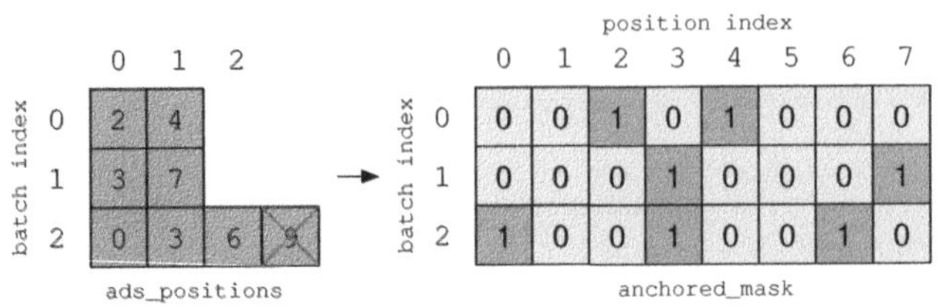

Figure 10.9 Illustration of converting `ads_positions` into ads `anchored_mask` binary indicators. Notice that the index value 9 is eliminated since the final output only returns up to eight items.

the anchor position 9 is eliminated from the `anchored_indices` so that we do not need to worry about out of scope/output index anchoring when we create the mask for the anchored positions in the next step.

- Lines 49–53 create a sparse indicator matrix, using the `anchored_indices` we determined from the previous steps. The shape of this matrix is `(batch_size, k)`, which corresponds to the final output matrix shape. This is essentially the logic of the `tf.sparse.to_indicator` operation that we have seen in Chapter 8. See Figure 10.9 for an illustration.
- Line 54 converts the sparse indicator matrix to a dense binary matrix.
- Line 57 calculates the anchored sorting positions using the `anchored_mask`. The nonzero values in this matrix are indices corresponding to the values `ads_items`. To illustrate the math operations implemented here, suppose the mask for a row is `[0, 0, 1, 0, 1, 0, 0, 0]` where `1`s indicate where the advertisements are. The first occurrence of `1` should correspond to the ads with the highest priority score, while the second occurrence of the `1` should correspond to the ads with the second highest priority score, etc. Therefore, we want to denote the positions of sorting in this array as `[0, 0, 1, 0, 2, 0, 0, 0]`, where `0`s are padding indices, while `1` and `2` indicate the priority of the ads that should be inserted at those positions. This transformation is achieved by `cumsum(mask) * mask`. More concretely, we can interpret this array as a form of ledgers in the following way: for the first two elements, the ledgers are `0` paddings, so we should retrieve the padding item from the sorted `ads_items` (i.e. `ads_items[0] = 0` which is a padding). The third element is a ledger of `1`, meaning we should take the ads with the highest priority, i.e. `ads_items[1]`. The fourth element is a padding value again. The fifth element `2` indicates that we should retrieve the ads with the second highest priority, i.e. `ads_items[2]`. The rest are all paddings again.
- Lines 59–61 use the `anchored_positions` derived from the previous step to look up the ads item indices. This will yield a dense matrix where the ads will be placed at the anchored positions. See Figure 10.10 for an illustration. Note that the same effect can also be achieved using `tf.scatter_nd` to directly

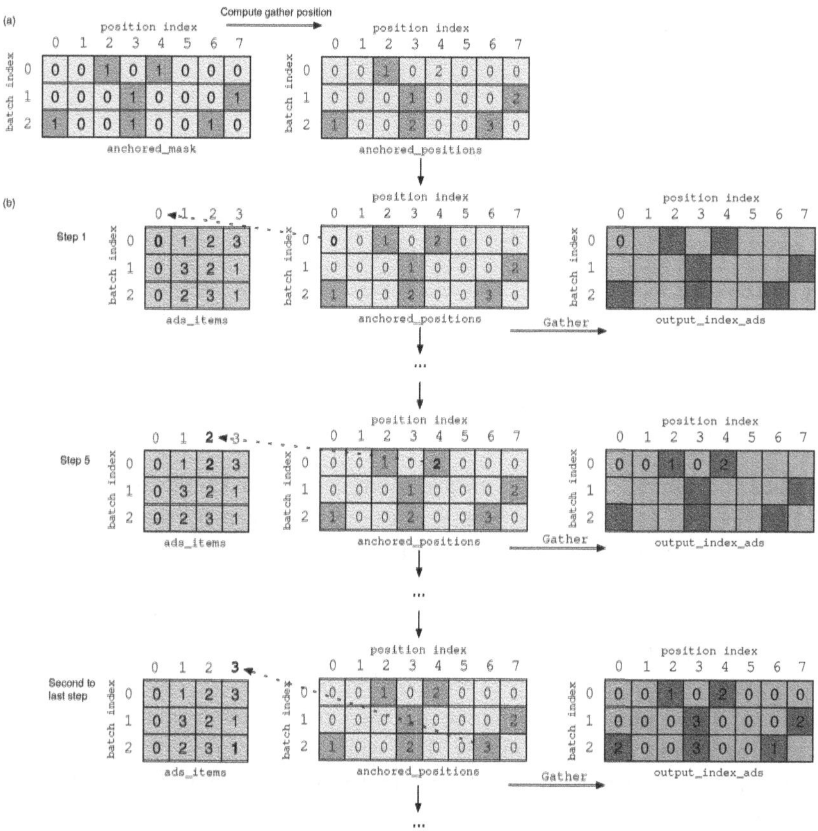

Figure 10.10 Illustration of creating the `output_index_ads` matrix. (a) Creating the ledger matrix to track which indices in the ads item index `ads_indices` should we gather. (b) Step-by-step, iterative illustration of using the sorted `ads_items` and the position ledger `anchored_positions` to gather the ads into the output at the corresponding locations. Let us walk through three examples. We start at location `(0, 0)` in Step 1. The ledger (middle matrix) points to index value `0`. We look for the value at index `0` for batch `0`, obtain the item `0` (which is a padding value) from the `ads_items` matrix (left), and put this item to the output `output_index_ads` (right). Skipping to Step 5 at location `(0, 4)`, the ledger (middle) indicates we should gather the item at index `2`, which has an item value of `2` in the `ads_items` matrix (left). We then put item `2` into the `output_index_ads` matrix (right) at the corresponding location. Skipping to the second to last step, at location `(2, 6)`, the ledger (middle) indicates we should gather the item at index `3`, which corresponds to the item value `1` in the `ads_items` matrix (left). Hence, we put item `1` to location `(2, 6)` in the `output_index_ads` matrix (right).

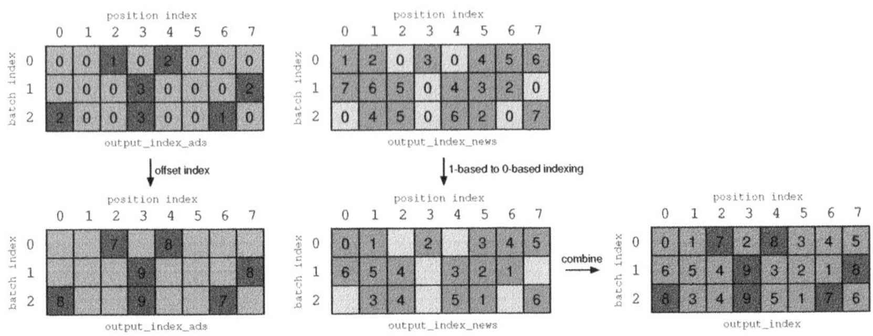

Figure 10.11 Combine ads and news output index into a single matrix. Notice the two matrices are complementary to each other. The values in `output_index_ads` are offset by the vocabulary size of the news. The values in `output_index_news` are converted back to 0-based index.

place the values, but it may involve more complicated indexing operations. This specific example illustrates that `tf.gather` can be used in place of `tf.scatter_nd` under certain scenarios.

- Lines 64–68 repeat the same logic but on news instead. Notice that `output_index_ads` and `output_index_news` complement each other. Figure 10.11: the "gaps"/"holes" in `output_index_ads` can be filled by `output_index_news` at the corresponding positions, and vice versa.
- Lines 71–78 attempt to combine the two output index matrices for ads and news into a single matrix. But, since index values for both news and ads start from 1 in those two matrices, if we naively combine the matrices together, we may end up having two indices `1s` in this `output_index` matrix, which can confuse news vs. ads items. So we offset the ads indices by the vocabulary size of the news. We also convert the indexing back to 0-based in the final `output_index` matrix. That way, news index values are `[0, 1, ..., <news_vocab_size - 1>]`, and ads index values are `[<news_vocab_size>, <news_vocab_size + 1>, ...., <news_vocab_size + ads_vocab_size - 1>]`. This is as if we are indexing using a combined vocabulary, `vocab = concat([news_vocab, ads_vocab])`. See Figure 10.11 for an illustration of the transformation.
- Lines 80–84 apply the `output_index` to look up the item names in the combined vocabulary if the string vocabulary of the two sets is given. This will return the actual item names in the final output.
- Lines 85–87 return the indices instead, since either `news_vocab` or `ads_vocab` is missing in the input.

We can test our implementation using the following example:

```
news_scores = tf.constant([
    [0.7, 0.6, 0.5, 0.4, 0.3, 0.2, 0.1],
    [0.1, 0.2, 0.3, 0.4, 0.5, 0.6, 0.7],
    [0.1, 0.4, 0.2, 0.7, 0.6, 0.5, 0.3],
])
ads_scores = tf.constant([
    [0.5, 0.3, 0.1],
    [0.1, 0.4, 0.8],
    [0.2, 0.6, 0.4],
])
ads_positions = tf.ragged.constant([
    [2, 4], [3, 7], [0, 3, 6, 9]
], dtype=tf.int64).to_sparse()
k = 8 # max number of contents to keep

news_vocab = tf.constant(["a", "b", "c", "d", "e", "f", "g"])
ads_vocab = tf.constant(["x", "y", "z"])

output, mask = arrange_ads_among_news(
    news_scores,
    ads_scores,
    ads_positions,
    news_vocab=news_vocab,
    ads_vocab=ads_vocab,
    k=k,
)

# ouput:
# <tf.Tensor: shape=(3, 8), dtype=string, numpy=
# array([[b'a', b'b', b'x', b'c', b'y', b'd', b'e', b'f'],
#        [b'g', b'f', b'e', b'z', b'd', b'c', b'b', b'y'],
#        [b'y', b'd', b'e', b'z', b'f', b'b', b'x', b'g']],
↪ dtype=object)>
#
# mask:
# <tf.Tensor: shape=(3, 8), dtype=int64, numpy=
# array([[1, 1, 0, 1, 0, 1, 1, 1],
#        [1, 1, 1, 0, 1, 1, 1, 0],
#        [0, 1, 1, 0, 1, 1, 0, 1]])>
```

Bibliography

Charles R. Harris, K. Jarrod Millman, Stéfan J. van der Walt, Ralf Gommers, Pauli Virtanen, David Cournapeau, Eric Wieser, Julian Taylor, Sebastian Berg, Nathaniel J. Smith, Robert Kern, Matti Picus, Stephan Hoyer, Marten H. van Kerkwijk, Matthew Brett, Allan Haldane, Jaime Fernández del Río, Mark Wiebe, Pearu Peterson, Pierre Gérard-Marchant, Kevin Sheppard, Tyler Reddy, Warren Weckesser, Hameer Abbasi, Christoph Gohlke, and Travis E. Oliphant.

Np.partition. https://numpy.org/doc/stable/reference/generated/numpy.partition .html, 2023.

Iris A. M. Huijben, Wouter Kool, Max B. Paulus, and Ruud J. G. van Sloun. A review of the gumbel-max trick and its extensions for discrete stochasticity in machine learning. *IEEE Transactions on Pattern Analysis and Machine Intelligence*, 45(2):1353–1371, March 2022.

11

Building a Language Model from Scratch

Throughout the book, we have illustrated examples of building deep learning models from scratch (e.g. Chapter 6, Case Study: Factorization Machine; Chapter 9 Case Study: Vision Transformer (ViT)). These solutions presented are the results of many years of research on the topic. In this chapter, let us follow the evolution of language modeling using the transformer architecture to illustrate how machine learning models, specifically, large language models (LLMs) like LLaMA (Large Language Model Meta AI), came to be. By the end of this chapter, I hope the readers can appreciate the large amount of effort dedicated to research in machine learning.

11.1 Language Modeling with Transformer

Prior to the introduction of the transformer architecture [Vaswani et al., 2017], deep learning-based natural language modeling has been relying on previously introduced sequence representation architectures such as recurrent neural networks (RNNs) and convolutional neural networks (CNNs). However, these architectures have a limitation: When learning from a long sequence of text/documents, they tend to remember only the most recently encountered tokens and forget those encountered earlier (e.g. the beginning of a paragraph). Deep learning researchers assess an architecture's capability to learn from input signals using a metric called **receptive field** size, a concept borrowed from vision neuroscience. If a model can memorize and learn from a long sequence (in 1D sequence modeling) or a large area (in 2D/3D computer vision problems), then it is said to have a large receptive field. Having a limited receptive field creates problems for tasks like language translation or question-answering, where the text input (i.e. prompt) can be very long (e.g. hundreds of words). Just imagine you were being asked an exceedingly long question, but you can only remember the last few words being said. How could you understand the question and,

Vectorization: A Practical Guide to Efficient Implementations of Machine Learning Algorithms,
First Edition. Edward DongBo Cui.

therefore, give a response without being able to recall what is being asked at the beginning? Attention, specifically the scaled-dot-product self-attention that we discussed in Chapter 5 is proven to resolve this issue in the original transformer paper [Vaswani et al., 2017]. By performing pair-wise dot-product between input tokens, the model learns to focus on what to pay "attention" to (pun intended here~) during sequence modeling. The transformer model was introduced in the context of neural machine translation. However, later research has shown that the transformer model is especially powerful in learning long-sequence data. It has been demonstrated that the transformer architecture allows researchers to build state-of-the-art models in domains not only in natural language processing (NLP), but also in areas including but not limited to text-to-speech (e.g. FastSpeech2 [Ren et al., 2020]), speech-to-text (e.g. Whisper [Radford et al., 2022]), computer vision (e.g. ViT [Dosovitskiy et al., 2021] model that we discussed in Chapter 9), drug discovery (e.g. SMILES [Honda et al., 2019; Grechishnikova, 2021]), and recommendation engines (e.g. SASRec [Kang and McAuley, 2018]).

11.1.1 Encoder and Decoder of the Transformer Architecture

The original transformer uses an encoder–decoder architecture, where the encoder compresses the inputs, and then the decoder decompresses the hidden states to generate outputs specific to the task of translation. As a side note, a recent study has demonstrated that the transformer model is a powerful data compressor for representation learning [Yu et al., 2023], with the interesting slogan "Compression is all there is." The simplest encoder–decoder architecture is the autoencoder, which we introduced in Chapter 8, where the encoder and decoder have the same architecture (i.e. a dense layer). Transformer instead uses a combination of attention, layer normalization, and feedforward layers to compress and decompress the data. The key differences between the encoder and the decoder blocks are:

- The encoder uses bidirectional attention. This means that past tokens can be used to predict future tokens, while future tokens can also be used to predict past tokens.
- The decoder uses masked or causal attention. Past tokens can be used to predict future tokens, but not vice versa. In Chapter 5, when we were discussing the scaled-dot-product attention mechanism, we have illustrated the two types of attention mechanism.
- Specific to the original transformer paper, the decoder block also receives inputs from the hidden state of the encoder layer with an additional attention layer. Therefore, the encoder layer only has one attention layer, while the decoder layer has two attention layers. However, in more recent architectures

such as generative pretrained transformer (GPT) [Radford et al., 2018], decoder architecture specifically refers to a transformer block that has only one causal attention layer.

More often than not, subsequent transformer-based models may use only either the encoder or decoder part of the transformer for specific tasks. For instance, bidirectional encoder representation from transformers (BERT) [Devlin et al., 2019] is an encoder-only pretrained language model from Google. Many sequence representers also tend to use encoder architecture, e.g. FastSpeech2 and Whisper, whose hidden states are then decoded by downstream architectures that may not necessarily be a transformer block, e.g. convolutional layers. By contrast, the GPT model family uses decoder-only architecture, where causal attention is used. The choice of specific architecture is tightly associated with the training tasks used by different models, which we will discuss in Section 11.5.

11.1.2 Training of Transformer Models

Self-supervised pretraining has been an effective technique in building models solving NLP tasks. Researchers would gather a large text corpus with billions to trillions of tokens to build models that "understand" text so that they can be fine-tuned downstream to solve NLP-specific tasks like text classification or sentiment analysis, which usually has less human-labeled data available. What they typically do is pretraining the model using the "Cloze" task (i.e. fill in the blank) [Devlin et al., 2019], in which a word in a sentence is blanked out, and the model learns to predict the word in the blank. It is said that the model learns via a self-supervised fashion, where the data used to generate both the input and the label come from the same sentence, without a human deciding what the inputs and the labels are. Indeed, text data are "self-labeling" (as explained by Yann Lecun). Similar to text data, time series or time-associated events data are also self-labeling. Let's say, we can use the last seven days of historical data to forecast the next event, e.g. whether or not it is raining tomorrow. At the inference time, we do not know what tomorrow's weather is going to be, so we do not have the label at the moment. But, we only need to wait to see the weather on the next day, and the label is automatically generated, without a human deliberately deciding what it is.

Going back to the context around NLP, the BERT model that uses encoder-only transformer architecture with bidirectional attention is trained using the Cloze task. The goal of BERT is to learn a sequence representation, such that it can be used in downstream tasks that would utilize the entirety of the sentence inputs. The Cloze task is an auxiliary task for the model to comprehend a full sentence. By contrast, the GPT model that uses decoder-only transformer architecture with causal attention is trained to predict the next token. The goal of GPT is, as the name suggests, to generate new tokens given an input prompt, in contrast to predicting a token in the middle of the sentence. Therefore, GPT will never see any future tokens during generation, so it needs to learn to predict the ordering of the tokens rather than representing a full sentence. Therefore, different training objectives can dictate different model architecture decisions in a model.

11.2 Pre-LN vs. Post-LN Transformer

Ever since the original transformer paper, various improvements to the architecture have been suggested to further enhance the learning capacity and training stability of the model. For the rest of the chapter, we will introduce various improvements made to the transformer model, which led to what is currently considered to be the strongest recipe for transformer, called Transformer++ [Gu and Dao, 2023] or LLaMA [Touvron et al., 2023a].

One significant caveat about the original transformer architecture is that the training of the model can become unstable, where the loss can diverge. To address this issue, the original transformer paper used a learning-rate schedule during model training, specifically

$$l_{\text{rate}} = d_{\text{model}}^{-0.5} \cdot \min\left(\text{step}^{-0.5}, \text{step} \cdot \text{warm-up\_steps}^{-1.5}\right)$$

where

- step is the number of training steps
- d_{model} is the embedding size of the tokens
- warm-up_steps is the number of steps used for a linear ramping-up period, which the original paper set to 4000.

The learning rate curve is plotted in Figure 11.1, with the accompanying implementation in NumPy below:

```
import numpy as np
import matplotlib.pyplot as plt

# learning rate schedule
d = 128 # d_model
step = np.arange(160_000)
warmup_steps = 4000
lr=(1/np.sqrt(d)) * np.minimum(1/np.sqrt(step), step * warmup_steps**-1.5)

fig, ax = plt.subplots(1, 1, figsize=(10, 4))
```

```
ax.plot(step, lr)
ax.set_xlabel("Step")
ax.set_ylabel("Learning Rate")
```

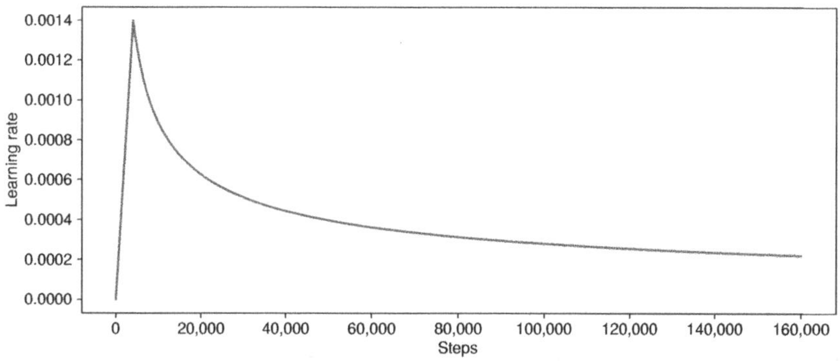

Figure 11.1 Learning rate schedule from the original transformer paper.

However, many researchers still found that they need to carefully tune the learning rate for their specific tasks. Techniques used in the transformer model such as normalization and skip connections are known to address the problem of either an exploding or vanishing gradient, which is crucial for model training stability and convergence, yet these problems still exist. Hence, Xiong et al., 2020 proposed an alternative arrangement of the layers within the transformer block [Xiong et al., 2020]. They call it Pre-LN architecture, to distinguish it from the original architecture which they call Post-LN. A comparison of the two architectures is illustrated in Figure 11.2. The key difference between the two architectures is that the original Post-LN architecture performs attention and feedforward operations and then normalizes their outputs, whereas Pre-LN architecture starts with layer normalization on the inputs. With this change, Xiong et al., 2020 demonstrated that the transformer can be trained more stably without the use of a learning-rate schedule with a warm-up period, hence significantly improving the training efficiency of the transformer model by allowing faster convergence. I have seen many domains of study still use Post-LN with a learning-rate schedule in their neural architecture design. If trained successfully, the Post-LN architecture almost always outperforms that of Pre-LN architecture, especially in certain areas such as text-to-speech.

Let us implement both the Pre-LN and Post-LN architecture of the transformer block.

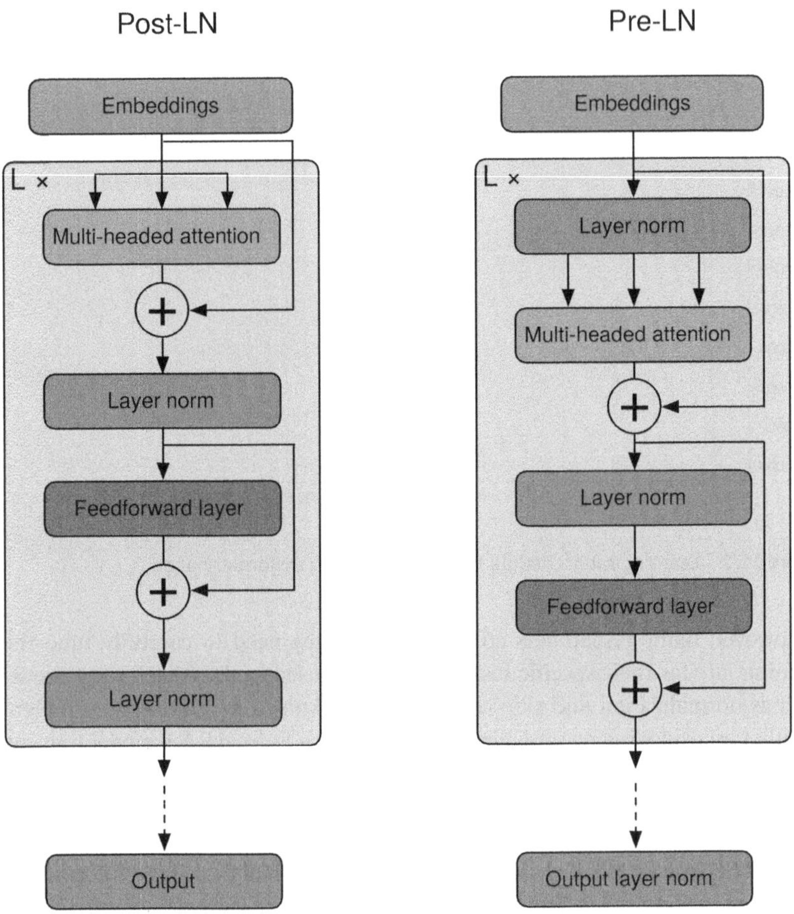

Figure 11.2 Comparison of Pre- and Post-layer normalization architecture. Post-LN architecture places the layer normalization between the residual blocks (indicated by "+" in the diagram), whereas Pre-LN architecture places layer normalization inside the residual blocks. Notice that Pre-LN architecture has a final layer normalization at the output, while Post-LN architecture does not. This final layer normalization in Pre-LN can be easily missed during implementation.

```
import torch
from torch import nn

class PreLNEncoderBlock(nn.Module):
    def __init__(self, embed_dim, ffn_size, num_heads=2, dropout=0.1):
        super().__init__()
        self.mhattn = nn.MultiheadAttention(
```

```
            embed_dim=embed_dim,
            num_heads=num_heads,
            dropout=dropout,
            batch_first=True,
        )
        self.ffn = nn.Sequential(
            nn.Linear(embed_dim, ffn_size),
            nn.GELU(),
            nn.Dropout(dropout),
            nn.Linear(ffn_size, embed_dim),
        )
        self.input_ln = nn.LayerNorm(embed_dim)
        self.ffn_ln = nn.LayerNorm(embed_dim)

    def forward(self, x):
        y = self.input_ln(x)
        y, _ = self.mhattn(y, y, y)
        y += x

        z = self.ffn_ln(y)
        z = self.ffn(z)
        z += y

        return z


class PostLNencoderBlock(nn.Module):
    def __init__(self, embed_dim, ffn_size, num_heads=2, dropout=0.1):
        super().__init__()
        self.mhattn = nn.MultiheadAttention(
            embed_dim=embed_dim,
            num_heads=num_heads,
            dropout=dropout,
            batch_first=True,
        )
        self.ffn = nn.Sequential(
            nn.Linear(embed_dim, ffn_size),
            nn.GELU(),
            nn.Dropout(dropout),
            nn.Linear(ffn_size, embed_dim),
        )
        self.input_ln = nn.LayerNorm(embed_dim)
        self.ffn_ln = nn.LayerNorm(embed_dim)

    def forward(self, x):
        y, _ = self.mhattn(x, x, x)
        y += x
        y = self.input_ln(y)

        z = self.ffn(y)
        z += y
        z = self.ffn_ln(z)

        return z
```

- `PreLNEncoderBlock` (**Lines 4–31**): This is identical to the implementation we had in Chapter 9 when we were looking at VisionTransformer.
  - `__init__` (**Lines 5–20**): This method initializes the necessary layers for the transformer, including `MultiHeadAttention`, the feedforward layer represented as an `nn.Sequential` layer, and the two layer normalization layers.
  - `forward` (**Lines 22–31**): This implements the Pre-LN architecture as follows:
    * Line 23: The input `x` is first normalized by the input layer normalization layer.
    * Line 24: The layer normalized values of `x` are then passed to multiheaded attention.
    * Line 25: Skip connection is implemented here, where we add the raw input `x` to the output of the multiheaded attention `y`.
    * Line 27: The result from the previous step is then passed to another layer of normalization layer.
    * Line 28: The normalized results are then passed to the feedforward layer.
    * Line 29: Another skip connection is implemented here, where we add the layer-normalized output from the multiheaded attention `y` with the output of the feedforward layer `z`. Note that after stacking a series of Pre-LN encoders together, the output requires a final layer normalization.
- `PostLNEncoderBlock` (**Lines 34–61**): This implements the Post-LN architecture from the original transformer paper.
  - `__init__` (**Lines 35–50**): This is identical to `PreLNEncoderBlock`
  - `forward` (**Lines 52–61**): This implements the Post-LN architecture as follows:
    * Line 53: The input `x` is first passed to multiheaded attention.
    * Line 54: Skip connection is implemented here, where we add the raw input `x` to the output of multiheaded attention `y`.
    * Line 55: Layer normalization is applied on the output `y`. Lines 54–55 are together referred to as "Add-and-Norm" in the original transformer paper.
    * Line 57: The result from the above normalization is passed to the feedforward layer.
    * Line 58: Another skip connection is implemented here, where we add the output of the normalized attention output `y` to the result of the feedforward layer `z`.
    * Line 59: Another layer normalization is applied here.

11.3 Layer Normalization

Normalization can help stabilize neural network training by reducing the likelihood of exploding or vanishing gradients and improving model generalization. Different normalization schemes have been explored previously, though most attempted to standardize the inputs or outputs of the network layer such that the mean of the values are centered around 0 and the variance becomes 1.

$$\tilde{x}_i = \frac{x_i - E[x_j]}{\sqrt{\mathrm{Var}[x_j] + \epsilon}}\gamma + \beta \tag{11.1}$$

where γ and β are learnable parameters for each dimension to be normalized; γ can be considered as the learned standard deviation, while β can be considered as the learned mean. ϵ is a smoothing value that prevents division by zero. Depending on which dimensions are normalized, the normalization schemes were given different names. Specifically, given an input tensor of shape `(batch_size, sequence_length, num_channels)`,

- **Batch normalization**: Standardizes each feature/channel/embedding value across the batches. Hence, γ and β are vectors of shape `(num_channels,)`.
- **Layer normalization**: Standardizes within the batch. It has two formulations: the input tensor is normalized either across both sequence and channel dimensions (Variant 1, see Figure 11.3b) or across only the channel dimensions within each sequence (Variant 2, see Figure 11.3c). Different from batch normalization which applies a scalar value for each channel, layer normalization uses an element-wise affine transformation (i.e. scale and shifting). Hence, β and γ are tensors of shape identical to the plane that the normalization is computed

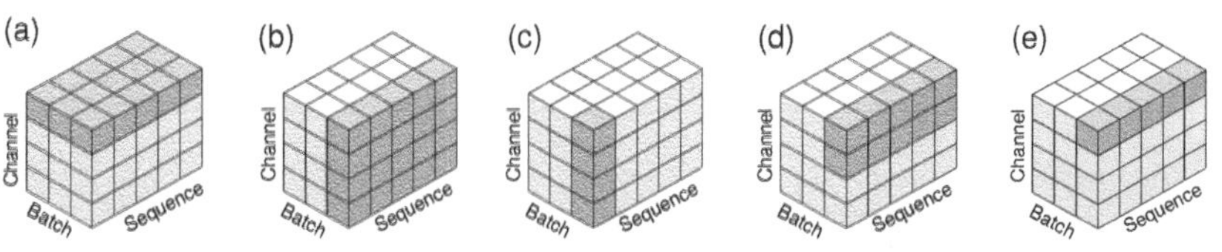

Figure 11.3 Comparison of normalization scheme. (a) Batch normalization normalizes across the batch within each channel/feature dimension, (b) layer normalization normalizes the channel by sequence within each batch, (c) an alternate formulation of layer normalization is to normalize within the batch and sequence across only the channels; this is the default behavior for the implementation of LayerNormalization in Keras and it is a popular usage pattern of layer normalization. This is also the normalization scheme for RMSNorm, (d) group normalization is similar to layer normalization, but further separates channel dimensions into groups and normalizing within the group, (e) instance normalization normalizes within each batch by channel instance, e.g. each sequence.

over. In the first variant of layer normalization, β and γ are matrices of shape (seq_len, num_channels); in the second variant, β and γ are vectors of shape (num_channels,).

- **Group normalization**: Divides the channel dimension into groups, and normalizes within the group. In this normalization scheme, γ and β are learned for each group of the channel.
- **Instance normalization**: Normalizes within each channel/feature. This can be seen as a special case of Group Normalization where the group size is 1.

Figure 11.3 illustrates the differences between various normalization schemes (which are inspired by Figure 2 of "Group Normalization" white paper from Facebook AI Research [Wu and He, 2018]).

RMSNorm, or root-mean-square normalization [Zhang and Sennrich, 2019], modifies the formulation in Equation 11.1; it no longer computes the mean, but only aims to normalize the variance of the inputs. The normalization formulation for RMSNorm is then

$$\tilde{x}_i = \frac{x_i}{\sqrt{\frac{1}{n}\sum_j x_j^2 + \epsilon}}\gamma \tag{11.2}$$

Let us implement both layer normalization and RMS normalization for comparison

```
import torch
from torch import nn


class LayerNorm(nn.Module):
    def __init__(self, nchannels, epsilon=1e-5):
        super().__init__()
        self.epsilon = epsilon
        self.gamma=nn.Parameter(torch.ones(nchannels), requires_grad=True)
        self.beta=nn.Parameter(torch.zeros(nchannels), requires_grad=True)

    def forward(self, x):
        num = x - x.mean(dim=-1, keepdims=True)
        denom = torch.var(x, dim=-1, keepdims=True)
        denom = torch.sqrt(denom + self.epsilon)
        return num / denom * self.gamma + self.beta


class RMSNorm(nn.Module):
    def __init__(self, nchannels, epsilon=1e-5):
        super().__init__()
        self.epsilon = epsilon
        self.gamma=nn.Parameter(torch.ones(nchannels), requires_grad=True)
```

```
24
25      def forward(self, x):
26          denom = (x ** 2).mean(dim=-1, keepdims=True)
27          denom = torch.sqrt(denom + self.epsilon)
28          return x / denom * self.gamma
```

- `LayerNorm` (**Lines 5–16**):
  - `__init__` (**Lines 6–10**): Initializes the layer. Here, we initialize two learnable parameters, `gamma` and `beta`, with a size equal to the number of channels for element-wise scaling and shifting.
  - `forward` (**Lines 12–16**): Implements the standardization formulation as described in Equation 11.1. Note that in the current implementation, we normalize over the last dimension, which is the channel/embedding dimension of the input.
- `RMSNorm` (**Lines 19–28**):
  - `__init__` (**Lines 20–23**): Initializes the layer. In contrast to `LayerNorm`, we only initialize the scaling parameter `gamma`.
  - `forward` (**Lines 25–28**): Implements the root-mean-square normalization formulation in Equation 11.2. Similar to the above `LayerNorm` implementation, we also normalize over the channel/embedding dimension of the input.

11.4 Positional Encoding and Embedding

11.4.1 Sinusoidal Positional Encoding

In the original transformer paper [Vaswani et al., 2017], the authors used position encodings added on top of the input embedding vectors of each word token to allow the attention model to disambiguate words from different positions along the sequence. Specifically, given a sequence of word tokens, the transformer model looks up the corresponding embeddings of each vector and then forms a tensor of shape `(batch_size, seq_len, emb_size)`. Then, for each sequence position by embedding position, an encoding value is added based on the formulation:

$$\begin{aligned} \mathrm{PE}_{(p,2i)} &= \sin\left(p/(10000^{2i/d_{\mathrm{model}}})\right) \\ \mathrm{PE}_{(p,2i+1)} &= \cos\left(p/(10000^{2i/d_{\mathrm{model}}})\right) \end{aligned} \tag{11.3}$$

where

- p is the position index starting from 0,

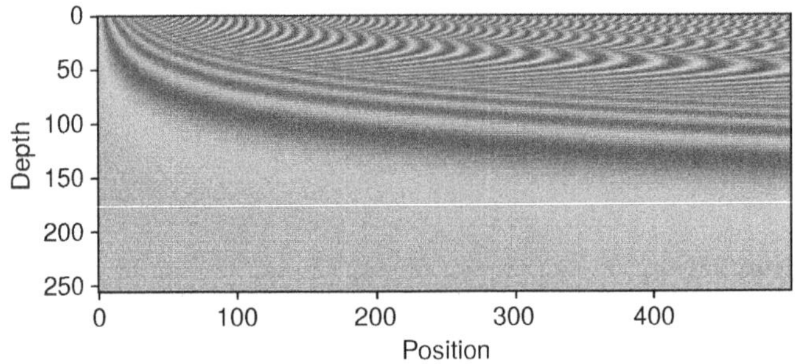

Figure 11.4 Sinusoidal positional encoding from the original transformer paper.

- i is the index along the embedding dimension, $i \in (0, 1, 2, \ldots, d_{\text{model}}//2)$,
- d_{model} is the embedding size.

This is typically referred to as sinusoidal positional encoding since it utilizes sinusoidal waves parameterized by position along the sequence and embedding dimensions. The following code block shows how to implement sinusoidal positional encoding from the original paper using PyTorch:

```
1  import torch
2
3  def sinusoidal_positional_encoding_vaswani(emb_size, seq_len):
4      position = torch.arange(seq_len)
5      timescales = torch.pow(
6          1 / 10000,
7          2 * (torch.arange(emb_size) // 2) / emb_size,
8      )
9      angles = position.unsqueeze(1) * timescales.unsqueeze(0)
10     out = torch.zeros((seq_len, emb_size))
11     # Interleaving sin and cos along embedding dimension
12     out[:, 0::2] = torch.sin(angles[:, 0::2])
13     out[:, 1::2] = torch.cos(angles[:, 1::2])
14     return out
```

- Lines 4–9 implement the term inside the sine and cosine function in Equation 11.3. This creates a sequence (e.g. time) by embedding a dimension matrix.
- Line 10 creates the output placeholder, initialized with zeros.
- Lines 12–13 assign the even position with sine values and odd positions with cosine values, respectively.

We can plot out the positional encoding given the embedding size of 256 and sequence length of 500, as shown in Figure 11.4.

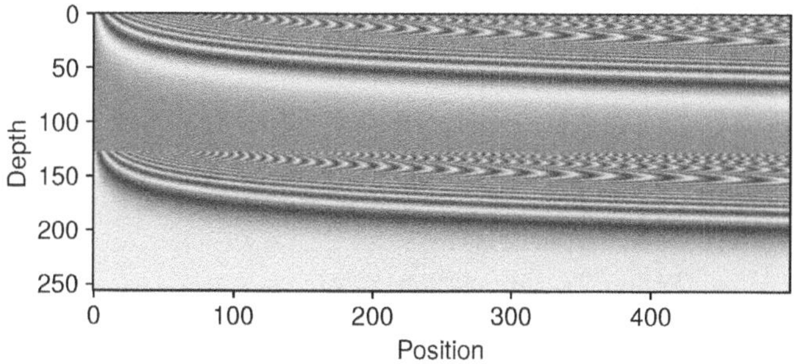

Figure 11.5 Tensor2Tensor variant of the sinusoidal positional encoding.

An alternative formulation of the sinusoidal positional encoding was introduced in the TransformerXL architecture implemented by Google's tensor2tensor [Vaswani et al., 2018] deep learning library [Vaswani et al., 2023]. (This is also the implementation in Google's tutorial on transformer [Abadi et al., 2023].) Instead of interleaving the sine and cosine terms along the depth dimension, the two pieces are simply concatenated together. Here is the implementation of this variant of sinusoidal positional encoding, and the encoding map is plotted in Figure 11.5.

```
import torch

def sinusoidal_positiona_encoding_tensor2tensor(emb_size, seq_len):
    position = torch.arange(seq_len)
    timescales = torch.pow(
        1 / 10000,
        2 * (torch.arange(emb_size) // 2) / emb_size,
    )
    angles = position.unsqueeze(1) * timescales.unsqueeze(0)
    out = torch.zeros((seq_len, emb_size))
    # Stacking the two partitions
    out = torch.concat([
        torch.sin(angles[:, 0::2]),
        torch.cos(angles[:, 1::2])
    ], axis=1)
    return out
```

11.4.2 Position as Categorical Embeddings

The sinusoidal position information is hard-coded. However, it is also possible to learn to represent the position information. An approach to learning to represent positional information is to treat positions as a categorical feature. Then, through back-propagation, the network learns to efficiently encode positional information. Well-known architectures that use learned categorical embeddings of position include BERT [Devlin et al., May 2019] and RoBERTa [Liu et al., 2019]. Let us implement a simple layer that adds positional embedding to an input.

```
1   import torch
2   from torch import nn
3
4   class PositionalEmbedding(nn.Module):
5       def __init__(self, seq_len, emb_size):
6           super().__init__()
7           self.embed = nn.Embedding(seq_len, emb_size)
8
9       def forward(self, x):
10          # x: (batch_size, seq_len, emb_size)
11          pos = torch.arange(x.shape[1], dtype=torch.int32)
12          # embedding lookup (batch_size, seq_len, emb_size)
13          pos_emb = self.embed(pos)
14          # add
15          x = x + pos_emb
16
17          return x
```

- `__init__` **(Lines 5–7)**: Initializes the positional embedding module and creates the embedding lookup table.
- `forward` **(Lines 9–17)**:
  - Line 11 creates the position vector for the current sequence inputs.
  - Line 13 looks up the corresponding embedding vector.
  - Line 15 adds the positional embedding vector to the input.

11.4.3 Relative Positional Encoding (RPE)

The sinusoidal positional encoding is sometimes also called absolute positional encoding. Thus, as a counterpart, relative positional encoding (RPE) is developed to encode only the relative distances between pairs of the sequence in the attention mechanism. Recall that self-attention is defined in the original transformer paper [Vaswani et al., 2017] as

$$\text{Attention}(Q, K, V) = \text{softmax}\left(\frac{QK^T}{\sqrt{d_k}}\right)V \tag{11.4}$$

where

- d_k is the embedding size of the key.

- $Q, K,$ and V are query, key, and value vectors that are obtained by passing inputs into a dense projection layer, e.g. $Q = W_q X_q$ with X_q being the embedding vector of the query. In the scenario of self-attention, $X_q, X_k,$ and X_v are identical, so we will denote it as X.

In the case of absolute sinusoidal positional encoding, it is typical to add positional encoding on top of the embedding output X, then project with a dense layer before computing the attention score. Mathematically, we can formulate it as $Q = (X + PE)W_q$ and $K = (X + PE)W_k$. Then,

$$QK^T = (X + PE)W_q \cdot \left((X + PE)W_k\right)^T$$

RPE, however, formulates this term as [Luo et al., 2022]

$$QK^T = (XW_q) \cdot (XW_k)^T + B$$

where B is a square matrix. That is, RPE adds the positional encoding information on top of the attention score map, instead of directly on the input token representation, where relative distances between each pair of tokens are computed. Well-known formulations of B include:

- Original RPE paper [Shaw et al., 2018]: $b_{ij} = X_i W_q r_{i-j}^T$ where r_{i-j} is a learnable parameter between the positions i and j in the sequence.
- Transformer-XL [Dai et al., 2019]: $b_{ij} = X_i W_q (r_{i-j} \tilde{W}_k)^T + u(X_j W_k)^T + v(r_{i-j} \tilde{W}_k)^T$, where $u, v,$ and $\tilde{W}_k$ are learnable vectors, while r_{i-j} is a sinusoidal positional encoding.
- T5 (Text-to-Text Transfer Transformer) [Raffel et al., 2020]: b_{ij} is a learnable parameter on its own.

Figure 11.6 illustrates the case for the T5 model. Pair-wise distances between the positions of the key and the query are computed. Then, the embedding of the relative distances is used to construct the position embedding tensor of shape `(q_len, k_len, emb_size)`, which is then added to the attention score [Wolf et al., 2020; HugginFace, 2023]. Note that the T5 model uses a strategy to map large relative distances to a bucket number in a logarithmic scale, such that these large relative distances may correspond to the same embedding vectors.

11.4.4 Rotary Positional Encoding (RoPE)

The RoFormer paper [Su et al., 2020] introduced the rotary positional encoding (RoPE), which is also a form of relative attention. Instead of adding a positional encoding to the input, the idea is to rotate the embedding vector of the sequence

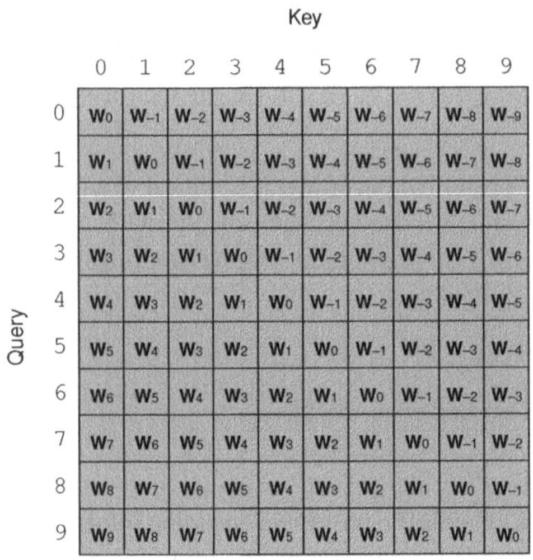

Figure 11.6 Relative positional embedding in T5 model.

by an angle parameterized by the position. Specifically, given a rotation matrix R, rotary positional encoding is incorporated into the input X by multiplication RX. Here, R has the form:

$$R^{(p)}_{\Theta,d} = \begin{bmatrix} \cos(p\theta_1) & -\sin(p\theta_1) & 0 & 0 & \dots & 0 & 0 \\ \sin(p\theta_1) & \cos(p\theta_1) & 0 & 0 & \dots & 0 & 0 \\ 0 & 0 & \cos(p\theta_2) & -\sin(p\theta_2) & \dots & 0 & 0 \\ 0 & 0 & \sin(p\theta_2) & \cos(p\theta_2) & \dots & 0 & 0 \\ \vdots & \vdots & \vdots & \vdots & \vdots & \vdots & \vdots \\ 0 & 0 & 0 & 0 & \dots & \cos(p\theta_{d/2}) & -\sin(p\theta_{d/2}) \\ 0 & 0 & 0 & 0 & \dots & \sin(p\theta_{d/2}) & \cos(p\theta_{d/2}) \end{bmatrix} \tag{11.5}$$

where,

- $R^{(p)}_{\Theta,d}$ is the pth slice of the rotation matrix for position p.
- Each slice is a $d \times d$ matrix, where d is the embedding size (i.e. d_{model}).
- Θ is an array with each element as $\theta_i = 10,000^{-2(i-1)/d}$, where $i \in [1, 2, 3, \dots, d/2]$.

Notice that each block is a 2D rotation matrix of the form

$$\begin{bmatrix} \cos(\theta) & -\sin(\theta) \\ \sin(\theta) & \cos(\theta) \end{bmatrix}$$

This means, we are rotating each pair of values along the embedding vector by an angle computed by $p\theta_i$. The following code block is an implementation of the rotation matrix R:

```
1  import torch
2
3  def rotary_positional_encoding(emb_size, seq_len):
4      pos = torch.arange(seq_len).unsqueeze(1)
5      depths = torch.arange(emb_size//2).unsqueeze(0)
6      theta = 10000**(-2*depths/emb_size)
7      p_theta = theta * pos
8      # Construct empty R matrix (l, d, d)
9      R = torch.zeros(seq_len, emb_size, emb_size)
10     # Fill the R matrix
11     depths_0 = torch.arange(0, emb_size, 2)
12     depths_1 = torch.arange(1, emb_size, 2)
13     # Computing rotations
14     R[:, depths_0, depths_0] = torch.cos(p_theta)
15     R[:, depths_1, depths_1] = R[:, depths_0, depths_0]
16     R[:, depths_0, depths_1] = -torch.sin(p_theta)
17     R[:, depths_1, depths_0] = -R[:, depths_0, depths_1]
18     return R
```

- Lines 4–7 compute $p\theta$ based on Equation 11.5.
- Line 9 initializes the placeholder tensor for the rotation matrix output.
- Lines 11–12 determine the positions to insert the sine and cosine values. Specifically, each block i has four positions, which are `(i, i)` for cos, `(i, i+1)` for – sin, `(i+1, i)` for sin, and `(i+1, i+1)` for cos.
- Lines 14–17 insert the sine and cosine values based on the positions.

We can visualize the first nine slices of the rotation matrix, given a sequence length of 500 and embedding size of 64 in Figure 11.7.

Although we can simply multiply together the rotation matrix with the input sequence of embedding based on the vanilla formulation, the sparsity of the matrix R suggests that there may be a more efficient way to formulate the encoding mechanism. The open-source repository for LLaMA from Meta implements a rather smart reformulation of rotary positional embedding [Facebook Research, 2023]. It can be seen that the matrix R is a block-diagonal matrix, where each block is a size of 2×2 rotation matrix. This means when multiplying together with the embedding sequence vector X, each rotation block at the position p, embedding positions k and $k+1$ is only multiplied with a pair of values along the same position p and embedding positions k and $k+1$ (see Figure 11.8 for illustration). We can denote the pair at embedding positions k and $k+1$ as group i, which corresponds to the index i for each θ in Equation 11.5. Then, we can write the multiplication within each group (p, i) as

$$\begin{bmatrix} \cos(p\theta_i) & -\sin(p\theta_i) \\ \sin(p\theta) & \cos(p\theta) \end{bmatrix} \begin{bmatrix} x_1^{(i)} \\ x_2^{(i)} \end{bmatrix} = \begin{bmatrix} x_1^{(i)}\cos(p\theta_i) - x_2^{(i)}\sin(p\theta_i) \\ x_1^{(i)}\sin(p\theta_i) + x_2^{(i)}\cos(p\theta_i) \end{bmatrix} \tag{11.6}$$

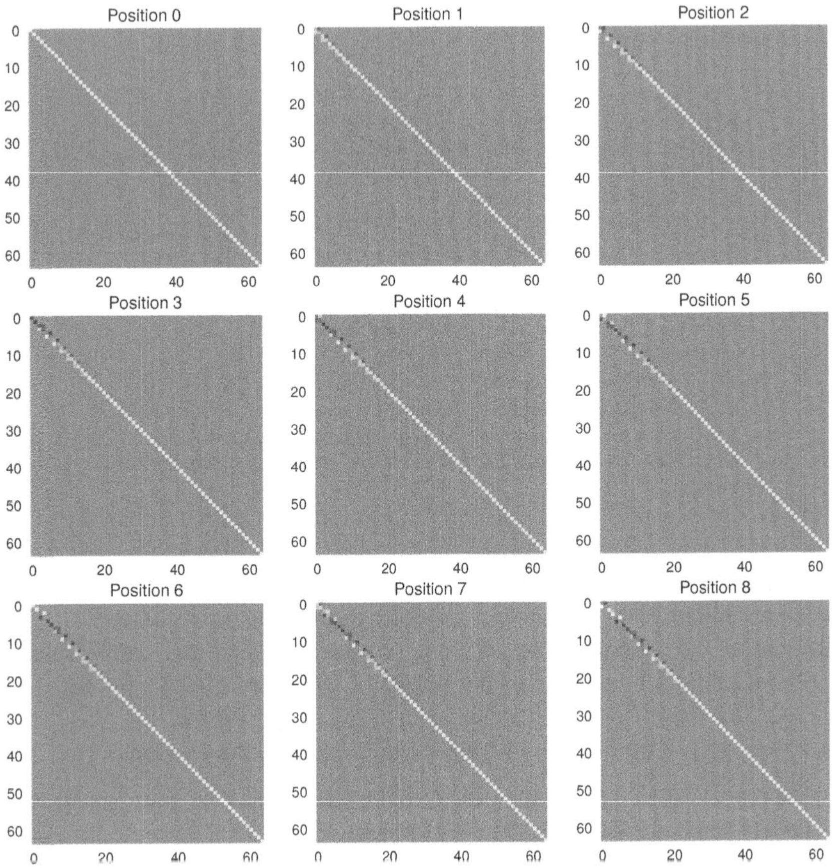

Figure 11.7 Rotation matrix of the first nine sequence positions, given embedding size of 64.

It turns out we can reformulate the rotation matrix using a complex value as

$$\phi_p^{(i)} = \cos(p\theta_i) + \sin(p\theta_i)j \tag{11.7}$$

where the imaginary number $\sqrt{-1}$ is denoted by the letter j following the engineering convention, rather than i under the mathematical convention. We can also treat the pair of values as a complex number, i.e.

$$x_p^{(i)} = x_1^{(i)} + x_2^{(i)}j \tag{11.8}$$

Then, it can be shown that

$$\begin{aligned} x_p^{(i)}\phi_p^{(i)} &= \left(x_1^{(i)} + x_2^{(i)}j\right)\left(\cos(p\theta_i) + \sin(p\theta_i)\right) \\ &= \left(x_1^{(i)}\cos(p\theta_i) - x_2^{(i)}\sin(p\theta_i)\right) + \left(x_1^{(i)}\sin(p\theta_i) + x_2^{(i)}\cos(p\theta_i)\right)j \end{aligned} \tag{11.9}$$

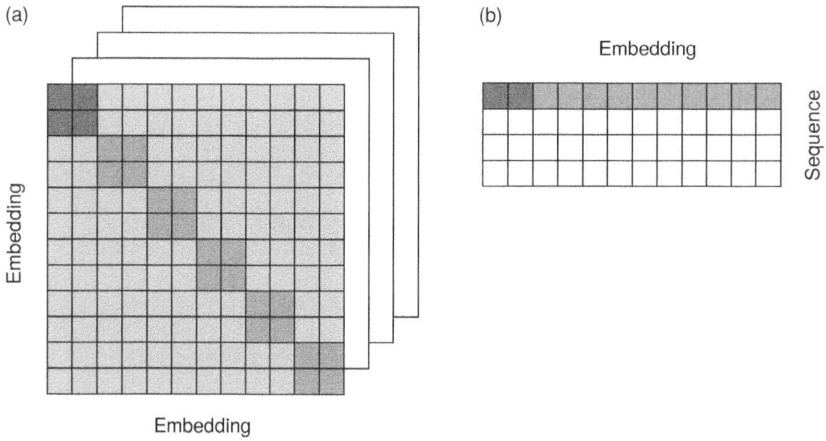

Figure 11.8 Corresponding values in R and X to be multiplied. Suppose we have an embedding size of 12 and a sequence length of 4. The R tensor (a) should have the shape $4 \times 12 \times 12$, and the sequence input tensor X (b) should have the shape 4×12. Due to the sparsity nature of the matrix R (a block diagonal matrix), the only effective multiplication is between the corresponding upper left corner regions in R (the upper left corner 4-grid square block) and X (the first two grids in the first row). In this example, the first block in R is multiplying together with the first pair in X, at the corresponding sequence position.

Notice that this is the same as representing the right-hand side column vector in Equation 11.6 (as the result of the multiplication) as a complex value.

This means if we represent the diagonal block as a vector of complex numbers and convert each pair of embedding values into a complex number as well, then we can simply do an element-wise product between the transformed vectors and then recover the result by converting the complex results back into a pair of numbers along the embedding dimension. Therefore, we can implement the following two functions which we can use to apply the rotary positional encoding on an input of shape `(batch_size, seq_len, ..., emb_size)`, where the elided dimension can be the additional `num_heads` dimension depending on the specific settings of the attention mechanism.

```
import torch

def create_phi(emb_size, seq_len):
    pos = torch.arange(seq_len).unsqueeze(1)
    depths = torch.arange(emb_size//2).unsqueeze(0)
    theta = 10000**(-2*depths/emb_size)
    # (seq_len, emb_size // 2)
    p_theta = theta * pos
    # cos(p_theta) + sin(p_theta) j
```

```
10         phi = torch.polar(torch.ones_like(p_theta), p_theta)
11         return phi
12 
13     def apply_rotary_position_encoding(phi, x):
14         """
15         Apply rotary positional encoding
16 
17         x has (batch_size, seq_len, <num_heads>, emb_size)
18         phi has shape (max_seq_len, emb_size)
19         """
20         # (batch_size, ..., emb_size // 2, 2)
21         x = x.reshape(*x.shape[:-1], -1, 2)
22         # (batch_size, ..., emb_size // 2), complex
23         x = torch.view_as_complex(x)
24         # Slice and reshape phi to match x
25         phi = phi[:x.shape[1], :] # same length as x
26         broadcast_shape = [
27             d if ii == 1 or ii == x.ndim - 1 else 1
28             for ii, d in enumerate(x.shape)
29         ]
30         phi = phi.view(broadcast_shape)
31         # Multiply: (batch_size, ..., emb_size // 2, 2)
32         out = torch.view_as_real(x * phi)
33         # (batch_size, ..., emb_size)
34         out = out.flatten(out.ndim - 2)
35         return out
```

- create_phi (**Lines 3–11**): The function creates the (seq_len, emb_size) matrix of phi based on the formulation we have in Equation 11.7.
- apply_rotary_position_encoding (**Lines 13–35**): The function takes an input tensor of the shape (batch_size, seq_len, <num_heads>, emb_size) (where num_heads dimension may be optional) and the reformulated rotation phi (a (seq_len, emb_size) matrix) and apply the rotary positional encoding.
  - Line 21: We first reshape the last dimension to create value pairs of the embeddings, converting the input tensor of shape (batch_size, seq_len, <num_heads>, emb_size) to (batch_size, seq_len, <num_heads>, emb_size // 2, 2), where the final dimension is a pair of embedding values.
  - Line 23: Following the formulation in Equation 11.8, the torch.view_as_complex operation converts the pair of values along the last dimension into complex representations, resulting in a tensor of shape (batch_size, seq_len, <num_heads>, emb_size // 2).

- Line 25: We slice the sequence dimension of phi such that phi and x have identical sequence lengths. In practice, it is possible that x can have variable size sequence lengths over different training or inference batches, we would need to precompute the matrix phi with an expected maximum sequence length for inference.
- Lines 26–29: This determines the broadcasting shape of the matrix phi. We would like to insert some singleton dimensions for batches, and optionally, the num_heads dimension (which may or may not exist depending on how we formulate attention).
- Line 30 applies the reshaping so that the singleton dimensions for broadcasting are inserted.
- Line 32: We first perform element-wise multiplication between the two complex matrices (with additional broadcasting). Then, we convert the complex-valued results back into real-value tensor, resulting in a shape of (batch_size, seq_len, <num_heads>, emb_size // 2, 2).
- Line 34: This reshapes the last two dimensions of the above tensor, giving back a result of shape (batch_size, seq_len, <num_heads>, emb_size).

11.5 Activation Functions in Feedforward Layer

Recall in Chapter 9, when we were implementing the Vision Transformer model, we implemented the feedforward layer that follows the multiheaded attention layer as

```
self.ffn = nn.Sequential(
    nn.Linear(embed_dim, ffn_size),
    nn.GELU(),
    nn.Dropout(dropout),
    nn.Linear(ffn_size, embed_dim),
)
```

Here, nn.GELU() is an activation function applied to the first dense layer of the feedforward layer. This is an improvement on top of the original transformer paper, which used a simple rectified linear unit (ReLU) activation function, and did not add a dropout layer before the second dense layer. Subsequent studies have, therefore, attempted to use more sophisticated activation functions to capture the nonlinearity in the feedforward layer. Specifically, a family of linear unit-based activation functions seemed to have various degrees of success, including but not limited to exponential linear unit (ELU) [Clevert et al., 2016],

Gaussian error linear unit (GELU) [Hendrycks and Gimpel, 2016], Mish [Misra, 2020], sigmoid linear unit (SiLU) [Hendrycks and Gimpel, 2016], and Swish [Ramachandran et al., 2017]. Their mathematical formulations are as follows:

$$\begin{aligned}
\mathrm{ReLU}(x) &= \max(x, 0) \\
\mathrm{ELU}(x) &= \begin{cases} \alpha \exp(x) - 1 & \text{if } x \leq 0 \\ x & \text{if } x > 0 \end{cases} \\
\mathrm{GELU}(x) &= x \cdot \frac{1}{2}\left[1 + \mathrm{erf}(x/\sqrt{2})\right] \\
\mathrm{Mish}(x) &= x \cdot \tanh(\mathrm{softplus}(x)) = x \cdot \tanh(\log(\exp(x) + 1)) \\
\mathrm{SiLU}(x) &= x\sigma(x) \\
\mathrm{Swish}_\beta(x) &= x\sigma(\beta x)
\end{aligned}$$

Notice that $\mathrm{Swish}_{\beta=1}(x) \equiv \mathrm{SiLU}(x)$.

The following code block implements these activation functions. Figure 11.9 illustrates several variants of activation functions.

```
import math
import torch
import torch.nn.functional as F

def relu(x):
    return torch.clamp(x, 0)

def elu(x, alpha=1.0):
    return torch.where(x > 0, x, alpha * torch.exp(x) - 1)

def gelu(x):
    return x * 0.5 * (1 + torch.erf(x / math.sqrt(2.0)))

def mish(x):
    return x * torch.tanh(F.softplus(x))

def swish(x, beta=1):
    return x * torch.sigmoid(beta * x)
```

Others have attempted to reformulate the linear layer completely to unlock more learning power. Specifically, linear gated units (GLUs) pass the input through two learnable dense layers and activate one with a nonlinear activation function, then multiply together with the other linear unit to generate the activated output. GLU is formulated as the following:

$$\mathrm{GLU}(x, W, V, b, c) = \sigma(xW + b) \otimes (xV + c)$$

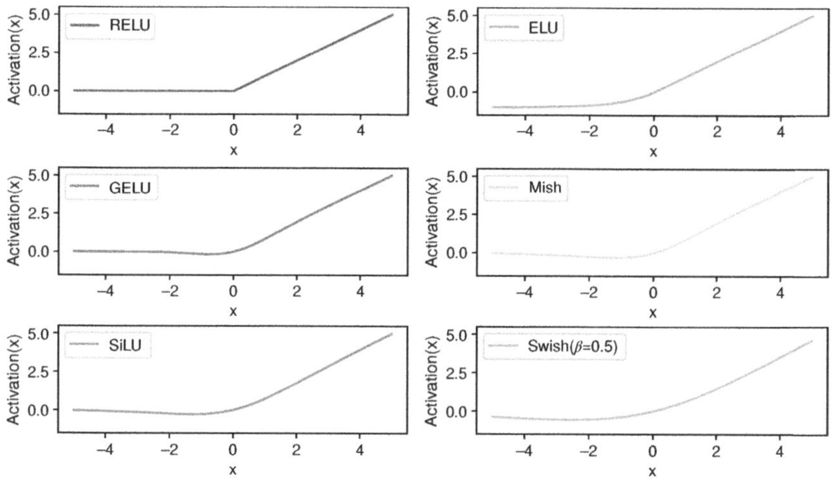

Figure 11.9 Comparison of nonlinear activation functions.

where $W, V, b,$ and c are learnable weights of the dense layer. A successful variant is to let the nonlinear activation function $\sigma =$ Swish, obtaining what is known as SwiGLU, i.e.

$$\text{SwiGLU}(x, W, V, b, c, \beta) = \text{Swish}_\beta(xW + b) \otimes (xV + c)$$

In practice, however, researchers have found that dropping the bias terms b and c and setting $\beta = 1$ (i.e. use SiLU) are more helpful for training the model. Therefore, the final form of the feedforward layer with SwiGLU becomes

$$\text{FFN}_{\text{SwiGLU}}(x, W, V, U) = (\text{SiLU}(xW) \otimes (xV))\, U$$

where $W, V,$ and U are learnable weights for feedforward layers. We can implement the new feedforward layer above as the following:

```
import torch
from torch import nn
import torch.nn.functional as F


class FFN(nn.Module):
    """
    Original feedforward layer with GeLU activation
    """
    def __init__(self, input_dim, hidden_dim):
        super().__init__()
        self.ffn1 = nn.Linear(input_dim, hidden_dim, bias=False)
        self.ffn2 = nn.Linear(hidden_dim, input_dim, bias=False)

```

```
    def forward(self, x):
        x = self.ffn1(x)
        x = F.gelu(x)
        x = self.ffn2(x)
        return x


class FFNSwiGLU(nn.Module):
    """
    SwiGLU feedforward layer.
    * input_dim: input dimension
    * hidden_dim: hidden dimension
    """
    def __init__(self, input_dim, hidden_dim):
        super().__init__()
        self.linear_gate = nn.Linear(input_dim, hidden_dim, bias=False)
        self.swi_gate = nn.Linear(input_dim, hidden_dim, bias=False)
        self.output = nn.Linear(hidden_dim, input_dim, bias=False)

    def forward(self, x):
        linear_term = self.linear_gate(x)
        swi_term = self.swi_gate(x)
        swi_term = F.silu(swi_term)
        return self.output(swi_term * linear_term)
```

11.6 Case Study: Training a Tiny LLaMA Model for Next Token Prediction

In this case study, we will implement a LLaMA [Touvron et al., 2023a]-like model using the improvements on the transformer architecture that we have discussed so far. This combination of transformer improvements forms the so-called Transformer++ recipe [Gu and Dao, 2023]. While this case study will not involve creating a full-scale LLM with billions of parameters, due to the challenges of acquiring a high-quality text corpus with trillions of tokens and the prohibitive costs of training such a model with extensive GPU time, it aims to provide a practical and accessible learning experience. Building a complete LLM would also require a dedicated team of research scientists and machine learning engineers, along with extensive fine-tuning and artificial intelligence (AI) safety alignment work to avoid generating toxic content. However, the goal of this book is to offer materials that are reproducible on a single machine or a Google Colab instance with limited computing resources. This approach ensures that every reader can execute the proposed solutions. By understanding the theory behind the transformer model and applying it to build a small-scale language model, readers can gain an appreciation for the effort and complexity involved in creating sophisticated AI models like ChatGPT.

Let's dive right into the code. We first implement all the supporting layers for the Transformer++ model based on our previous discussions:

```python
import torch
from torch import nn
import torch.nn.functional as F
import math

class RMSNorm(nn.Module):
    """Root-Mean-Square Layer Normalization."""
    def __init__(self, nchannels, epsilon=1e-5):
        super().__init__()
        self.epsilon = epsilon
        self.gamma=nn.Parameter(torch.ones(nchannels), requires_grad=True)

    def forward(self, x):
        denom = (x ** 2).mean(dim=-1, keepdims=True)
        denom = torch.sqrt(denom + self.epsilon)
        return x / denom * self.gamma


class RotaryPositionalEncoding(nn.Module):
    """Layer to apply rotary positional encoding."""
    def __init__(self, emb_size, seq_len):
        super().__init__()
        phi = self.create_phi(emb_size, seq_len)
        self.phi = nn.Parameter(phi, requires_grad=False)

    @staticmethod
    def create_phi(emb_size, seq_len):
        pos = torch.arange(seq_len).unsqueeze(1)
        depths = torch.arange(emb_size//2).unsqueeze(0)
        theta = 10000**(-2*depths/emb_size)
        # (seq_len, emb_size // 2)
        p_theta = theta * pos
        # cos(p_theta) + sin(p_theta) j
        phi = torch.polar(torch.ones_like(p_theta), p_theta)
        return phi

    @staticmethod
    def apply_rotary_position_encoding(phi, x):
        """
        Apply rotary positional encoding

        x has (batch_size, seq_len, <num_heads>, emb_size)
        phi has shape (max_seq_len, emb_size)
        """
        # (batch_size, ..., emb_size // 2, 2)
        x = x.reshape(*x.shape[:-1], -1, 2)
        # (batch_size, ..., emb_size // 2), complex
        x = torch.view_as_complex(x)
        # Slice and reshape phi to match x
        phi = phi[:x.shape[1], :] # same length as x
```

```
        broadcast_shape = [
            d if ii == 1 or ii == x.ndim - 1 else 1
            for ii, d in enumerate(x.shape)
        ]
        phi = phi.view(broadcast_shape)

        # Multiply: (batch_size, ..., emb_size // 2, 2)
        def complex_multiply_to_real(a, b):
            # Element-wise multiplication of 2 complex numbers
            a = torch.view_as_real(a) # (..., 2)
            b = torch.view_as_real(b) # (..., 2)
            out_real = a[..., 0] * b[..., 0] - a[..., 1] * b[..., 1]
            out_img  = a[..., 0] * b[..., 1] + a[..., 1] * b[..., 0]
            out = torch.stack([out_real, out_img], axis=-1)
            return out

        # out = torch.view_as_real(x * phi)
        out = complex_multiply_to_real(x, phi) # MPS device workaround!
        # (batch_size, ..., emb_size)
        out = out.flatten(out.ndim - 2)
        return out

    def forward(self, x):
        return self.apply_rotary_position_encoding(self.phi, x)


class FFNSwiGLU(nn.Module):
    """
    SwiGLU feedfoward layer.
    * input_dim: input dimension
    * hidden_dim: hidden dimension
    """
    def __init__(self, input_dim, hidden_dim):
        super().__init__()
        self.linear_gate = nn.Linear(input_dim, hidden_dim, bias=False)
        self.swi_gate = nn.Linear(input_dim, hidden_dim, bias=False)
        self.output = nn.Linear(hidden_dim, input_dim, bias=False)

    def forward(self, x):
        linear_term = self.linear_gate(x)
        swi_term = self.swi_gate(x)
        swi_term = F.silu(swi_term)
        return self.output(swi_term * linear_term)


class RoPEMultiHeadedAttention(nn.Module):
    """RoPE supported multi-headed attention."""
    def __init__(self, emb_dim, num_heads, causal=False):
        super().__init__()
        self.emb_dim = emb_dim
        self.num_heads = num_heads
        self.head_dim = emb_dim // num_heads
        self.causal = causal

```

```
        # Layers
        self.rope = RotaryPositionalEncoding(self.head_dim, 1024)
        self.project_q = nn.Linear(emb_dim, emb_dim, bias=False)
        self.project_k = nn.Linear(emb_dim, emb_dim, bias=False)
        self.project_v = nn.Linear(emb_dim, emb_dim, bias=False)
        self.project_o = nn.Linear(emb_dim, emb_dim, bias=False)

    def attention(self, query, key, value, mask=None):
        # Apply rotary embeddings
        query = self.rope(query)
        key = self.rope(key)

        # Compute attention score
        scores = torch.matmul(
            query.permute(0, 2, 1, 3),
            key.permute(0, 2, 3, 1)
        )
        scores = scores / math.sqrt(self.emb_dim)

        # Masking
        if mask is not None:
            # true indicate the entries to keep
            mask = mask.unsqueeze(0).unsqueeze(0)
            scores += (1 - mask) * -1e9

        if self.causal:
            causal_mask = torch.triu(torch.ones(scores.shape[-2:]))
            scores += causal_mask.to(scores.device) * -1e9

        attn = F.softmax(scores, dim=-1)
        output = torch.matmul(attn, value.permute(0, 2, 1, 3))
        output = output.permute(0, 2, 1, 3)

        return output, attn

    def forward(self, q, k, v, mask=None):
        # projecting each input
        x_q = self.project_q(q)
        x_k = self.project_k(k)
        x_v = self.project_v(v)

        # (batch_size, seq_len, emb_size)
        b, t, d = q.shape
        shape = (b, t, self.num_heads, self.head_dim)
        # reshaping ->
        # (batch_size, seq_len, num_heads, emb_size // num_heads)
        query = x_q.view(*shape)
        key = x_k.view(*shape)
        value = x_v.view(*shape)

        # Compute attention
        x_attn, attn_weight = self.attention(query, key, value, mask=mask)

        # Reshaping back to (batch_size, seq_len, emb_size)
```

```
        x_attn = x_attn.reshape(b, t, d)

        # Final projection layer
        x_attn = self.project_o(x_attn)

        return x_attn, attn_weight


class DecoderBlock(nn.Module):
    def __init__(self, embed_dim, ffn_size, num_heads=4, causal=False):
        super().__init__()
        self.mhattn = RoPEMultiHeadedAttention(
            emb_dim=embed_dim,
            num_heads=num_heads,
            causal=causal,
        )
        self.ffn = FFNSwiGLU(embed_dim, ffn_size)
        self.input_ln = RMSNorm(embed_dim)
        self.ffn_ln = RMSNorm(embed_dim)

    def forward(self, x):
        y = self.input_ln(x)
        y, _ = self.mhattn(y, y, y)
        y += x

        z = self.ffn_ln(y)
        z = self.ffn(z)
        z += y

        return z


class TransformerPlusPlus(nn.Module):
    def __init__(self,n_layers,vocab_size,embed_dim,ffn_size,num_heads):
        super().__init__()
        self.embed = nn.Embedding(vocab_size+1, embed_dim, padding_idx=0)
        self.encoder = nn.ModuleList([
            DecoderBlock(embed_dim, ffn_size, num_heads, causal=True)
            for _ in range(n_layers)
        ])
        self.final_ln = RMSNorm(embed_dim)
        self.output_layer = nn.Linear(embed_dim, vocab_size)

    def forward(self, tokens):
        # (batch_size, seq_len, emb_size)
        x = self.embed(tokens)
        # passing through encoder layers
        for enc in self.encoder:
            x = enc(x)

        # Final layer normalization
        x = self.final_ln(x)

        # output layer
```

```
213         logits = self.output_layer(x[:, -1, :])
214
215         return logits
```

- `RMSNorm` (**Lines 6–16**): This implements the RMSNorm layer. This is a just a copy of the previous implementation without modifications.
- `RotaryPostionalEncoding` (**Lines 19–74**): This incorporates the two functions we implemented previously, `create_phi` and `apply_rotary_position_encoding` as part of a module.
  - The `__init__` method (**Lines 21–24**): Calls the `create_phi` static method to pre-compute the `phi` matrix.
  - The `forward` method (**Lines 73–74**): Calls the `apply_rotary_position_encoding` method.
  - Lines 58–68 of the `apply_rotary_position_encoding` method: this is a workaround for training on Apple Silicon devices ("MPS"), where multiplication between two complex tensors is not implemented. Instead, we implement a subroutine `complex_multiply_to_real` (Lines 58–65) according to Equation 11.9 to replace Line 67, `torch.view_as_real(x * phi)`. See also PyTorch GitHub issue #105665 [Paszke et al., 2023].
- `FFNSwiGLU` (**Lines 77–93**): This is a copy of the previous implementation of the SwiGLU-activated feedforward layer in Section 11.1.2.
- `RoPEMultiHeadedAttention` (**Lines 96–164**): This is our attention layer with RoPE to encode positional information.
  - `__init__` (**Lines 98–106**): This method initializes:
    * The rotary positional encoding layer (Line 106) is used to apply positional encodings before computing attention scores.
    * The input projection layers for query, key, and value (Lines 107–109).
    * The output projection layer (Line 110) for an additional projection.
  - `attention` (**Lines 112–136**): This method implements the scaled-dot-product attention logic as described by Equation 11.4.
    * Lines 114 and 115 add rotary positional encoding to query and key, which are both of shape `(batch_size, seq_len, num_heads, head_dim)`, where `num_heads * head_dim = emb_size`.
    * Lines 118–120 compute the attention scores by performing dot product between query and key. This results in a tensor of shape `(batch_size, num_heads, query_seq_len, key_seq_len)`.
    * Line 122 normalizes the scores by $\sqrt{d_{\text{model}}}$.
    * Lines 125–129 apply predefined masking to the attention scores. The trick here is to add a large negative number to scores to be ignored such that after softmax activation, the scores are near zero.

Lines 130 and 131 apply causal masking to the attention scores. It creates an upper triangular mask, which will indicate which part of the score will be masked out.

  * Line 134 activates the scores with softmax.
  * Line 135 completes the attention formula, by multiplying the attention weight with value.
  * Line 136 rearranges the indices such that the output is of shape `(batch_size, seq_len, num_heads, head_dim)`.

- `forward` (**Lines 140–164**): This method implements the main logic of the multiheaded attention layer.
  * Lines 142–144 perform a series of projections for query, key, and value.
  * Lines 147–148 determine the shape of the multiheaded attention inputs, i.e. `(batch_size, seq_len, num_heads, head_dim)`.
  * Lines 151–153 apply the shape above on query, key, and value.
  * Line 156 calls the `attention` method to compute output after attention `x_attn`, as well as attention score `attn_weight`.
  * Line 159: The output from the `attention` method still has the shape `(batch_size, seq_len, num_heads, head_dim)`, so we would like to reshape the output back to `batch_size, seq_len, emb_size)` by flattening the last two dimensions together.
  * Line 162 performs a final linear projection for the output.

• `DecoderBlock` (**Lines 167–188**): This is an adaption of the `PreLNEncoderBlock` layer that we implemented previously.
  - `__init__` (**Lines 168–177**): The `DecoderBlock` layer replaces the following components:
    * Lines 170–174: Instead of using the original `nn.MultiHeadAttention` layer from PyTorch, we use our own `RoPEMultiHeadedAttention` for the attention mechanism. Note that we are using causal attention by setting `causal=True`. In the original transformer paper's terminology, this is actually the decoder layer which uses causal attention. Hence, GPT and LLaMA models for text generation are decoder-only architectures.
    * Line 175: Instead of using two feedforward layers with GELU activations that we implemented previously, using `nn.Sequential`, we use the `FFNSwiGLU` feedforward layer that we implemented above.
    * Lines 176–177: Instead of using `nn.LayerNorm`, we use the `RMSNorm` that we implemented above.
  - `forward` (**Lines 179–188**): This is identical to our previous implementation in `PreLNEncoderBlock`.

• `TransformerPlusPlus` (**Lines 191–215**): This module constructs our final Transformer++ (a.k.a. LLaMA) model.
  - `__init__` (**Lines 193–200**):

  * Line 194: We initialize the token embedding based on the `vocab_size` and `embed_dim`. Here, we assume index `0` is used for padding.
  * Lines 195–198: We initialize a series of decoder blocks.
  * Line 199 initializes a final `RMSNorm` layer which will be called at the transformer blocks.
  * Line 200 initializes a final output layer which projects the hidden state to assign a logit value on each token.
- `forward` (**Lines 202–215**):
  * Line 204 looks up the embedding of each token in the input.
  * Lines 206–207 pass the embedding representation of the sequence of tokens through the transformer blocks.
  * Line 210 adds a final layer normalization via `RMSNorm`.
  * Line 213 computes the logit value of each token in the vocabulary via linear projection. Here, we use the last hidden state of the sequence as input to this layer, since we would like to predict the next token.

We can visualize the model structure via `torchsummary`.

```
# View model architecture
from torchsummary import summary
model = TransformerPlusPlus(
    n_layers=4,vocab_size=vocab_size,embed_dim=256,ffn_size=512,num_heads=4
)
summary(model)
```

which gives us

```
=================================================================
Layer (type:depth-idx)                                  Param #
=================================================================
Embedding: 1-1                                        7,813,888
ModuleList: 1-2                                       --
|    DecoderBlock: 2-1                                --
|    |    RoPEMultiHeadedAttention: 3-1               (294,912)
|    |    FFNSwiGLU: 3-2                              393,216
|    |    RMSNorm: 3-3                                256
|    |    RMSNorm: 3-4                                256
|    DecoderBlock: 2-2                                --
|    |    RoPEMultiHeadedAttention: 3-5               (294,912)
|    |    FFNSwiGLU: 3-6                              393,216
|    |    RMSNorm: 3-7                                256
|    |    RMSNorm: 3-8                                256
|    DecoderBlock: 2-3                                --
|    |    RoPEMultiHeadedAttention: 3-9               (294,912)
|    |    FFNSwiGLU: 3-10                             393,216
|    |    RMSNorm: 3-11                               256
|    |    RMSNorm: 3-12                               256
```

```
|    DecoderBlock: 2-4                                         --
|    |    RoPEMultiHeadedAttention: 3-13                       (294,912)
|    |    FFNSwiGLU: 3-14                                      393,216
|    |    RMSNorm: 3-15                                        256
|    |    RMSNorm: 3-16                                        256
RMSNorm: 1-3                                                   256
Linear: 1-4                                                    7,844,154
=========================================================================
Total params: 18,412,858
Trainable params: 17,233,210
Non-trainable params: 1,179,648
=========================================================================
```

With the current settings, our model has only ~18 million parameters, which is several magnitudes smaller than the smallest 7B parameter LLaMA model [Touvron et al., 2023a].

Let us train the model using WikiText2 [Merity et al., 2016], which is accessible through `torchtext.datasets.WikiText2` (a copy of the dataset is available on the book supplement repository: https://github.com/sapphire008/vectorization-book-supplement/tree/main/wikitext2, in case `torchtext` fails to get this dataset from a remote source). This corpus has about 2.3 million tokens in the training set. Though it is several magnitudes smaller than the 1 billion+ tokens used to train a full-sized LLaMA [Touvron et al., 2023a], it is sufficient for our purpose of illustrating the concepts. We will also use `BERTTokenizer` to map the string values of each token from this text corpus to integer values. The data preparation steps are implemented as below:

```
import numpy as np
# !pip install torchtext==0.16.0 portalocker==2.8.2
# Make sure to restart the python shell after installing portalocker
import torchtext
from torchtext.transforms import BERTTokenizer
from tqdm import tqdm

def sample_sentence(element, max_context_size=64):
    """Take a sentence, and subset the tokens into input and next word."""
    l = len(element)
    start = np.random.randint(0, l-3)
    end = np.random.randint(start, l-1)
    if end - start > max_context_size:
        start = end - max_context_size
    sentence = element[start:end]
    sentence = np.array(sentence).astype(int).tolist()
    next_word = element[end]
    next_word = int(next_word)
    return sentence, next_word

```

```
def collate_fn(elements):
    """Collect data into padded batches."""
    sentence, next_word = zip(*elements)
    sentence = torch.nested.nested_tensor(
        list(sentence)).to_padded_tensor(padding=-1)
    sentence = sentence.long() + 1
    next_word = torch.tensor(next_word) + 1 # index offset by 1
    return sentence, next_word


VOCAB_FILE="https://huggingface.co/bert-base-uncased/resolve/main/vocab.txt"
vocab_size = 30522
tokenizer = BERTTokenizer(
    vocab_path=VOCAB_FILE, do_lower_case=True, return_tokens=False
)
train_dataset = torchtext.datasets.WikiText2(
    "./wikitext2", split="train"
).map(tokenizer).filter(lambda x: len(x) > 5).map(sample_sentence)
eval_dataset = torchtext.datasets.WikiText2(
    "./wikitext2", split="valid"
).map(tokenizer).filter(lambda x: len(x) > 5).map(sample_sentence)

train_data = torch.utils.data.DataLoader(
    train_dataset, batch_size=32, collate_fn=collate_fn
)
eval_data = torch.utils.data.DataLoader(
    eval_dataset, batch_size=32, collate_fn=collate_fn
)
```

Note that to use `torchtext==0.16.0`, we need to install `portalocker ==2.8.1`. Make sure to restart the kernel before continuing.

Let us train the model. This is a similar routine to the one we implemented for the VisionTransformer training in Chapter 9.

```
if torch.cuda.is_available():  # NVIDIA GPU
    device = torch.device("cuda")
elif torch.backends.mps.is_available():  # Apple Silicon
    device = torch.device("mps")
else:
    device = torch.device("cpu")

# prepare model
model = TransformerPlusPlus(
   n_layers=4,vocab_size=vocab_size,embed_dim=256,ffn_size=512,num_heads=4
)
model = model.to(device)
optimizer = torch.optim.AdamW(model.parameters(), lr=0.001)
num_epochs = 10

def train_epoch(model, device, train_data, optimizer):
```

```
    model.train()
    loss_train = 0.0
    accuracy_train = 0.0
    n_examples = 0

    # iterate over data
    for X, y in tqdm(train_data):
        # Ship data to training device
        X, y = X.to(device), y.to(device)
        n_examples += len(y)
        # Forward pass
        optimizer.zero_grad()
        y_pred = model(X)
        loss = F.cross_entropy(y_pred, y, reduction="sum")
        loss_train += loss # accumulating
        loss = loss / len(y) # average
        loss.backward()
        optimizer.step()

        # Compute accuracy
        accuracy_train += (y_pred.argmax(dim=1) == y).sum()

    # Gather losses
    avg_loss = loss_train / n_examples
    avg_acc = accuracy_train / n_examples

    return avg_loss, avg_acc


def eval_epoch(model, device, eval_data):
    loss_eval = 0.0
    accuracy_eval = 0.0
    n_examples = 0
    with torch.no_grad():
        for X, y in tqdm(eval_data):
            X, y = X.to(device), y.to(device)
            n_examples += len(y)
            y_pred = model(X)
            loss = F.cross_entropy(y_pred, y, reduction="sum")
            loss_eval += loss.item()

            # Compute accuracy
            accuracy_eval += (y_pred.argmax(dim=1) == y).sum()

    # Gather losses
    avg_loss = loss_eval / n_examples
    avg_acc = accuracy_eval / n_examples
    return avg_loss, avg_acc


# train
history = {
    "train_loss": [],
    "train_acc": [],
```

```
        "eval_loss": [],
        "eval_acc": [],
    }
    for epoch in range(num_epochs):
        train_loss, train_acc=train_epoch(model,device,train_data,optimizer)
        eval_loss, eval_acc = eval_epoch(model, device, eval_data)
        print(
            f"Epoch {epoch+1}/{num_epochs}:",
            f"train_loss - {train_loss:.2f},",
            f"train_acc - {train_acc:.2f}",
            f"eval_loss - {eval_loss:.2f},",
            f"eval_acc - {eval_acc:.2f}",
        )
        history["train_loss"].append(float(train_loss))
        history["eval_loss"].append(float(eval_loss))
        history["train_acc"].append(float(train_acc))
        history["eval_acc"].append(float(eval_acc))
        # saving checkpoint
        if (epoch + 1) % 20 == 0:
            torch.save({
                'epoch': epoch + 1,
                'model_state_dict': model.state_dict(),
                'optimizer_state_dict': optimizer.state_dict(),
                'loss': eval_loss,
                }, f"./checkpoints/tiny_llm_checkpoint_{epoch+1:03d}.pt")

    # Saving metrics
    with open("./training_history.json", "w") as fid:
        json.dump(history, fid)
    # Saving the final model
    torch.save(model.state_dict(), "./final_model_weights.pt")
```

Over 10 epochs, we reached the best evaluation accuracy at 0.27.

```
Epoch 1/10: train_loss - 6.96, train_acc - 0.16 eval_loss - 5.85, eval_acc - 0.25
Epoch 2/10: train_loss - 6.46, train_acc - 0.20 eval_loss - 5.83, eval_acc - 0.22
Epoch 3/10: train_loss - 6.27, train_acc - 0.21 eval_loss - 5.65, eval_acc - 0.26
Epoch 4/10: train_loss - 6.23, train_acc - 0.21 eval_loss - 5.67, eval_acc - 0.25
Epoch 5/10: train_loss - 6.18, train_acc - 0.21 eval_loss - 5.73, eval_acc - 0.26
Epoch 6/10: train_loss - 6.14, train_acc - 0.21 eval_loss - 5.67, eval_acc - 0.25
Epoch 7/10: train_loss - 6.13, train_acc - 0.21 eval_loss - 5.73, eval_acc - 0.24
Epoch 8/10: train_loss - 6.03, train_acc - 0.21 eval_loss - 5.40, eval_acc - 0.27
Epoch 9/10: train_loss - 6.10, train_acc - 0.21 eval_loss - 5.56, eval_acc - 0.26
Epoch 10/10: train_loss - 6.00, train_acc - 0.21 eval_loss - 5.54, eval_acc - 0.26
```

However, if we train for more epochs, we can obtain even better results. Figure 11.10 plots the training and evaluation loss as well as prediction accuracy for this model trained over 300 epochs.

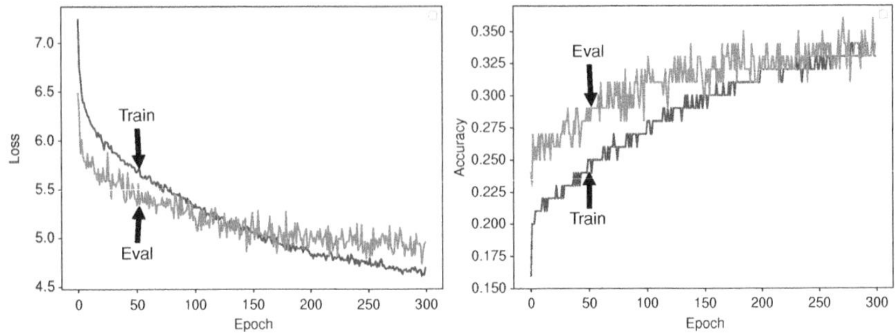

Figure 11.10 Training our tiny LLaMA model using WikiText2.

11.7 A Word on AI Safety and Alignment

Before we close the chapter, I would like to add some thoughts on AI safety and alignment. The advent of LLMs such as LLaMA and GPT, along with other generative models such as DALL-E and Stable Diffusion, has intensified the debate on AI safety among both academics and practitioners in the industry. These technologies, while groundbreaking, have raised significant concerns. Associated topics of the debate include but are not limited to:

- **Demographic biases**: One of the critical issues is demographic bias, where AI models may exhibit prejudices based on gender, race, or ethnicity. This is often (though not only) a result of training data reflecting existing societal biases. For instance, a language model might generate stereotypical or discriminatory content if it is trained on biased data, leading to concerns about reinforcing negative stereotypes. Perhaps the most infamous example was that in 2015, Google's photo service used AI to mistakenly label a photo of a black person as "gorilla" [BBC News, 2015].
- **Content toxicity**: Another area of concern is content toxicity. AI models can inadvertently generate harmful or offensive content, especially if they are exposed to toxic data during training. This raises questions about the responsibility of AI developers in filtering and moderating training data and the mechanisms needed to prevent the propagation of harmful content. The hate-speech generating GPT-4Chan model, a GPT-3 model trained on toxic comments from the internet forum 4Chan, sparked much controversy in 2022 [Gault, 2022].
- **Truthfulness and misinformation**: The ability of AI models to generate realistic yet potentially false information poses a challenge in terms of truthfulness and misinformation. As these models become more sophisticated, distinguishing between AI-generated and human-generated content becomes more and

more difficult, potentially leading to the spread of misinformation. Scams that use sophisticated AI technologies, such as face- and voice-swapping DeepFakes, have also been on the rise recently. Scammers can pretend to be the victim's friends and family who are in an emergency and ask for financial assistance, making it more difficult for the victims to recognize the deception [Butts, 2024].

- **Ethical and societal considerations**: Ethical concerns encompass a broad range of issues, from the potential misuse of AI for deceptive purposes to the ethical implications of AI-generated art and literature. Societal considerations involve the broader impact of AI on society, including its influence on employment, privacy, and social dynamics. Most notably, data sources used to train the large-scale models are neither "scraped" from the internet, without first asking for permission from the creators, nor would the model provide appropriate credit to the creators. Whether or not using scraped content to train AI models that can regenerate such content constitutes fair use is still under debate. More recently, the New York Times has sued OpenAI over using copyrighted materials to train LLM [Grynbaum and Mac, 2023].
- **Risk mitigation**: Addressing these challenges involves comprehensive risk mitigation strategies. This includes developing robust algorithms that can detect and counteract biases, implementing strict guidelines for data handling and model training, and continuously monitoring AI systems for unintended behaviors. Major players of the landscape including Meta [Touvron et al., 2023b], Google [Chowdhery et al., 2022], and OpenAI [OpenAI et al., 2023] are investing heavily in the research of this topic. Their efforts focus not only on further enhancing the technical capabilities of generative models but also on ensuring the downstream applications of the models are carefully evaluated to align with AI safety standards.
- **Policy and regulation**: The role of policy and regulation is paramount in shaping the safe development and deployment of AI technologies. Governments and regulatory bodies are increasingly focusing on developing frameworks to ensure that AI advances in a manner that is safe, ethical, and beneficial for society. This includes regulations around data privacy, AI transparency, and accountability. Government entities across the world, including the European Union [European Parliament, 2023], the United States [iapp, 2024; Klobuchar, 2024], China [Costigan, 2024], and Japan [AP, 2024] are actively developing new legislation and guidelines regarding AI safety and regulation.

Addressing these issues is critical to ensure the new AI technologies are safe and beneficial for society. As the technology continues to innovate, we will see more and more collaborations between researchers, industry leaders, and policymakers to navigate the complex landscape of AI safety and alignment.

11.8 Concluding Remarks

Congratulations! You have learned how to translate various mathematical theories into code. You have acquired the skills to implement data processing methods and machine learning models using many standard frameworks in Python. You have also mastered how to vectorize your code for efficiency and scalability. Using these skills and knowledge, you have also successfully built many practical machine learning solutions, solving problems in computer vision, audio processing, and recommendation systems, and even making your own mini-LLM! I hope you enjoy this journey!

Bibliography

Martín Abadi, Ashish Agarwal, Paul Barham, Eugene Brevdo, Zhifeng Chen, Craig Citro, Greg S. Corrado, Andy Davis, Jeffrey Dean, Matthieu Devin, Sanjay Ghemawat, Ian Goodfellow, Andrew Harp, Geoffrey Irving, Michael Isard, Yangqing Jia, Rafal Jozefowicz, Lukasz Kaiser, Manjunath Kudlur, Josh Levenberg, Dan Mane, Rajat Monga, Sherry Moore, Derek Murray, Chris Olah, Mike Schuster, Jonathon Shlens, Benoit Steiner, Ilya Sutskever, Kunal Talwar, Paul Tucker, Vincent Vanhoucke, Vijay Vasudevan, Fernanda Viegas, Oriol Vinyals, Pete Warden, Martin Wattenberg, Martin Wicke, Yuan Yu, and Xiaoqiang Zheng. Google's tutorial on transformer. https://www.tensorflow.org/text/tutorials/transformer, 2023.

AP. Japan's Kishida unveils a framework for global regulation of generative AI. https://apnews.com/article/oecd-ai-japan-kishida-artificial-intelligence-023ac08e04db5a2109cf35f8b8c9b102, May 2024.

BBC News. Google apologises for Photos app's racist blunder. https://www.bbc.com/news/technology-33347866, July 2015.

Dylan Butts. DeepFake scams have robbed companies of millions. Experts warn it could get worse. https://www.cnbc.com/2024/05/28/deepfake-scams-have-looted-millions-experts-warn-it-could-get-worse.html, May 2024.

Aakanksha Chowdhery, Sharan Narang, Jacob Devlin, Maarten Bosma, Gaurav Mishra, Adam Roberts, Paul Barham, Hyung Won Chung, Charles Sutton, Sebastian Gehrmann, Parker Schuh, Kensen Shi, Sasha Tsvyashchenko, Joshua Maynez, Abhishek Rao, Parker Barnes, Yi Tay, Noam Shazeer, Vinodkumar Prabhakaran, Emily Reif, Nan Du, Ben Hutchinson, Reiner Pope, James Bradbury, Jacob Austin, Michael Isard, Guy Gur-Ari, Pengcheng Yin, Toju Duke, Anselm Levskaya, Sanjay Ghemawat, Sunipa Dev, Henryk Michalewski, Xavier Garcia, Vedant Misra, Kevin Robinson, Liam Fedus, Denny Zhou, Daphne Ippolito, David Luan, Hyeontaek Lim, Barret Zoph, Alexander Spiridonov, Ryan Sepassi, David

Dohan, Shivani Agrawal, Mark Omernick, Andrew M. Dai, Thanumalayan Sankaranarayana Pillai, Marie Pellat, Aitor Lewkowycz, Erica Moreira, Rewon Child, Oleksandr Polozov, Katherine Lee, Zongwei Zhou, Xuezhi Wang, Brennan Saeta, Mark Diaz, Orhan Firat, Michele Catasta, Jason Wei, Kathy Meier-Hellstern, Douglas Eck, Jeff Dean, Slav Petrov, and Noah Fiedel. PaLM: Scaling language modeling with pathways, October 2022.

Djork-Arné Clevert, Thomas Unterthiner, and Sepp Hochreiter. Fast and accurate deep network learning by exponential linear units (ELUs), February 2016.

Johanna Costigan. China's New Draft AI Law Prioritizes Industry Development. https://www.forbes.com/sites/johannacostigan/2024/03/22/chinas-new-draft-ai-law-prioritizes-industry-development/?sh=3ba2ac836095, March 2024.

Zihang Dai, Zhilin Yang, Yiming Yang, Jaime Carbonell, Quoc V. Le, and Ruslan Salakhutdinov. Transformer-XL: Attentive language models beyond a fixed-length context, June 2019.

Jacob Devlin, Ming-Wei Chang, Kenton Lee, and Kristina Toutanova. BERT: Pre-training of deep bidirectional transformers for language understanding, May 2019.

Alexey Dosovitskiy, Lucas Beyer, Alexander Kolesnikov, Dirk Weissenborn, Xiaohua Zhai, Thomas Unterthiner, Mostafa Dehghani, Matthias Minderer, Georg Heigold, Sylvain Gelly, Jakob Uszkoreit, and Neil Houlsby. An image is worth 16x16 words: Transformers for image recognition at scale, June 2021.

European Parliament. EU AI Act: First regulation on artificial intelligence. https://www.europarl.europa.eu/topics/en/article/20230601STO93804/eu-ai-act-first-regulation-on-artificial-intelligence, December 2023.

Facebook Research. LLaMA reformaulation of RoPE. https://github.com/facebookresearch/llama/blob/main/llama/model.py\LY1\textbackslash#L80-L161, 2023.

Matthew Gault. AI Trained on 4Chan Becomes 'Hate Speech Machine'. https://www.vice.com/en/article/7k8zwx/ai-trained-on-4chan-becomes-hate-speech-machine, June 2022.

Daria Grechishnikova. Transformer neural network for protein-specific de novo drug generation as a machine translation problem. *Scientific Reports*, 11(1):321, January 2021. ISSN 2045-2322. doi: 10.1038/s41598-020-79682-4.

Michael M. Grynbaum and Ryan Mac. The Times Sues OpenAI and Microsoft Over A.I. Use of Copyrighted Work. https://www.nytimes.com/2023/12/27/business/media/new-york-times-open-ai-microsoft-lawsuit.html, December 2023.

Albert Gu and Tri Dao. Mamba: Linear-time sequence modeling with selective state spaces, December 2023.

Dan Hendrycks and Kevin Gimpel. Gaussian error linear units (GELUs), June 2016.

Shion Honda, Shoi Shi, and Hiroki R. Ueda. SMILES transformer: Pre-trained molecular fingerprint for low data drug discovery, November 2019.

HugginFace. HuggingFace T5 Model. https://huggingface.co/transformers/v3.2.0/_modules/transformers/modeling_t5.html, 2023.

iapp. Checking in on proposed California privacy and AI legislation. https://iapp.org/news/a/checking-in-on-proposed-california-privacy-and-ai-legislation, March 2024.

Wang Cheng Kang and Julian McAuley. Self-attentive sequential recommendation. *Proceedings – IEEE International Conference on Data Mining, ICDM*, 2018-November, pages 197–206, 2018. ISSN 15504786. doi: 10.1109/ICDM.2018.00035.

Klobuchar. Klobuchar Statement on Rules Committee Passage of Three Bipartisan AI and Elections Bills. https://www.klobuchar.senate.gov/public/index.cfm/2024/5/klobuchar-statement-on-rules-committee-passage-of-three-bipartisan-ai-and-elections-bills, May 2024.

Yinhan Liu, Myle Ott, Naman Goyal, Jingfei Du, Mandar Joshi, Danqi Chen, Omer Levy, Mike Lewis, Luke Zettlemoyer, and Veselin Stoyanov. RoBERTa: A robustly optimized BERT pretraining approach, July 2019.

Shengjie Luo, Shanda Li, Shuxin Zheng, Tie-Yan Liu, Liwei Wang, and Di He. Your transformer may not be as powerful as you expect, October 2022.

Stephen Merity, Caiming Xiong, James Bradbury, and Richard Socher. Pointer sentinel mixture models, September 2016.

Diganta Misra. *Mish: A self regularized non-monotonic activation function*, August 2020.

OpenAI, Josh Achiam, Steven Adler, Sandhini Agarwal, Lama Ahmad, Ilge Akkaya, Florencia Leoni Aleman, Diogo Almeida, Janko Altenschmidt, Sam Altman, Shyamal Anadkat, Red Avila, Igor Babuschkin, Suchir Balaji, Valerie Balcom, Paul Baltescu, Haiming Bao, Mo Bavarian, Jeff Belgum, Irwan Bello, Jake Berdine, Gabriel Bernadett-Shapiro, Christopher Berner, Lenny Bogdonoff, Oleg Boiko, Madelaine Boyd, Anna-Luisa Brakman, Greg Brockman, Tim Brooks, Miles Brundage, Kevin Button, Trevor Cai, Rosie Campbell, Andrew Cann, Brittany Carey, Chelsea Carlson, Rory Carmichael, Brooke Chan, Che Chang, Fotis Chantzis, Derek Chen, Sully Chen, Ruby Chen, Jason Chen, Mark Chen, Ben Chess, Chester Cho, Casey Chu, Hyung Won Chung, Dave Cummings, Jeremiah Currier, Yunxing Dai, Cory Decareaux, Thomas Degry, Noah Deutsch, Damien Deville, Arka Dhar, David Dohan, Steve Dowling, Sheila Dunning, Adrien Ecoffet, Atty Eleti, Tyna Eloundou, David Farhi, Liam Fedus, Niko Felix, Simón Posada Fishman, Juston Forte, Isabella Fulford, Leo Gao, Elie Georges, Christian Gibson, Vik Goel, Tarun Gogineni, Gabriel Goh, Rapha Gontijo-Lopes, Jonathan Gordon, Morgan Grafstein, Scott Gray, Ryan Greene, Joshua Gross, Shixiang Shane Gu, Yufei Guo, Chris Hallacy, Jesse Han, Jeff Harris, Yuchen He, Mike Heaton, Johannes Heidecke, Chris Hesse, Alan Hickey, Wade Hickey, Peter Hoeschele, Brandon Houghton, Kenny Hsu, Shengli Hu, Xin Hu, Joost Huizinga, Shantanu

Jain, Shawn Jain, Joanne Jang, Angela Jiang, Roger Jiang, Haozhun Jin, Denny Jin, Shino Jomoto, Billie Jonn, Heewoo Jun, Tomer Kaftan, Lukasz Kaiser, Ali Kamali, Ingmar Kanitscheider, Nitish Shirish Keskar, Tabarak Khan, Logan Kilpatrick, Jong Wook Kim, Christina Kim, Yongjik Kim, Hendrik Kirchner, Jamie Kiros, Matt Knight, Daniel Kokotajlo, Lukasz Kondraciuk, Andrew Kondrich, Aris Konstantinidis, Kyle Kosic, Gretchen Krueger, Vishal Kuo, Michael Lampe, Ikai Lan, Teddy Lee, Jan Leike, Jade Leung, Daniel Levy, Chak Ming Li, Rachel Lim, Molly Lin, Stephanie Lin, Mateusz Litwin, Theresa Lopez, Ryan Lowe, Patricia Lue, Anna Makanju, Kim Malfacini, Sam Manning, Todor Markov, Yaniv Markovski, Bianca Martin, Katie Mayer, Andrew Mayne, Bob McGrew, Scott Mayer McKinney, Christine McLeavey, Paul McMillan, Jake McNeil, David Medina, Aalok Mehta, Jacob Menick, Luke Metz, Andrey Mishchenko, Pamela Mishkin, Vinnie Monaco, Evan Morikawa, Daniel Mossing, Tong Mu, Mira Murati, Oleg Murk, David Mély, Ashvin Nair, Reiichiro Nakano, Rajeev Nayak, Arvind Neelakantan, Richard Ngo, Hyeonwoo Noh, Long Ouyang, Cullen O'Keefe, Jakub Pachocki, Alex Paino, Joe Palermo, Ashley Pantuliano, Giambattista Parascandolo, Joel Parish, Emy Parparita, Alex Passos, Mikhail Pavlov, Andrew Peng, Adam Perelman, Filipe de Avila Belbute Peres, Michael Petrov, Henrique Ponde de Oliveira Pinto, Michael (Rai)Pokorny, Michelle Pokrass, Vitchyr Pong, Tolly Powell, Alethea Power, Boris Power, Elizabeth Proehl, Raul Puri, Alec Radford, Jack Rae, Aditya Ramesh, Cameron Raymond, Francis Real, Kendra Rimbach, Carl Ross, Bob Rotsted, Henri Roussez, Nick Ryder, Mario Saltarelli, Ted Sanders, Shibani Santurkar, Girish Sastry, Heather Schmidt, David Schnurr, John Schulman, Daniel Selsam, Kyla Sheppard, Toki Sherbakov, Jessica Shieh, Sarah Shoker, Pranav Shyam, Szymon Sidor, Eric Sigler, Maddie Simens, Jordan Sitkin, Katarina Slama, Ian Sohl, Benjamin Sokolowsky, Yang Song, Natalie Staudacher, Felipe Petroski Such, Natalie Summers, Ilya Sutskever, Jie Tang, Nikolas Tezak, Madeleine Thompson, Phil Tillet, Amin Tootoonchian, Elizabeth Tseng, Preston Tuggle, Nick Turley, Jerry Tworek, Juan Felipe Cerón Uribe, Andrea Vallone, Arun Vijayvergiya, Chelsea Voss, Carroll Wainwright, Justin Jay Wang, Alvin Wang, Ben Wang, Jonathan Ward, Jason Wei, C. J. Weinmann, Akila Welihinda, Peter Welinder, Jiayi Weng, Lilian Weng, Matt Wiethoff, Dave Willner, Clemens Winter, Samuel Wolrich, Hannah Wong, Lauren Workman, Sherwin Wu, Jeff Wu, Michael Wu, Kai Xiao, Tao Xu, Sarah Yoo, Kevin Yu, Qiming Yuan, Wojciech Zaremba, Rowan Zellers, Chong Zhang, Marvin Zhang, Shengjia Zhao, Tianhao Zheng, Juntang Zhuang, William Zhuk, and Barret Zoph. GPT-4 Technical Report, December 2023.

Adam Paszke, Sam Gross, Francisco Massa, Adam Lerer, James Bradbury, Gregory Chanan, Trevor Killeen, Zeming Lin, Natalia Gimelshein, Luca Antiga, Alban Desmaison, Andreas Köpf, Edward Yang, Zach DeVito, Martin Raison, Alykhan Tejani, Sasank Chilamkurthy, Benoit Steiner, Lu Fang, Junjie Bai, Soumith Chintala, and PyTorch GitHub Issue #105665. Running Llama 2 on Apple Silicon

GPUs – missing MPS types and operators #105665. https://github.com/pytorch/pytorch/issues/105665, July 2023.

Alec Radford, Karthik Narasimhan, Tim Salimans, and Ilya Sutskever. Improving language understanding by generative pre-training. June 2018.

Alec Radford, Jong Wook Kim, Tao Xu, Greg Brockman, Christine McLeavey, and Ilya Sutskever. Robust speech recognition via large-scale weak supervision, December 2022.

Colin Raffel, Noam Shazeer, Adam Roberts, Katherine Lee, Sharan Narang, Michael Matena, Yanqi Zhou, Wei Li, and Peter J. Liu. Exploring the Limits of Transfer Learning with a Unified Text-to-Text Transformer, September 2020.

Prajit Ramachandran, Barret Zoph, and Quoc V. Le. Searching for activation functions, October 2017.

Yi Ren, Chenxu Hu, Xu Tan, Tao Qin, Sheng Zhao, Zhou Zhao, and Tie-Yan Liu. FastSpeech 2: Fast and high-quality end-to-end text to speech. pages 1–15, 2020.

Peter Shaw, Jakob Uszkoreit, and Ashish Vaswani. Self-attention with relative position representations, April 2018.

Jianlin Su, Yu Lu, Shengfeng Pan, Ahmed Murtadha, Bo Wen, and Yunfeng Liu. RoFormer: Enhanced transformer with rotary position embedding. September 2020.

Hugo Touvron, Thibaut Lavril, Gautier Izacard, Xavier Martinet, Marie-Anne Lachaux, Timothée Lacroix, Baptiste Rozière, Naman Goyal, Eric Hambro, Faisal Azhar, Aurelien Rodriguez, Armand Joulin, Edouard Grave, and Guillaume Lample. LLaMA: Open and efficient foundation language models, February 2023a.

Hugo Touvron, Louis Martin, Kevin Stone, Peter Albert, Amjad Almahairi, Yasmine Babaei, Nikolay Bashlykov, Soumya Batra, Prajjwal Bhargava, Shruti Bhosale, Dan Bikel, Lukas Blecher, Cristian Canton Ferrer, Moya Chen, Guillem Cucurull, David Esiobu, Jude Fernandes, Jeremy Fu, Wenyin Fu, Brian Fuller, Cynthia Gao, Vedanuj Goswami, Naman Goyal, Anthony Hartshorn, Saghar Hosseini, Rui Hou, Hakan Inan, Marcin Kardas, Viktor Kerkez, Madian Khabsa, Isabel Kloumann, Artem Korenev, Punit Singh Koura, Marie-Anne Lachaux, Thibaut Lavril, Jenya Lee, Diana Liskovich, Yinghai Lu, Yuning Mao, Xavier Martinet, Todor Mihaylov, Pushkar Mishra, Igor Molybog, Yixin Nie, Andrew Poulton, Jeremy Reizenstein, Rashi Rungta, Kalyan Saladi, Alan Schelten, Ruan Silva, Eric Michael Smith, Ranjan Subramanian, Xiaoqing Ellen Tan, Binh Tang, Ross Taylor, Adina Williams, Jian Xiang Kuan, Puxin Xu, Zheng Yan, Iliyan Zarov, Yuchen Zhang, Angela Fan, Melanie Kambadur, Sharan Narang, Aurelien Rodriguez, Robert Stojnic, Sergey Edunov, and Thomas Scialom. Llama 2: Open foundation and fine-tuned chat models, July 2023b.

Ashish Vaswani, Noam Shazeer, Niki Parmar, Jakob Uszkoreit, Llion Jones, Aidan N. Gomez, Lukasz Kaiser, and Illia Polosukhin. Attention is all you need. *Advances in*

Neural Information Processing Systems, 2017-Decemember (NIPS), pages 5999–6009, 2017. ISSN 10495258.

Ashish Vaswani, Samy Bengio, Eugene Brevdo, Francois Chollet, Aidan N. Gomez, Stephan Gouws, Llion Jones, Lukasz Kaiser, Nal Kalchbrenner, Niki Parmar, Ryan Sepassi, Noam Shazeer, and Jakob Uszkoreit. Tensor2Tensor for Neural Machine Translation, March 2018.

Ashish Vaswani, Samy Bengio, Eugene Brevdo, Francois Chollet, Aidan N. Gomez, Stephan Gouws, Llion Jones, Lukasz Kaiser, Nal Kalchbrenner, Niki Parmar, Ryan Sepassi, Noam Shazeer, and Jakob Uszkoreit. Tensor2Tensor Sinusoidal Positional Encoding. https://github.com/tensorflow/tensor2tensor/blob/23bd23b9830059fbc349381b70d9429b5c40a139/tensor2tensor/layers/common_attention.py\LY1\textbackslash#L387-L433, 2023.

Wolf Thomas, Lysandre Debut, Victor Sanh, Julien Chaumond, Clement Delangue, Anthony Moi, Pierric Cistac et al. "Transformers: State-of-the-art natural language processing." In *Proceedings of the 2020 conference on empirical methods in natural language processing: system demonstrations*, pp. 38–45. 2020.

Yuxin Wu and Kaiming He. Group normalization, June 2018.

Ruibin Xiong, Yunchang Yang, Di He, Kai Zheng, Shuxin Zheng, Chen Xing, Huishuai Zhang, Yanyan Lan, Liwei Wang, and Tie-Yan Liu. On layer normalization in the transformer architecture, June 2020.

Yaodong Yu, Sam Buchanan, Druv Pai, Tianzhe Chu, Ziyang Wu, Shengbang Tong, Hao Bai, Yuexiang Zhai, Benjamin D. Haeffele, and Yi Ma. White-box transformers via sparse rate reduction: Compression is all there is?, November 2023.

Biao Zhang and Rico Sennrich. Root mean square layer normalization, October 2019.

Index

Vectorization: A Practical Guide to Efficient Implementations of Machine Learning Algorithms, First Edition. Edward DongBo Cui.